T0406617

New Trends in Soil Micromorphology

Selim Kapur · Ahmet Mermut ·
Georges Stoops (Eds.)

New Trends in Soil Micromorphology

 Springer

Selim Kapur
University of Cukorova
Department of Soil and Archaeometry
Adana
Turkey

Dr. Ahmet Mermut
University of Saskatchewan
Dept. Soil Science
Saskatoon SK S7N 5A8
Canada

Prof. Dr. Georges Stoops
Ghent University
Dept. Geology & Soil Sciences
Krijgslaan 281 S8
9000 Gent
Belgium

ISBN: 978-3-540-79133-1   e-ISBN: 978-3-540-79134-8

Library of Congress Control Number: 2008932572

Cover design: deblik, Berlin

Printed on acid-free paper

9 8 7 6 5 4 3 2 1

springer.com

Contents

List of Contributors

Hema Achyuthan
Department of Geology, Anna University, Chennai 600 025, India,
e-mail: hachyuthan@yahoo.com

Montserrat Antúnez
Department of Environment and Soil Sciences, University of Lleida Av. Rovira
Roure 191, 25198 Lleida, Catalonia

O. Arnalds
Agricultural University of Iceland, Reykjavik, Iceland

J. M. Arocena
College of Science and Management; Canada Research Chair – Soil and
Environmental Sciences University of Northern British Columbia 3333 University
Way, Prince George, BC Canada V2N 4Z9, e-mail: arocenaj@unbc.ca

S. Blok
Institute for Biodiversity and Ecosystem Dynamics (IBED), University of
Amsterdam, Nieuwe Achtergracht 166, 1024 AD Amsterdam, The Netherlands

W.E.H. Blum
Institute of Soil Research, Department of Forest and Soil Sciences, University of
Natural Resources and Applied Life Sciences (BOKU), Vienna, Austria,
e-mail: herma.exner@boku.ac.at

Jaume Boixadera
Department of Environment and Soil Sciences, University of Lleida Av. Rovira
Roure 191, 25198 Lleida, Catalonia; Secció d'Avaluació de Recursos Agraris,
DARP, Generalitat de Catalunya

Z. Chen
Environmental Monitoring and Evaluation Branch Alberta Environment, 9820-106
Street, Edmonton, AB CANADA T5K 2J6, e-mail: Chi.Chen@gov.ab.ca

Marie-Agnès Courty
UMR 5198, CNRS-IPH, Centre Européen de Recherches Préhistoriques,
Avenue Léon-Jean Grégory, 66720 Tautavel, France,
e-mail: courty@tautavel.univ-perp.fr

Alex Crisci
CMTC INP, 1260 rue de la Piscine, BP 75, 38402 Saint Martin d'Hères, France,
e-mail: alexandre.crisci@cmtc.inpg.fr

María de Lourdes Flores-Delgadillo
Departamento de Edafología, Instituto de Geología, UNAM

Scott L. Fedick
Department of Anthropology, University of California, Riverside,
e-mail: scott.fedick@ucr.edu

Michel Fedoroff
Laboratoriy Electrochimie & Chimie analytique, ENSCP, 11 rue Pierre et Marie
Curie, 75231 Paris, France, e-mail: michel-fedoroff@enscp.jussieur.fr

N. Fedoroff
Institut National Agronomique, 78850 Thiverval-Grignon, France,
e-mail: nicolas.fedoroff@wanadoo.fr

T. S. Gendler
Institute of Physics of the Earth, Russian Academy of Sciences, Bolshaya
Gruzinskaya 10, Moscow 123810, Russia, e-mail: gendler06@mail.ru

M. Gérard
UR GEOTROPE, Institut de Recherche pour le Développement, Bondy, France

Paul Greenwood
Centre for Land Rehabilitation, School of Earth and Geographical Sciences
University of WA, 35 Stirling Hwy, Crawley, 6009, Australia,
e-mail: Paul.Greenwood@uwa.edu.au

Kliti Grice
Department of Applied Chemistry, Curtin University of Technology, GPO Box
1987, 6845 Perth, Australia, e-mail: K.Grice@curtin.edu.au

A. Heidari
Department of Soil Science, Agriculture College, University of Tehran,
Daneshkadeh St. Karaj, Iran, e-mail: aheidari82@yahoo.com,

F. Heller
Institut für Geophysik, ETH Hönggerberg, CH-8093 Zürich, Switzerland,
e-mail: heller@mag.ig.erdw.ethz.ch

G.S. Humphreys
Department of Physical Geography, School of Environmental and Life Sciences,
Macquarie University, NSW, Australia

Tang Keli
Institute of Soil and Water Conservation, CAS, Yangling, Shaanxi 712100,
China, e-mail: kltang@ms.iswc.ac.cn

Miroslav Kutílek
Emeritus Professor, Nad Patankou 34, 160 00 Prague 6, Czech Rep.,
e-mail: miroslav.kutilek@volny.cz

Hua Lizhong
Institute of Mountain Hazards and Environment, CAS, Chengdu, Sichuan 610041;
Graduate University of Chinese Academy of Science, Beijing 100039,
China, e-mail: hualizhong2008@yahoo.com.cn

S. Mahmoodi
Department of Soil Science, Agriculture College, University of Tehran,
Daneshkadeh St. Karaj, Tran, e-mail: smahmodi@chamran.ut.ac.ir,

Maria Gerasimova
Moscow Lomonosov University, Faculty of Geography Leninskie Gory 119992,
Russia, e-mail: etingof@glasnet.ru

Marina Lebedeva
Dokuchaev Soil Science Institute, Pyzhevskiy pereulok, 7, Moscow 109017,
Russia, e-mail: etingof@glasnet.ru

Michel Mermoux
LEPMI-ENSEEG, Domaine Universitaire, BP75, 38042 Saint-Martin d'Hères,
France, e-mail: Michel.Mermoux@enseeg.inpg.fr

Marcello Pagliai
CRA – Centro di ricerca per l'agrobiologia e la pedologia, Piazza M. D'Azeglio
30–50121 Firenze, Italy, e-mail: marcello.pagliai@entecra.it

Teresa Pi-Puig
Departamento de Geoquímica, Instituto de Geología, UNAM,
e-mail: tpuig@geologia.unam.mx

Rosa Maria Poch
Department of Environment and Soil Sciences, University of Lleida Av. Rovira
Roure 191, 25198 Lleida, Catalonia, e-mail: rosa.poch@macs.udl.cat

P. Sanborn
Ecosystem Science and Management, University of Northern British Columbia,
3333 University Way, Prince George, BC Canada V2N4Z9, e-mail: sanborn@unbc.ca

David Smith
Lab. LEME/Nanoanalysis, MNHN, Laboratoire de Minéralogie, 61 rue Buffon,
75005 Paris, France, e-mail: davsmith@mnhn.fr

Sergey Sedov
Departamento de Edáfología, Instituto de Geología, UNAM,
e-mail: sergey@geologia.unam.mx

Elizabeth Solleiro-Rebolledo
Departamento de Edáfología, Instituto de Geología, UNAM,
e-mail: solleiro@geologi a.unam.mx

G. Stoops
Laboratorium voor Mineralogie, Petrologie en Micropedologie, Universiteit Gent;
Department of Geology and Soil Science, Ghent University, Krijgslaan 281, S8,
B-9000 Ghent, Belgium, e-mail: georges.stoops@ugent.be,

Mark Thiemens
Department of Chemistry and Biochemistry, University of California, San Diego
92093-0352, USA, e-mail: mht@chem.ucsd.edu

A. Tsatskin
Zinman Institute of Archaeology, University of Haifa, Haifa 31905, Israel,
e-mail: tsatskin@research.haifa.ac.il

Ernestina Vallejo-Gómez
Departamento de Edáfología, Instituto de Geología, UNAM

J. van Mourik
Institute for Biodiversity and Ecosystem Dynamics (IBED), University of
Amsterdam. Nieuwe Achtergracht 166, 1024 AD Amsterdam

P. G. Walsh
Department of Physical Geography, School of Environmental and Life Sciences,
Macquarie University; Forests NSW, Department of Primary Industries, NSW,
Australia, e-mail: peterwa@sf.nsw.gov.au

He Xiubin
Institute of Mountain Hazards and Environment, CAS, Chengdu, Sichuan 610041;
Institute of Soil and Water Conservation, CAS, Yangling, Shaanxi 712100, China,
e-mail: xiubinh@imde.ac.cn

Bao Yuhai
Institute of Mountain Hazards and Environment, CAS, Chengdu, Sichuan 610041;
Graduate University of Chinese Academy of Science, Beijing 100039,
China, e-mail: byhcw@126.com

Geoffrey Steel Humphreys 1953–2007

Geoff Humphreys died suddenly and unexpectedly in Sydney on August 12th, 2007, tragically early at 54 but surrounded by the bush he loved so much. As recently as two days before his death, Geoff was in the field having a great time being outdoors, digging holes, talking about his research and engaging with people interested in understanding how landscapes function. Geoff's interest in the formation of soil and its management, and his breadth of experience and interests, brought him into contact with many in the geomorphological, Quaternary and ecological communities.

Geoff enrolled at Macquarie University as an undergraduate in the early 1970s initially studying economics but soon transferring to the earth sciences where his real interests lay. He took full advantage of the flexibility Macquarie offered by studying a mix of geography, geology and biology. He graduated with first class Honours in 1977 and was encouraged to move directly onto a PhD. Although Geoff worked in a range of different areas, what he regarded as the core of his research was his work on bioturbation of soils. He started this work in his honours research in 1975/76 when he began investigating the effects of ants, termites, worms and insects on podsol soils around Sydney. Geoff expanded this into the topic of his PhD research: how soil-dwelling organisms affect the formation of soils (specifically texture-contrast soils, typical of hillslopes around Sydney). The subject arose at the suggestion of Ron Paton, his supervisor, but Geoff insisted on adopting a quantitative approach. Geoff knew that to convince people of the importance of these small animals he would need to measure volumes, areas, rates and depths so that there could be no argument about their significance in soil formation. In this Geoff succeeded: the quantitative work forms the backbone of both journal papers and the bioturbation chapter in 'Soils: a new global view' which is still some of the best work published on the topic. This multi-disciplinary approach stayed with

Geoff throughout his career as he devised new ways of understanding the role of the biosphere in soil formation.

Geoff met his wife Janelle at Macquarie and together they moved to Papua New Guinea in 1979, first in Chimbu Province in the Highlands and later at the University of Papua New Guinea from 1983 to 1987. Geoff worked on his PhD as a student of Macquarie University before and during his time in PNG, somehow managing to work up the Sydney-based field data while living in PNG, teaching at the University of Papua New Guinea and raising a family. After graduating in 1985, Geoff took up a position at the University of New South Wales in 1987 and then, in 1989, an appointment in the Land Management Project in the Research School of Pacific Studies at ANU which saw him return to PNG and also to other parts of Asia and the Pacific and even Africa for long field seasons. Geoff was fascinated by the spectacular and highly dynamic landscapes of these countries, which he investigated with boundless enthusiasm.

Geoff began lecturing at Macquarie University in mid-1994 just as he and his Macquarie colleagues Ron Paton and Peter Mitchell were completing their book 'Soils: A new global view'. One of Geoff's proudest achievements was to be awarded (with his co-authors) the G.K. Gilbert Award for excellence in geomorphological research by the Geomorphology Specialty Group of the Association of American Geographers in March 1999 for this book and for the impact the book created. In an article titled *Shock the World (and then some)*, Randall Schaetzl included it within the four most groundbreaking and influential treatises on geomorphology and pedology of the 20th Century. Other reviewers put the book at the front of a paradigm shift in the understanding of soil genesis, although it is fair to say that views were wide-ranging. He was sometimes frustrated by the entrenched and intransigent positions within pedology. Partly as a response, and partly just recognising a good scientific opportunity, in recent years Geoff in collaboration with others, was very innovative in bending new techniques in earth sciences to his task of measuring soil processes including single-grain optically stimulated luminescence dating to measure soil turnover rates, terrestrial cosmogenic nuclides to measure rates of soil production (conversion of rock to soil) and, uranium series disequilibrium methods to measure soil formation age. In addition to those innovative directions, he continued to collaborate with Ron Paton on soil genesis issues right up to his death, with two papers critically evaluating the zonalistic foundations of soil science in the USA published in *Geoderma* in 2007.

Geoff was a highly respected and valued member of the soil science community in Australia and around the world. Geoff was instrumental in the establishment of the Soil Morphology and Micromorphology Commission of the IUSS, which he chaired from 2002 to 2006 and at the time of his death was 2nd Vice Chair. In this role he is said to have breathed new life into the morphological study of soils. He was active in ASSSI (Australian Soil Science Society Inc), representing the NSW Branch on the organising Committee of the Brisbane 2010 World Congress.

In addition to his fine research contributions, Geoff will be remembered as a great teacher and advocate of soil science and scientific research in general. As an Associate Dean of Research at Macquarie University, he was an energetic contributor on several post graduate and research guiding committees. For 11 years he was co-editor of the Australian Geographer.

In his time at Macquarie Geoff revelled in the supervision of many students in a range of subjects. Somewhere in each of those projects were a link back to understanding how soils form and an original and imaginative approach to tackling intractable problems. Along the way he made important contributions to studies of soil erosion and land degradation, ecology, geomorphology and the Quaternary.

Geoff recognised the critical importance of detailed quantitative observations of soil morphology at the macro and micro scales. In collaboration with others in Australia and internationally, he sought to unlock secrets of pedology revealed by soil morphological features, generating an impressive publication output along the way. More generally, Geoff has been credited with paving the way for a truly modern, interdisciplinary approach to pedology, one that effectively incorporates geomorphological and ecological principles, and this is perhaps the primary legacy of Geoff's career.

The outpouring of sadness at the news of his death was the greatest demonstration of the warmth with which Geoff was regarded by hundreds of friends, colleagues and students, past and present. Geoff will be greatly missed by his many friends, colleagues (past and present) and former students as well as his family: Janelle, Sheridan, Lachlan, Rowan, William and grandson Max.

Paul Hesse, Jonathan Gray and others

The Role of Soil Micromorphology in the Light of the European Thematic Strategy for Soil Protection

W. E. H. Blum

Abstract The role of soil micromorphology within new soil research concepts developed for the European Thematic Strategy for Soil Protection is explained. Soil micromorphology has a central function in the concept of integrated research and is able to support inter-disciplinary approaches for soil protection and management (Table 1). This could be one of the main assets for the future development of soil micromorphology and its survival in the medium or long term.

Keywords Micromorphology · soil protection strategy · integrated soil research

1 Introduction

A communication from the European Commission to the Council and the European Parliament, entitled: "Towards a thematic strategy for soil protection", ratified by the 15 ministers of environment of the European Union in 2002 (European Commission 2002), defines five main functions of soil for human societies and the environment:

1. the production of food and other biomass,
2. the capacity for storing, filtering and transformation,
3. the soil as a habitat and a gene pool,
4. the soil as a physical and cultural environment for humankind, and
5. as a source of raw materials.

In addition, it devices 8 main threats to soil: erosion, decline in organic matter, soil contamination (local and diffuse), soil sealing, soil compaction, decline in soil bio-diversity, salinisation, and floods and landslides.

W. E. H. Blum
Institute of Soil Research, Department of Forest and Soil Sciences,
University of Natural Resources and Applied Life Sciences (BOKU),
Vienna, Austria, e-mail: herma.exner@boku.ac.at

S. Kapur et al. (eds.), *New Trends in Soil Micromorphology*,
© Springer-Verlag Berlin Heidelberg 2008

Table 1 Concept for integrated soil research, based on DPSIR

	Main research goals	Research clusters	Sciences involved
1	To understand the main processes in the eco-subsystem soil, underlying soil quality and soil functions, in relation to land uses and soil.	Analysis of processes related to the threats to soil and their interdependency: erosion, loss of organic matter, contamination, sealing, compaction, decline in biodiversity, salinisation, floods and landslides.	Inter-disciplinary research through co-operation of soil micromorphology, of soil physics, soil chemistry, soil mineralogy and soil biology.
2	To know where these processes occur and how they develop with time.	Development and harmonization and standardisation of methods for the analysis of the **State (S)** of the threats to soil and their changes with time = **soil monitoring**.	Multi-disciplinary research through co-operation of soil sciences with -geographical sciences, -geo-statistics, geo-information sciences (e.g. GIS)
3	To know the driving forces and pressures behind these processes, as related to policy and decision making on a local, regional or global basis.	Relating the 8 threats to **Driving forces (D)** and **Pressures (P)** = cross linking with cultural, social and economic drivers, such as policies (agriculture, transport, energy, environment etc.) as well as with technical and ecological drivers, e.g. global and climate change.	Multi-disciplinary research through co-operation of soil sciences with political sciences, legal sciences, social sciences, economic sciences, historical sciences, philosophical sciences and others.
4	To know the impacts on the eco services provided by the sub-system soil to other environmental compartments (eco-subsystems).	Analysis of the **Impacts (I)** of the threats, relating them to soil eco-services for other environmental compartments: air, water (open and ground water), biomass production, human health, biodiversity, culture.	Multi-disciplinary research through co-operation of soil sciences with geological sciences, biological sciences, toxicological sciences, hydrological sciences, physio-geographical sciences, sedimentological sciences and others.
5	To have operational tools (technologies) at one's disposal for the mitigation of threats and impacts.	Development of operational procedures for the mitigation of the threats = **Responses (R)**.	Multi-disciplinary research through co-operation of natural sciences with engineering sciences, technical sciences, physical sciences, mathematical sciences and others.

From 2002–2005, about 400 scientists from all over Europe worked together in an operational set-up of 5 working groups. Their reports are available on the soil internet site (http://europa.eu.int/comm/environment/soil/index.htm) or at the soil electronic library and discussion site CIRCA. In the following, the outcome of the working group on research is of major importance, because it develops new concepts and directions (Blum et al. 2004a,b).

The specific task was to use the DPSIR approach, distinguishing between driving forces (D), which develop pressures (P), resulting in a state (S), which by itself creates impacts (I) and for which responses (R) are needed (European Environment Agency 1999).

This approach allows for key questions to be answered in the understanding of complex soil and environmental systems, such as: (1) what is the D behind a problem?, (2) what are the Ps deriving from the Ds?, (3) what is the S, which the P creates?, (4) what are the Is that result from the S?, and (5) it also allows Rs to change the Ds in order to alleviate or reverse a problem, developing solutions through the implementation of operational measures.

Based on this approach, a new concept for integrated research in soil protection and soil resource management was developed (Table 1). From this table, the 5 main research goals as well as the 5 main research clusters, which are needed to reach these goals, and the sciences which have to be involved, can be identified. Special importance for micromorphology is the first research cluster, targeting at the analysis of soil processes which can only be performed by inter-disciplinary research.

2 The Role of Soil Micromorphology

The primary role of soil micromorphology within this new concept could be to link the different soil disciplines, like soil physics, chemistry, mineralogy, soil biology together by providing a basis on which they can co-operate. This means that soil micromorphology acts as an integrating tool for all soil disciplines involved.

The comparative advantage of micromorphology is its capacity to develop three-dimensional models for describing the complexity of soil, especially regarding the pore system which is the spatial basis for all physical, chemical and biological soil processes. In this context, the walls of the pores, which contain humic substances, clay minerals, oxides and others, and the pore space itself, in which, besides air and water, living organisms such as fungi, bacteria and others are actively participating in soil processes, are of main importance. Soil micromorphology has a central function in the concept of integrated research and is able to support inter-disciplinary approaches for soil protection and management (Table 1). This could be one of the main assets for the future development of soil micromorphology and its survival in the medium or long term.

It can be summarized that the new research concepts, developed within the European Soil Thematic Strategy, underline the importance of soil micromorphology as an integrating soil discipline.

References

Blum WEH, Büsing J, Montanarella L (2004a) Research needs in support of the European thematic strategy for soil protection. Trends in Analytical Chemistry 23: 680–685

Blum WEH, Barcelo D, Büsing J, Ertel T, Imeson A, Vegter J (2004b) Scientific Basis for the Management of European Soil Resources – Research Agenda. Verlag Guthmann-Peterson, Wien

European Commission (2002) Towards a Thematic Strategy for Soil Protection. COM (2002) 179 final (eur-lex.europa.eu/LexUriServ/site/en/com/2002/com2002_0179en01.pdf)

European Environment Agency (1999) Environment in the European Union at the turn of the century. Environmental assessment report No. 2. EEA, Luxembourg (ISBN 92-9157-202-0)

Soil Micromorphology and Soil Hydraulics

Marcello Pagliai and Miroslav Kutilek

Abstract The characterization of soil porosity by micromorphological approach is largely used to evaluate the modification of soil structure induced by the impact of agricultural activity. On the contrary, few studies are addressed to the characterisation of soil porosity to evaluate water movements in soils in spite of the fact that soil hydraulic functions are strongly dependent on the soil porous system. The physical interpretation of one of these functions, the saturated hydraulic conductivity, by soil micromorphological parameters have been studied in a loam soil, representative of the hilly environment of Italy, cultivated to maize.

Besides the confirmation that the continuous conventional tillage induced soil structure degradation in terms of reduction of soil porosity and particularly elongated pores, this paper showed a significant correlation between the elongated continuous transmission pores and the saturated hydraulic conductivity. Results clearly showed that the shape, the size, the orientation and the continuity of pores regulated the flux of the saturated hydraulic conductivity. The micromorphological research also showed that the walls of pores plays an important role on the stability of pores. Formation and existence of the vesicular pores, combined with the orientation of elongated pores parallel to the soil surface is the main factor of the substantial decrease in hydraulic conductivity.

Further research should include all the existing information on pore micromorphology into physically based soil hydraulic functions.

Keywords Soil thin sections · soil porous system · elongated transmission pores · image analysis · saturated hydraulic conductivity

Marcello Pagliai
CRA – Centro di ricerca per l'agrobiologia e la pedologia,
Piazza M. D'Azeglio 30 – 50121 Firenze, Italy, e-mail: marcello.pagliai@entecra.it

Miroslav Kutilek
Emeritus Professor, Nad Patankou 34, 160 00 Prague 6, Czech Rep.
e-mail: miroslav.kutilek@volny.cz

S. Kapur et al. (eds.), *New Trends in Soil Micromorphology*,
© Springer-Verlag Berlin Heidelberg 2008

1 Introduction

The soil water retention curve, the saturated hydraulic conductivity and the unsaturated hydraulic conductivity function are basic soil hydraulic functions and parameters. Ample apprehension of the soil hydraulic functions and parameters is required for a successful formulation of the principles leading to sustainable soil management, agricultural production and environmental protection. From these, all the other parameters, required in the solution of the practical tasks, are derived.

The basic soil hydraulic functions are strongly dependent upon the soil porous system. The development of models is characteristic by the gradual transition from the simplest concepts up to the sophisticated approaches, which should correspond to the visual reality studied by soil micromorphology.

2 Soil Porous System and Soil Micromorphometry

2.1 An Overview on the Quantification of the Soil Porous System

Quantification of the soil porous system consists of classification of soil pores, characterization of the soil pores shapes and the estimation of the pore size distribution function.

When the hydraulic functions of the soil pores are considered, the following laws of hydrostatics and hydrodynamics are applied as best fitting to the classification criteria of the size of the pores (Kutilek and Nielsen 1994, p. 20, Kutilek 2004):

A. Submicroscopic pores that are so small that they preclude clusters of water molecules from forming fluid particles or continuous water flow paths.
B. Micropores, or capillary pores where the shape of the interface between air and water is determined by the configuration of the pores and by the forces on the interface. The resulting air-water interface is the capillary meniscus. The unsaturated flow of water is described by the Darcy-Buckingham equation. The category of micropores is further subdivided into two sub-categories:

   B1. Matrix (intra-aggregate, intrapedal) pores within soil aggregates or within blocks of soil, if aggregates are not present. The shape and size of pores in aggregates as well as coating of the particles, cutans and nodules depend on soil genesis. Aggregates may or may not be stable during the transport of water and thus the porous system may change. Due to the cutaneous film-like forms, which cover the surface of the majority of aggregates, the saturated conductivity at the surface of stable aggregates is usually strongly reduced when compared with that inside the matrix of aggregates (Horn 1994).

B2. Structural (inter-aggregate, interpedal) pores between the aggregates. Since the shape and size of aggregates depend on soil genesis and soil use, the morphology of pores between aggregates depends upon the soil genesis and soil use. The eventual aggregate instability influences the configuration of structural pores when the soil water content is changed destroying even a certain portion of the structural pores. On the other hand side the porous system has a high stability when the aggregates are formed by the pedo-edaphon. Structural pores are sometimes interpreted as macropores with capillarity, or macropores where Richards equation is applicable. The equivalent pore radius of the boundary separating the two subcategories is in broad ranges from 2 to 50 m, being dependent on the soil taxon and the type of soil use (Kutilek et al., 2006). It is not appropriate to consider a fixed value of the boundary between the two subcategories. Its value is determined either by soil micromorphology (Pagliai and Vignozzi 2003), or from the derivative curve to soil water retention curve, where two or three peaks appear. One peak is characteristic for matrix pores (Sect. 2.1) and one or two peaks for structural pores (Sect. 2.2) (Othmer et al. 1991, Durner 1992).

C. Macropores, or non-capillary pores of such a size, that capillary menisci are not formed across the pore and the shape of air-water interface across the pore is planar. The boundary between micropores and macropores is approximated by the equivalent pore radius 1–1.5 mm. The flow in macropores is described either by a modified Chézy equation or by the kinematic wave equation (Germann and Beven 1985). A more detailed classification of macropores is related to their stability and persistence in time:

C1. Macropores formed by the activity of pedo-edaphon such as decayed roots, earthworm channels etc. Their main characteristic is their high stability and persistence in time.

C2. Fissures and cracks occurring as a consequence of volumetric changes of swelling-shrinking soils. They have planar form and they close when the soil matrix is saturated with water.

C3. Macropores originating due to soil tillage. The depth of their occurrence is limited and they disappear usually in less than one vegetation season. Their persistence depends on the genetic evolution of the soil, meteorological conditions and the type of plants being grown.

The accelerated flux in macropores and structural micropores is usually denoted as preferential flow. With the technique of image analysis it is now possible to characterize soil structure by the quantification of soil pore shape, size, distribution, irregularity, orientation, continuity, etc. in thin sections, prepared from undisturbed soil samples (Bouma et al. 1977, Pagliai et al., 1983, 1984, Pagliai 1988).

The pore size distribution is usually approximated by the lognormal distribution function, but the gamma distribution function was proposed, too (Brutsaert 1966). Pachepsky et al. (1992) and Kosugi (1994, 1999) used the analytical form of the soil hydraulic functions for pore size lognormal distribution and Kutilek (2004) applied this procedure for bi-modal soils distinguishing between the structural and matrix domains of pores.

The pore size distribution is a dynamic property. It is dependent on the water content especially in fine textured soils that are subjected to swelling and shrinkage (Kutilek and Nielsen 1994). Large number of studies exist that show the changes of pore size distribution in Vertisols at different water contents, i.e. when they are water saturated after the rain and the cracks are closed due to saturation and when they are dried and the cracks are open (Schweikle 1982, Kutilek 1983, 1996, Bui et al. 1989). The changes of pore size distribution are strictly correlated with the changes of hydraulic conductivity (Kutilek 1996).

2.2 An Overview on the Soil Micromorphometry

The morphometric technique has the advantage that the measurement and the characterization of pore space can be combined with a visual appreciation of the type and distribution of pores in soil at a particular moment in its dynamic evolution. In soil micromorphological studies the classification of pores according to their size and functional characteristics is restricted to fixed constant boundaries and this is the main difference from the hydraulic approach. Table 1 reports the most frequently used classification scheme of pores proposed by Brewer (1964). Values used by Greenland (1977) are given in Table 2.

The very fine pores less than $0.005\,\mu m$, called "bonding spaces", are critically important in terms of the forces holding domains and aggregates of primary particles together; pores of less than $0.5\,\mu m$ are the "residual pores" for the chemical interactions at the molecular level; pores which have an equivalent pore diameter ranging between 0.5 and $50\,\mu m$ are the "storage pores", i.e. the pores that store water for plants and for micro-organisms; and the pores ranging from 50 to $500\,\mu m$ are those called "transmission pores" in which the movement of water is important for plants, and, moreover, they are the pores needed by feeding roots to grow into. The water content, when pores larger than $50\,\mu m$ have drained, corresponds approximately to the field capacity of the soil. The wilting point commences when most pores larger than approximately $0.5\,\mu m$ are emptied.

Table 1 Morphologic pore size classification according to Brewer (1964)

Class	Subclass	Class limits (Equivalent diameter $\mu m - 10^{-6}\,m$)
Macropores	Coarse	above 5000
	Medium	2000–5000
	Fine	1000–2000
	Very Fine	75–1000
Mesopores		30–75
Micropores		5–30
Ultramicropores		0.1–5
Cryptopores		less than 0.1

Table 2 Classification of soil pores according to their size. Modified from Greenland (1977)

Equivalent diameter μm (10^{-6} m)	Water potential (bar)	Name
<0.005	>−600	Bonding space
0.005–0.5	−600/−6	Residual pores
0.5–50	−6/−0.06	Storage pores
50–500	−0.06/−0.006	Transmission pores
>500	<−0.006	Fissures

Pores larger than 500 μm can have some useful effects on root penetration and water movement (drainage), especially in fine-textured soils. However, a high percentage of this type of pore (above 70−80% of the total porosity) in soils is usually an index of poor soil structure, especially in relation to plant growth. This is because surface cracks, which develop after the rainfall, when the stability of soil aggregates is poor, belong to this size class (Pagliai et al. 1983). Until now, the necessary proportion of large pores for air and water transmission and easy root growth has generally been inadequately defined. In fact, adequate storage pores (0.5−50 μm) as well as adequate transmission pores (50−500 μm) are necessary for plant growth.

Using the image analysis, the shape factors allow division of pores into different shape groups such as, more or less rounded (regular), irregular, and elongated pores (Bouma et al. 1977, Pagliai et al. 1983). Pores of each shape group can be further subdivided into a selected number of size classes according to either the equivalent pore diameter for rounded and irregular pores or the width for elongated pores.

The regular pores are obviously those of a rounded shape and can be separated in two types according to their origin: the spherical pores formed by entrapped air during soil drying and the channels and chambers formed by biological activity (root growth and movement of soil fauna). Their distinction on soil thin sections is very evident, because spherical pores (vesicles, according to Brewer 1964) have very smooth walls, while channels, even though cut in a transversal mode on thin section, have rough walls with deposits of insect excrements or root exudates. The presence of many spherical pores of the first type (vesicles) creates a vesicular structure typical of soils with the evidence of degradation.

The irregular pores are the common soil voids that have irregular walls (vughs, according to the micromorphological terminology of Brewer 1964) and can be isolated as packing voids or interconnected. The presence of these pores produces the typical vughy structure (Bullock et al. 1985). In cultivated soils these pores can be produced by soil tillage implements as suggested by Kutilek (2004).

Two types of elongated pores can be distinguished, i.e., cracks and thin fissures (planes). The former are typical of clay soils with a depleted soil organic matter content and they are visible at the surface when the soil is dry. The thin fissures are the most important, especially from an agronomic point of view. They are the typical elongated transmission pores (Greenland 1977, Pagliai and Vignozzi 2003) being one part of the structural pores according to Kutilek (2004).

Using the image analysis each pore shape group can be further subdivided into a selected number of size classes according to either the equivalent pore diameter for rounded and irregular pores or the width for elongated pores. The equivalent pore diameters are calculated from the area of the regular and irregular pores, while the width of elongated pores is calculated from their area and perimeter data using a quadratic equation because it is assumed that elongated pores are long narrow rectangles (Pagliai et al. 1984).

The development of micromorphological techniques together with image analysis allows the improvement of hydraulic models. For example, Bouma et al. (1977) developed a method based on the preparation of undisturbed soil columns, saturated and then percolated with a 0.1% solution of methylene-blue that is adsorbed by the soil particles on the pore walls followed by the preparation of vertical and horizontal thin sections. Pores are divided into three shape groups as already explained above and then the pore size distribution is determined. For the planar elongated pores the total area, the area of the blue-stained pore walls, and their lengths, and the spatial distribution of the widths and lengths of the pores with blue-stained walls are determined. Particular attention should be paid to the measurement of the width of the necks of elongated pores because the hydraulic conductivity is determined by the necks in the flow system.

Following the above mentioned procedure the hydraulic conductivity (K_{sat}) can be calculated as proposed by Bouma et al. (1979). Further studies of Bouma (1992) confirmed that morphological information on the soil pore system is essential for the realisation of water flux models. The evolution of software for image analysis, that enables the acquisition of precise information about shape, size, continuity, orientation, and arrangement of pores in soil, permits the simplification of the modelling approach.

3 A Case of Study

3.1 Aim of the Study

The aim of this study was to emphasize the importance of the complete characterization of soil porosity, by micromorphological methods, in order to evaluate the hydraulic conductivity in a loam soil, representative of the hilly environment of Italy, cultivated by maize. Such a correlation is fundamental to improve the actual models of water movements in soils.

4 Materials and Methods

4.1 Soil

The soil is located in the field experiments at the Fagna Agricultural Experimental Centre (Scarperia – Firenze) of the Research Institute for Soil Study and Conservation

Table 3 Main physical and chemical characteristics of the soil

Sand (g kg^{-1})	400
Silt (g kg^{-1})	422
Clay (g kg^{-1})	178
CEC (me/100 g)	14.6
pH (1:2.5) H$_2$O	8.1
Organic matter (%)	1.4
CaCO^3 (%)	5.2
Total N (Kjieldahl) (g kg^{-1})	1.1
C/N	7.4

(Firenze, Italy). It is on a loam soil classified as Typic Haplustept (USDA 1999) or Lamellic Calcaric Cambisol (FAO-IUSS-ISRIC 1998). Some major characteristics of the soil are reported in Table 3.

The field experiment was established in 1994 and three replicates of each of three management practices were tested in 50 m × 10 m plots. The tillage treatments were: (1) minimum tillage (harrowing with a disc harrow to a depth of 10 cm); (2) conventional deep tillage (mouldboard ploughing to a depth of 40 cm) and (3) ripper subsoiling to a depth of 50 cm.

The soil had been cultivated with maize since 1970 adopting the same traditional management practices and, since 1980, the fertilisation has been mineral alone without any addition of farmyard manure or other organic materials.

4.2 Soil Porosity Measurements

The pore system was characterised by image analysis on thin sections from undisturbed soil samples to measure pores >50 μm (macroporosity). Six series of six replicate undisturbed samples were collected in the surface layer (0−100 mm) of each plot at the ripening time of the maize in September 2004. Due to the high variability in conventionally tilled plots an additional series of six replicate undisturbed samples were collected in one of these plots. In total 114 thin sections were prepared and each value reported represents the mean of six thin sections.

Samples were dried by acetone replacement of water (Murphy 1986), impregnated with a polyester resin and made into 60 3 70 mm, vertically oriented thin sections of 30 μm thickness (Murphy 1986). IMAGE PRO-PLUS software produced by Media Cybernetics (Silver Spring, MD, USA) calculated pore structure features from digital images of the thin-sections, using the approach described by Pagliai et al. (1984). The analysed image covered 45 3 55 mm of the thin section, avoiding the edges where disruption can occur. Total porosity and pore distribution were measured according to pore shape and size, the instrument being set to measure pores larger than 50 μm. Pore shape was expressed by a shape factor [perimeter^2/ (4π.area)] so that pores could be divided into regular (more or less rounded) (shape factor 122), irregular (shape factor 225) and elongated (shape factor .5). These classes correspond approximately to those used by Bouma et al. (1977). Pores of each shape group were further subdivided into size classes according to either their equivalent pore diameter (regular and irregular pores), or their width (elongated pores) (Pagliai et al. 1983, 1984). Thin sections were also examined using a Zeiss "R POL" microscope at 25 3 magnification to observe soil structure, i.e. to gain a qualitative assessment of the structure.

4.3 Saturated Hydraulic Conductivity

To measure saturated hydraulic conductivity, in areas adjacent to those sampled for thin section preparation, the same series of six undisturbed cores for each plot (57 mm diameter and 95 mm high) were collected in the surface layer (0−100 mm). The samples were slowly saturated and the saturated hydraulic conductivity was measured using the falling-head technique (Klute and Dirksen 1986).

5 Results and Discussion

Previous studies (e.g., Pagliai et al. 2004) reported that conventional ploughing induced the more relevant modification of soil physical properties resulting in damage to soil structure. The negative aspects associated with this management system are the formation of surface crusts. The formation of the crusts and the decrease of porosity, in particular the continuous elongated pores, in the surface layers of conventionally tilled soil, besides a reduction of water movement, may also hamper root growth. Minimum tillage and ripper subsoiling could be a good alternative to conventional ploughing.

The combination image analysis-micromorphological observations on thin section prepared from undisturbed soil samples allows the complete characterization of the soil porous system and the quantification of the above mentioned aspects of soil degradation and can also help to understand and to explain differences in water movement.

Figure 1 shows a linear correlation between porosity, represented by pores larger than 50 μm measured on soil thin sections, and saturated hydraulic conductivity,

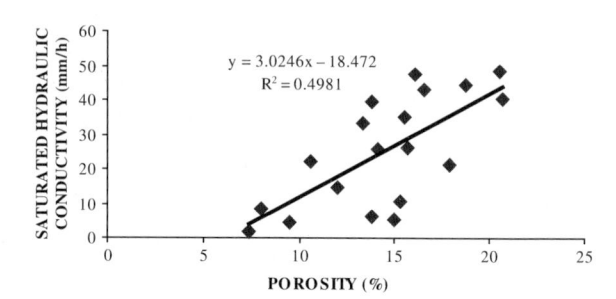

Fig. 1 Correlation between soil porosity, formed by pores larger than 50 μm, and saturated hydraulic conductivity in the surface layer (0–10 cm) of a loam soil (Lamelli-Calcaric Cambisol, according to FAO-IUSS-ISRIC classification, 1998) cropped with maize

$y = 3.0246x - 18.472$
$R^2 = 0.4981$

SATURATED HYDRAULIC CONDUCTIVITY (mm/h)

POROSITY (%)

Fig. 2 Correlation between elongated pores larger than 50 μm, and saturated hydraulic conductivity in the surface layer (0–10 cm) of a loam soil (Lamelli-Calcaric Cambisol, according to FAO-IUSS-ISRIC classification, 1998) cropped with maize

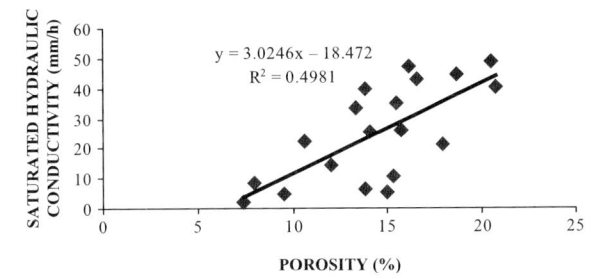

measured on undisturbed cores (57 mm × 95 mm) collected in the surface layer (0−100 mm), in areas adjacent to those sampled for thin section studies.

As already mentioned, the hydrological flux is regulated by elongated and continuous pores, therefore, such a correlation is strictly connected with the proportion of these pores. The correlation increased only when the elongated pores were taken into consideration, as reported in Fig. 2, while it was completely absent with regular and irregular pores. However, in Fig. 2 it is evident how some values diverged from the linear correlation. The microscopic examination of soil thin sections revealed that in the surface layer of samples collected in conventionally tilled plots a surface crust was present and the elongated pores were parallel oriented to the soil

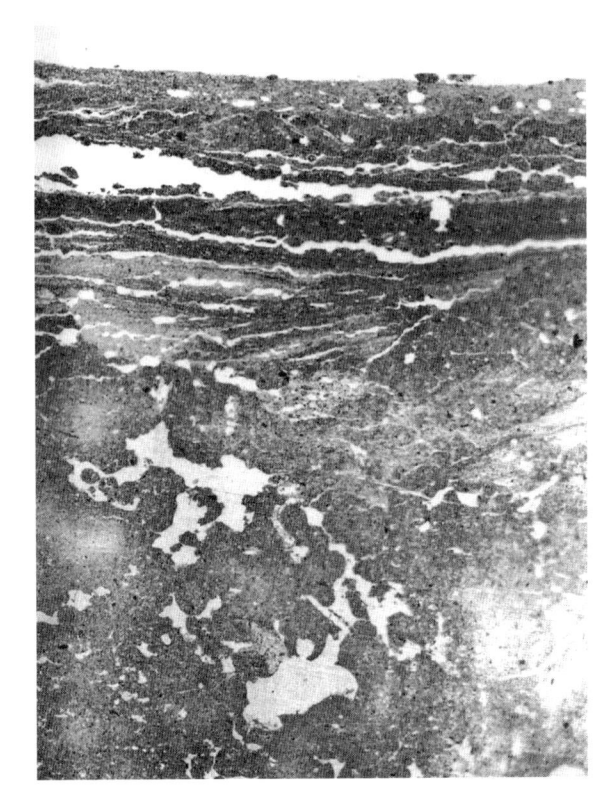

Fig. 3 Microphotograph of a vertically oriented soil thin section prepared from undisturbed samples from the surface layer (0–100 mm) of a loam soil (Lamelli-Calcaric Cambisol, according to FAO-IUSS-ISRIC classification, 1998). Surface crust formation is very evident and the elongated pores are parallel oriented to the soil surface. The white areas represent the pores. Frame length 3 × 5 mm

surface without continuity in a vertical sense (Fig. 3), with a disadvantage for water infiltration (Bui and Mermut 1989). Besides the elongated pores parallel to the soil surface, the indicators of soil degradation, i.e., the regular pores-vesicles-formed by entrapped air during the drying processes that do not conduct water, were also observed in the thin sections.

On the contrary, Fig. 4 shows a subangular blocky structure representatives of samples collected in plots under ripper subsoiling. The soil aggregates are separated by elongated continuous pores (planes), are of different sizes and can be rather porous inside. The walls of these elongated pores are moderately regular which do not perfectly accommodate each other. Therefore, these pores permit water movement even when the soil is wet and fully swollen (Pagliai et al. 1984). In contrast, the very regular elongated pores are flat and smooth pores with accommodating faces, which tend to seal when the soil is wet, thus preventing water movement. From an agronomic point of view, the subangular blocky structure is the best type of soil structure because the continuity of elongated pores allows good water movement and facilitates root growth. Moreover, it is a rather stable soil structure.

The visual appreciation of the differences between Figs. 3 and 4 can be quantified as reported in Fig. 5. This Figure represents the pore shape and size distribution of samples taken in the surface layer (0−100 mm) of conventionally tilled plots, in

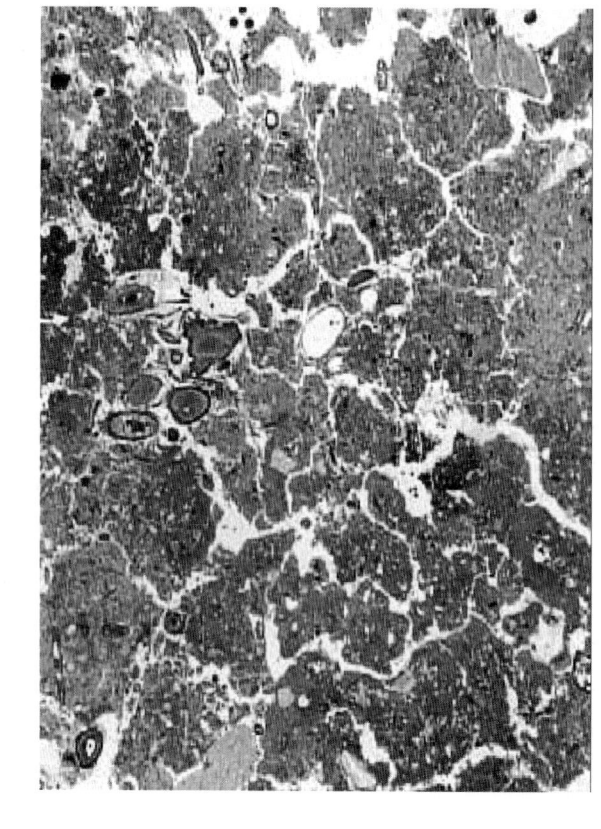

Fig. 4 Microphotograph of a vertically oriented soil thin section prepared from undisturbed samples from the surface layer (0–100 mm) of a loam soil (Lamelli-Calcaric Cambisol, according to FAO-IUSS-ISRIC classification, 1998) cropped with maize, showing an example of subangular blocky structure, where the aggregates are separated and surrounded by elongated continuous pores. The white areas represent the pores. Frame length 3 × 5 mm

Fig. 5 Correlation between elongated continuous pores larger than 50 μm, and saturated hydraulic conductivity in the surface layer (0–10 cm) of a loam soil (Lamelli-Calcaric Cambisol, according to FAO-IUSS-ISRIC classification, 1998) cropped with maize

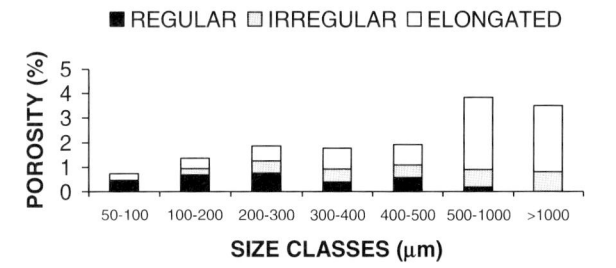

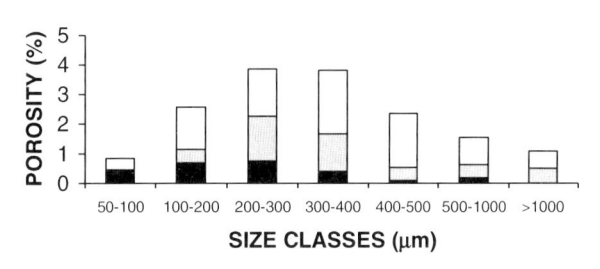

which a surface crust was pronounced, as reported in Fig. 3 and in the surface layer of plots under ripper subsoiling, showing a type of soil structure like those represented in Fig. 4. The total macroporosity was statistically similar (around 15% of area occupied by pores larger than 50 μm per thin section), but the pore shape and size distribution were quite different.

Pore shape and size distribution analysis revealed that the proportion of elongated transmission pores (50−500 μm) was lower in conventionally tilled plots than in plots under ripper subsoiling. In the conventionally tilled plots the great part of elongated pores were larger than 500 μm and generally parallel oriented to the soil surface as shown in Fig. 4. Also the irregular pores ("vughs") larger than 500 μm were higher in these plots. Many of these pores are the macropores originated by the tillage with mouldboard plough. The proportion of rounded pores in the range 50−500 μm was higher in conventionally tilled plots and these pores were represented by vesicles formed by entrapped air during soil drying (Fig. 3), so isolated in the soil matrix and, therefore, not conducting water. On the contrary, the regular pores in samples from plots under ripper subsoiling were mainly represented by channels, particularly root channels, conducting water. The higher proportion of elongated transmission pores (50−500 μm), that are important for water movement and for maintaining a good soil structure (Greenland 1977), associate to the presence of channels in soils under ripper subsoiling explained the higher value of saturated hydraulic conductivity, in comparison with conventionally tilled plots.

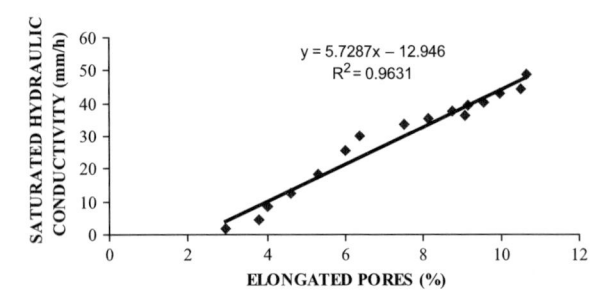

Fig. 6 Correlation between elongated continuous pores larger than 50 μm, and saturated hydraulic conductivity in the surface layer (0–10 cm) of a loam soil cropped with maize

The pore shape and size distribution in the surface layer of the minimum tilled plots showed, approximately, the same trend of that of soils under ripper subsoiling.

When the elongated pores were continuous in a vertical sense, as shown in Fig. 4, the correlation between the elongated pores and hydraulic conductivity was highly significant (Fig. 6).

Combined with image analysis, the use of fractal and fractal fragmentation models can help to characterize the geometry of a porous medium in relation to the transport process (Kutilek and Nielsen 1994). For example, the model of fractal fragmentation leads to a better understanding of the relationships between aggregation, n-modal porosity and soil hydraulic properties.

6 Conclusions

Results of this study indicated that a significant correlation exists between elongated continuous transmission pores and saturated hydraulic conductivity. The shape of the walls of pores plays an important role upon the stability of pores. The formation and existence of the vesicular pores is the main indicator of soil degradation and, combined with the orientation of elongated pores parallel to the soil surface, is the main factor of substantial decrease in hydraulic conductivity.

To preserve a good soil structure, preventing soil degradation, it is important to adopt those management practices that, in the long term, are able to promote a sub-angular blocky structure with a good proportion of elongated transmission pores, like the ripper subsoiling or minimum tillage.

This study also demonstrated the importance of pore characterisation to understand water movements in soils. Despite the information that can be obtained, the micromorphological research of the soil porous system did not proceed beyond the correlation to saturated conductivity. The next step in the future study on relations between soil micromorphology and hydraulic functions should be focused to the unsaturated conductivity function and mainly to the role of elongated pores upon this function in the structural domain.

Physical models of? soil hydraulic functions (soil water retention function and unsaturated hydraulic conductivity function) fitting to a certain type of morphological structure are still missing. The first step in linking soil hydraulics to soil micromorphology was performed when soil hydraulic functions were based upon the log-normal pore size distribution.

Acknowledgments This research was supported by the Czech Grant Agency, GACR 103/05/2143 and the Special Project "SUOLO" (D.M. 359/7303/01, 30 October 2001) of the Italian Ministry of Agriculture.

References

Bouma J (1992) Influence of soil macroporosity on environmental quality. Advances in Agronomy, 46: 1–37.

Bouma J, Jongerius A, Boersma OH, Jager A, Schoonderbeek D, (1977) The function of different types of macropores during saturated flow through four swelling soil horizons. Soil Sci. Soc. Am. J., 41, 945–950.

Bouma J, Jongerius A, Schoonderbeek D (1979) Calculation of saturated hydraulic conductivity of some pedal clay soils using micromorphometric data. Soil Sci. Soc. Am. J., 43, 261–264.

Brewer R (1964) Fabric and Mineral Analysis of Soils. John Wiley, New York, 470 pp.

Brutsaert W (1966) Probability laws for pore size distribution. Soil Sci., 101, 85–92.

Bui EN, Mermut AR, Santos MCD (1989) Microscopic and Ultra Microscopic Porosity of an Oxisol as Determined by Image Analysis and Water Retention. Soil Sci. Soc. Am. J., 53, 661–665.

Bui EN, Mermut AR (1989) Study of Orientation of Planar Voids in Vertisols and Soils with Vertic Properties Soil Sci. Soc. Am. J., 53, 171–178.

Bullock P, Fedoroff N, Jongerius A, Stoops G, Tursina T (1985) Handbook for soil thin section description. Waine Res. Pub., Wolverhampton.

Durner W (1992) Predicting the unsaturated hydraulic conductivity using multi-porosity water retention curves. In: van Genuchten, M.Th., Leij, F.J. and Lund, L.J. (Eds.), Indirect Methods for Estimating the Hydraulic Properties of Unsaturated Soils. University of California, Riverside, CA, USA, pp.185–202.

FAO-IUSS-ISRIC (1998) World Reference Base for Soil Survey. FAO report n° 84.

Germann PF, Beven K (1985) Kinematic wave approximation to infiltration into soils with sorbing macropores. Water Resour. Res., 21, 990–996.

Greenland DJ (1977) Soil damage by intensive arable cultivation: temporary or permanent? Philos. Trans. R. Soc. Lond., 281, 193–208.

Horn R (1994) The effect of aggregation of soils on water, gas and heat transport. In: Schulze, E.D. (Ed.), Flux Control in Biological Systems, Academic Press, 10, pp. 335–361.

Klute A, Dirksen C (1986) Hydraulic conductivity and diffusivity: laboratory methods. In: Klute, A. (Ed.), Methods of Soil Analysis, Part 1, Second Edn. Am. Soc. Agron. Publ., Madison, WI, pp. 687–734.

Kosugi K (1994) Three-parameter lognormal distribution model for soil water retention. Water Resour. Res., 30, 891–901.

Kosugi K (1999) General model for unsaturated hydraulic conductivity for soils with lognormal pore-size distribution. Soil Sci. Soc. Am. J., 63, 270–277.

Kutilek M (1983) Soil physical properties of saline and alkali Vertisols. In: Isotope and Radiation Techniques in Soil Physics and Irrigation Studies. IAEA, Vienna, pp. 179–190.

Kutilek M, Nielsen DR (1994) Soil Hydrology. Catena Verlag, Cremlingen Destedt, Germany.

Kutilek M (1996) Water relations and water management in Vertisols. In: Ahmad, N. and Mermut, A. (Eds.), Vertisosls and Technoloégies for their Management. Elsevier, Amsterdam, pp. 201–230.

Kutilek M (2004) Soil hydraulic properties as related to soil structure. Soil Tillage Res., 79, 175–184.

Kutilek M, Jendele L, Panayiotopolos KP (2006) The influence of uniaxial compression upon pore size distribution in bi-modal soils. Soil Tillage Res., 86, 27–37.

Murphy CP (1986) Thin section preparation of soils and sediments. A B Academic Publishers, Herts, U.K., 149 pp.

Othmer H, Diekkrüger B, Kutilek M (1991) Bimodal porosity and unsaturated hydraulic conductivity. Soil Sci., 152, 139–150.

Pachepsky YA, Mironenko EV, Shcherbakov RA (1992) Prediction and use of soil hydraulic properties. In: van Genuchten, M.Th., Leij, F.J. and Lund, L.J. (Eds.), Indirect Methods for Estimating the Hydraulic Properties of Unsaturated Soils. University of California, Riverside, CA. pp. 203–213.

Pagliai M (1988) Soil porosity aspects. Int. Agrophysics, 4, 215–232.

Pagliai M, La Marca M, Lucamante G (1983) Micromorphometric and micromorphological investigations of a clay loam soil in viticulture under zero and conventional tillage. J. Soil Sci., 34, 391–403.

Pagliai M, La Marca M, Lucamante G, Genovese L (1984) Effects of zero and conventional tillage on the length and irregularity of elongated pores in a clay loam soil under viticulture. Soil Tillage Res., 4: 433–444.

Pagliai M, Vignozzi N (2003) Image analysis and microscopic techniques to characterize soil pore system. In: Blahovec, J. and Kutilek, M. (Eds.), Physical Methods in Agriculture. Kluwe Academic Publishers, London, pp. 13–38.

Pagliai M, Vignozzi N, Pellegrini S (2004) Soil structure and the effect of management practices. Soil and Tillage Res., 79, 131–143.

Schweikle V (1982) Gefügeeigenschaften von Tonböden. Verlag Eugen Ulmer, Stuttgart.

USDA-NRCS (1999) Soil Taxonomy, a basic system of soil classification for making and interpreting soil surveys. 2nd ed. Agriculture handbook N° 436, Washington D.C., 869 pp.

Micromorphology of a Soil Catena in Yucatán: Pedogenesis and Geomorphological Processes in a Tropical Karst Landscape

Sergey Sedov, Elizabeth Solleiro-Rebolledo, Scott L. Fedick, Teresa Pi-Puig, Ernestina Vallejo-Gómez, and María de Lourdes Flores-Delgadillo

Abstract Development of the soil mantle in karst geosystems of the tropics is still poorly understood. We studied a typical soil toposequence formed over limestone in the northeastern Yucatán Peninsula of Mexico, to assess the pedogenetic and geomorphological processes which control soil formation and distribution, as well as to understand their relation to landscape development and their influence in ancient Maya agriculture. The soil cover is dominated by thin Leptic Phaeozems and Rendzic Leptosols in the uplands, and Leptic Calcisols in the wetlands. Upland soils have weathered groundmass containing abundant vermiculitic clay and iron oxides. The combination of thinness and high weathering status is explained by interaction between the intensive pedogenesis and vertical transport of soil material towards karst sinkholes. In wetlands, biochemical secondary calcite precipitation occurs, accompanied by surface accumulation of algal residues (periphyton crust). In the transitional area, a polygenetic profile (Calcisol over Cambisol) was developed, indicating recent advance of wetlands. Because of specific pedogenesis, the upland soils lack many disadvantages of other soils of humid tropics, such as acidity,

---

Sergey Sedov
Departamento de Edafología, Instituto de Geología, UNAM,
e-mail: sergey@geologia.unam.mx

Elizabeth Solleiro-Rebolledo
Departamento de Edafología, Instituto de Geología, UNAM

Scott L. Fedick
Department of Anthropology, University of California, Riverside

Teresa Pi-Puig
Departamento de Geoquímica, Instituto de Geología, UNAM

Ernestina Vallejo-Gómez
Departamento de Edafología, Instituto de Geología, UNAM

María de Lourdes Flores-Delgadillo
Departamento de Edafología, Instituto de Geología, UNAM

low humus content, and poor structure. However, ancient land-use practices had to be adjusted to thin soils, low P availability and soil loss due to karst erosion.

Keywords Pedogenesis · karst erosion · soil toposequence · micromorphology · Maya civilization

1 Introduction

Pedogenesis on calcareous parent materials is known to differ significantly from that of soils developed on silicate minerals. While soils formed on carbonate rocks in Mediterranean and Temperate regions (in particular, Terra Rossa, Terra Fusca, and Rendzinas) have been studied extensively, knowledge about tropical soils derived from carbonates is rather limited. Singer (1988), who studied soil diversity in South-Eastern China, describes shallow profiles formed on limestone, with low weathering status and mollic epipedon, quite different from profound, red, deeply weathered "latheritic" soils on shales and Pleistocene sediments. However other studies in the tropical islands of the Pacific and Caribbean demonstrated vast variety of soil types formed on limestones. Besides relatively "young" Entisols and Mollisols, those in an advanced stage of development like Alfisols and Ultisols were reported (Bruce 1983). Even deeply weathered ferrallitic profiles are known to form on limetones, e.g. kaolinitic "Red Ferrallitic" or "Latosolic" soils in Cuba (Ortega Sastriques 1984) and Oxisols associated with the karstic bauxites in Jamaica (Scholten and Andriesse 1986). The factors and processes controlling this high soil diversity as well as the origin of the carbonate-free soil materials are still poorly understood.

Ahmad and Jones (1969) favor the hypothesis of "residual" origin of soils on the Pleistocene limestones of Barbados, however Borg and Banner (1996) point to domination of "non-regolith eolian components" in their parent material, relying on neodymium and strontium isotopic compositions and Sm/Nd ratio. Muhs et al. (1990) state that transatlantic transport of Saharan dust, with minor input of volcanic ash deposition from local volcanoes provided the substrate for soils on calcareous rocks on Carribean islands. Brückner and Schnütgen (1995) demonstrated volcanic origin of soil parent material on the coral reef terraces of New Guinea.

Soil age is thought to be an important, but not unique factor controlling diversity. Study of a chronosequence of reef terraces on Barbados, with a reliable time scale provided, has shown that some pedogenic properties, in particular weathering status, depend upon landform age (Muhs 2001). At the same time, soil diversity within a single terrace was found to be rather high, especially regarding solum, and A and B horizon thicknesses.

The Yucatán Peninsula in southeastern Mexico presents a perfect area for studying tropical pedogenesis on calcareous parent materials. This extensive, flat, slightly uplifted limestone platform is occupied by humid to subhumid tropical forest ecosystems, in many cases with little contemporary human impact. The surface and subsurface karst forms are abundant and variable, indicating intensive and recent karstification processes.

The study of soil mantle formation and evolution in Yucatán also has an important archaeological significance. Soils of this region were once involved in long-term agroecosystems of ancient Maya civilization, perhaps the most developed prehispanic society of the Americas, which flourished for over a thousand years before declining for still unknown reasons around A.D. 900 (Sharer 1994). The high population of the region could only have been supported through intensive agriculture and resource management (Culbert and Rice 1990, Fedick 1996).

Many studies of ancient Maya intensive agricultural practices have now been conducted (Harrison and Turner 1978, Pohl 1985, Flannery 1982, Fedick 1996, White 1999), however, there has been relatively little research on how soil fertility was maintained under such intensive cultivation systems. Assessment of soil properties and soil fertility characteristics in Yucatán was conducted by Aguilera-Herrera (1963) and by Hernández et al. (1985) in relation to traditional slash-and burn agriculture. Recently Bautista-Zúñiga et al. (2003, 2004) studied soil diversity in Yucatán state, linking it to microrelief development and building up soil evolutionary schemes.

In this work we performed a pedogenetic study of most wide spread soils in the northeastern Yucatán Peninsula, based mainly in micromorphology, in order to (1) understand how different soil formation and geomorphological processes interact in generating the soil mantle in a tropical karst landscape and (2) characterize this mantle as a resource for productive and sustainable agroecosystems of the ancient Maya.

2 Environmental Conditions

We studied the soil cover distribution along a topographic transect (toposequence) in an area of northern Quintana Roo, Mexico, where ongoing archaeological survey by the Yalahau Regional Human Ecology Project is documenting dense Maya settlement dating primarily to portions of the Late Preclassic and Early Classic periods (ca. 100 B.C.–A.D. 350), with a widespread reoccupation of the region dating to the Late Postclassic period (A.D. 1250–1520) (see Fedick et al. 2000). The soil study focuses on land within and around the El Eden Ecological Reserve, which is located 38 km WNW of Cancun (21°3'N, 87°11' W) (Fig. 1).

Geologically the northern peninsula consists of Cretaceous age uplifted fossiliferous limestone. These rocks, mainly included in the Yucatán Evaporite Formation (López-Ramos 1975), reach a depth of 3,500 m and rest over a Paleozoic basement. Overlying evaporates, a sequence of limestones, sandstones and evaporitic deposits of Paleocene-Eocene age are found. The landscape is characterized by a relatively smooth plain broken through numerous karstic sinkholes (known locally as "cenotes"). The altitude of the platform averages 25–35 m (Lugo-Hupb et al. 1992). Pool (1980) described (using Yucatán Mayan terminology) four geomorphic elements of the relief: heights or ho´lu´um, plains or kan kab, depressions or k´op, and cenotes or dzoonot. Climate is hot and humid. Mean annual temperature is 25°C and the

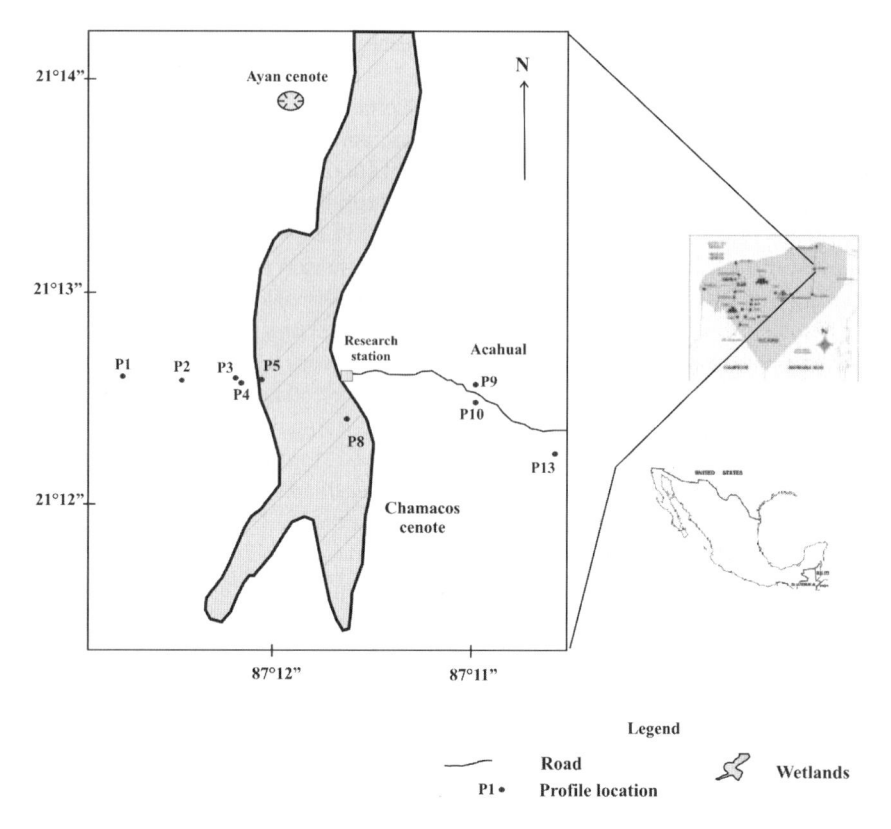

Fig. 1 Location of study area and soil profiles

annual precipitation is 1126 mm. The highest temperature, during May, is 28°C. The maximum precipitation is concentrated in summer, from May to October, reaching 915 mm (García 1988).

Natural vegetation is a moist, medium-high broadleaf, evergreen forest (Miranda 1959); the areas recently (less than 10 years ago) affected by fire are occupied by pioneer associations dominated by ferns. Tropical savanna vegetation dominates the wetlands.

3 Materials and Methods

This study was conducted along an E-W transect in order to identify soil variety and distribution across the landscape. This transect crossed the main geomorphological elements and related vegetation types: uplands with tropical forests, small karst pits, and extensive flat depressions with wetlands. Additional profiles were studied and sampled in the areas, affected by forest fires.

Soils were classified according to the World Reference Base (WRB 1998). Bulk samples for physical and chemical analyses, as well as undisturbed samples for thin sections, were collected from genetic horizons. Soil colors were determined according to the Munsell Soil Color Charts (1975). Thin sections were prepared from undisturbed soil samples impregnated with resin, and studied under a petrographic microscope with descriptions following the terminology of Bullock et al. (1985). Selected chemical properties, including organic carbon, available P, and pH were evaluated according to guidelines of the United States Department of Agriculture (USDA 1993).

We quantitatively separated particle size fractions: sand (2–0.05 mm) by sieving, silt (0.05–0.002 mm) and clay by gravity sedimentation, with previous destruction of micro-aggregation agents: carbonates (treatment with diluted HCl with pH 3) humus (10% H_2O_2) and iron oxides (DCB). We separated particle size fractions: sand (2–0.05 mm) by sieving, silt (0.05–0.002 mm) and clay by gravity sedimentation, with previous destruction of microaggregation agents: carbonates (treatment with diluted HCl with pH 3), humus (10% H_2O_2) and iron oxides (DCB). X-ray diffraction patterns of the oriented specimens of clay fraction were obtained with a diffractometer (Philips 1130/96) utilizing Cu K_radiation following pretreatments of air drying, heating to 400°C, and saturation with ethylene glycol. To identify neoformed crystalline components in the soil affected by forest fire, we studied soil materials under electron microscope equipped with an EDX microprobe, and obtained powder X-ray diffracto-gramms from the material enriched with the neoformed component to be identified.

4 Results

4.1 Soil Morphology

In uplands, the soil cover is discontinuous; rock outcrops are rather frequent and limestone looks strongly weathered, with dissolution pits and etched zones. In all geomorphological positions, including wetlands, we observed typical results of karstic alteration, forming fractures and sinkholes. In the uplands (Ho´lu´um) under mainly undisturbed forest, thin soils dominate (less than 30 cm depth), with an Ah/AC/C profile. They have a dark reddish brown (5YR 3/3, dry) Ah horizon, 10–15 cm thick. This horizon has a very well developed granular structure, and a high root density. The AC horizon is reddish brown (5YR 4/4, dry), with a fine granular structure, and is softer than the previous horizon.

Weathered limestone fragments are common; however, there is no reaction to HCl, indicating that soil matrix is free of carbonates. The C horizon is composed of 90% fractured limestone, infilled by soil material. It was classified as Leptic Calcaric Phaeo-zem (profiles 1 and 2). Other upland soils are thinner and more stony (profile 3). The Ah horizon (less than 10 cm thick), with a less developed granular structure, rests directly on limestone. In this case, they were classified as Rendzic Leptosols.

We associate both upland soil units with the traditional concept of "Rendzina" meaning shallow humus-rich soils on calcareous parent material. Similar soils in areas recently affected by fire and currently occupied by ferns, have signs of disturbance and turbation (profiles 9 and 10). Dark grey humus-rich material, typical for Rendzina Ah horizon, are mixed with red mottlesb and charcoal Powdery white material, showing clear reaction with HCl, is distributed unevenly over all above mentioned morphological elements. Another soil type was identified in a similar upland position, to the north-west of the transect (Fig. 1) where it occupies a small area about 10 m². In this location, soils are thicker (35 cm), and more clayey (profile 13). The Ah horizon (15 cm thick) is dark reddish brown (5YR 3/3, dry) and has a granular structure with a high root density and moderate contents of organic carbon. The B horizon, 15 cm thick, has a subangular blocky structure, is very hard when dry, and very clayey. The C horizon is rich in weathered limestone fragments, although the complete profile has no reaction to HCl. It was classified as Chromic Cambisol.

Within the relatively flat upland terrain, small depressions can be found (k´op in Yucatec Mayan). Within these depressions, soils are somewhat deeper and clayey. They show an A/Bg/C profile (profile 4). The Ah horizon is very dark grayish brown (10YR 3/2, dry), 11 cm thick, with a granular to subangular blocky structure and the highest content of organic carbon (15.3%). The Bg horizon is grayish brown (10YR 5/2, dry), clayey, with thin clay cutans over aggregates. Bioturbation is strong and the presence of Fe concretions is common. Some peds break into angular blocks with 60° angles, indicating vertic characteristics, although not very well expressed.

Weathered limestone fragments are found at a depth of 24 cm, but horizons have no reaction to HCl. The soil was classified as Leptic Gleyic Phaeozem. In the wetlands, periphyton appears, forming a soft patchy crust on the surface. Periphyton is a complex community of algae and other aquatic microorganisms that form in seasonal bodies of water such as wetlands (see Wetzel 1983) and are deposited on the land surface in the dry period. Surprisingly, soils in some portions of the wetlands are also thin (profile 8), with 1 cm of a very dark gray (10YR 3/1, dry) horizon and rich, highly decomposed organic material. Below we found a rather thin (less than 10 cm thick) gray Bk horizon, consisting mainly of fine secondary calcite and containing mollusk shells. It was classified as Leptic Calcisol. In the wet season this soil is flooded, however in the dry period (when the description was made) it was saturated with water.

In intermediate positions, between uplands and the lower areas of wetland, where forest is restricted to gentle slopes, periphyton appears again. Under the periphyton, a 5-cm pale Bk horizon was found with light gray color when dry (10YR 7/1) that becomes brown when humid (7.5YR 4/2). It is sandy and rich in carbonates. Below, a well developed soil profile was found. The buried A horizon is dark brown (7.5YR 3/4, dry), 10 cm thick, with a well expressed granular structure and with abundant charcoal fragments. Its organic carbon content is low (1.04%), and it does not react with HCl. The Bw horizon, 10 cm thick, is similar in color (7.5YR 3/4, dry), but coarser in texture (more silty), with abundant content of carbonates. The underlying C horizon has many weathered limestone fragments. This soil can be regarded as polygenetic, because two different cycles of pedogenesis and sedimentation were recognized.

4.2 Soil Micromorphology

Groundmass of Rendzic Leptosols (profile 3) is mainly formed by dark-colored (brown or black) isotropic organic material. Under higher magnification, a large part is dominated by fine detritus. Larger fragments of plant tissues in different stages of decomposition are also abundant. This material is organized in rather coarse subangular blocks or in small crumbs of irregular shape. Carbonate particles of various sizes are frequent in the groundmass. On the contrary, all kinds of silicate materials, and in particular clay, are scarce. We found only few clay-rich rounded aggregates with lighter yellow-brown color and speckled to granostri-ated b-fabric, embedded in the dark isotropic predominantly organic groundmass (Fig. 2a,b). Despite the macromorphological similarity with Rendzic Leptosols, the micromorphological properties of the two Leptic Calcaric Phaeozem profiles (1 and 2) are quite different. Only very few large (>2 mm) fragments of limestone with biogenic structure (numerous mollusk shells) could be found. Groundmass is completely free of carbonates and very rich in fine clay material with scarce silt-size quartz grains embedded in it. Clay has stipple-speckled b-fabric and rather low birefrigence, groundmass has a reddish-brown color; due to staining with iron oxides. Organic materials are present in different forms. Fragments of plant tissues are rather frequent, with signs of decomposition by microbes or mesofauna.

Mottles of dark gray-brown organic pigment and tiny black specks are abundant within groundmass. The material is remarkably well aggregated: granules of various size and small subangular blocks, all of coprogenic origin, are partly welded, to form the porous aggregates of higher order, and all together produce spongy fabric with a dense void system (Fig. 2c). Soils formed in the small karst depression have a quite specific set of micromorphological features of the Bg horizon. It is compact, and its structure is dom-inated by subangular to angular blocks separated by fissures. The color of soil material is brown-yellow, with dark- brown areas corresponding to frequent ferruginous nodules and mottles (Fig. 2d). The groundmass is rich in clay, having porostriated, granostriated (thick envelopes of oriented clay round the ferruginous nodules) and mosaic-speckled b-fabric. Stress cutans are rather frequent along open and closed fissures (Fig. 2e). A few deformed clay illuvial pedofeatures was found only in this horizon. Few charcoal fragments are embedded in the groundmass (Fig. 2f).

The micromorphology of the Leptic Calcisol in the wetland implies a quite distinct mode of pedogenesis. The major part of groundmass consists of carbonates – in con-trast to highland Phaeozems and Cambisols, in which groundmass is dominated by silicate clay. However, these carbonates are represented not by fragmented parent limestone, as in Rendzic Leptosol, but exclusively by neoformed micritic calcite. Micrite (with admixture of microsparite) forms blocky and granular aggregates, often welded to build up spongy fabric with high porosity. Frequently the micrite accumulations have specific biogenic microstructure: ooids (Fig. 3a), channeled clusters, related to the activity of algae (Fig. 3b) (Vogt 1987). Granules are often col-ored with dark-brown organic pigment – the presence of which lowers the material birefrigence. Partly decomposed plant tissues, as well as shell fragments (Fig. 3c) of terrestrial mollusks, are common. Soils on the gentle slope towards the wetland

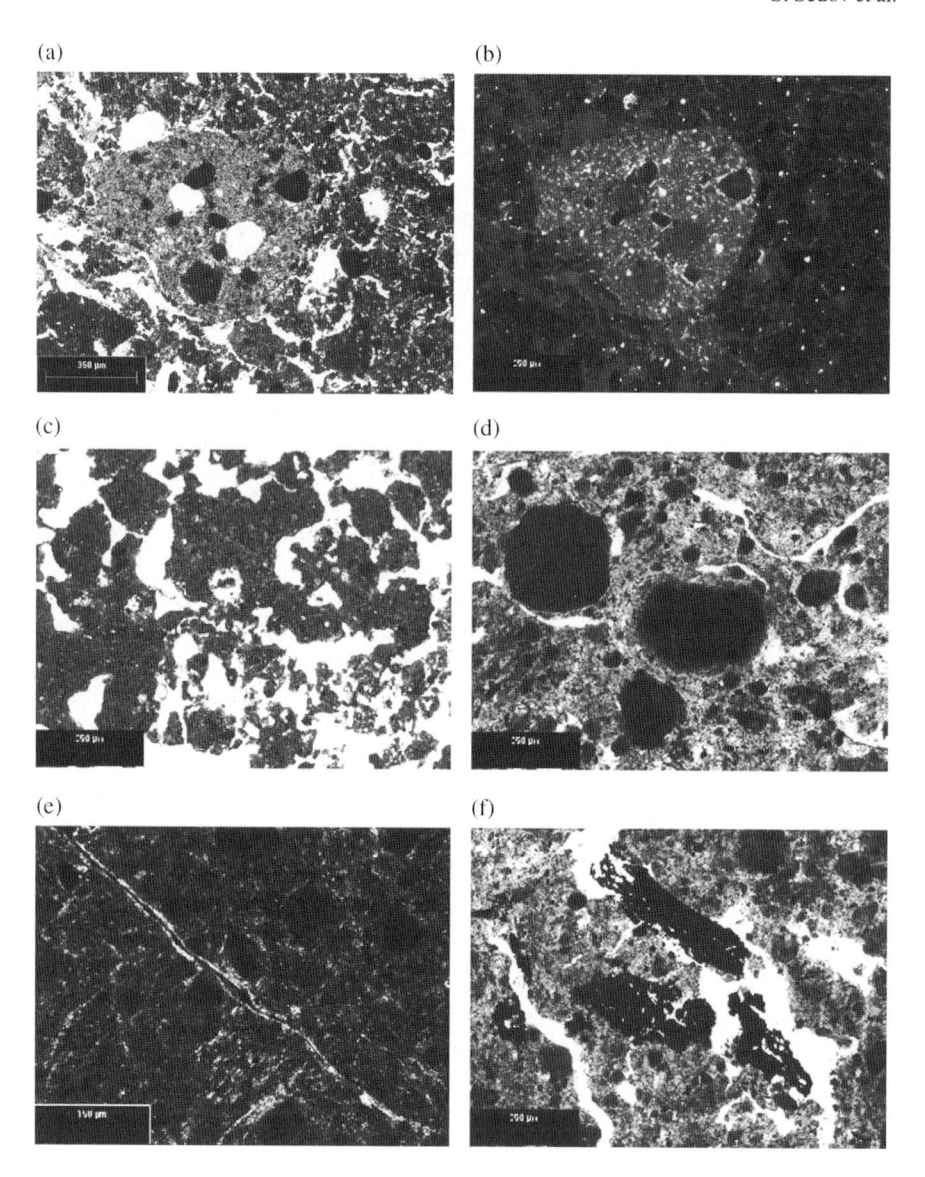

Fig. 2 Micromorphology of the upland soils from El Eden toposequence. PPL – plane polarized light, XPL – crossed polarizers. (**a**) Clay aggregate, Ah horizon of Rendzic Leptosol, PPL, (**b**) Same as (**a**), XPL, (**c**) Complex structure (subangular blocks and granules) and spongy fabric of Ah horizon of Leptic Phaeozem, PPL, (**d**) Compact material with fey fissures, frequent ferruginous nodules; Bg horizon of Gleyic Phaeozem, PPL, (**e**) Stress cutan, Bg horizon of Gleyic Phaeozem, XPL, (**f**) Charcoal fragments, Bg horizon of Gleyic Phaeozem, PPL

(profile 5) demonstrate a peculiar adn contradictory combination of micromorphological properties. According to earlier observations, some of the soils (profile 5) are typical for the wetlands, and others, for the upland soils. The surface carbonate

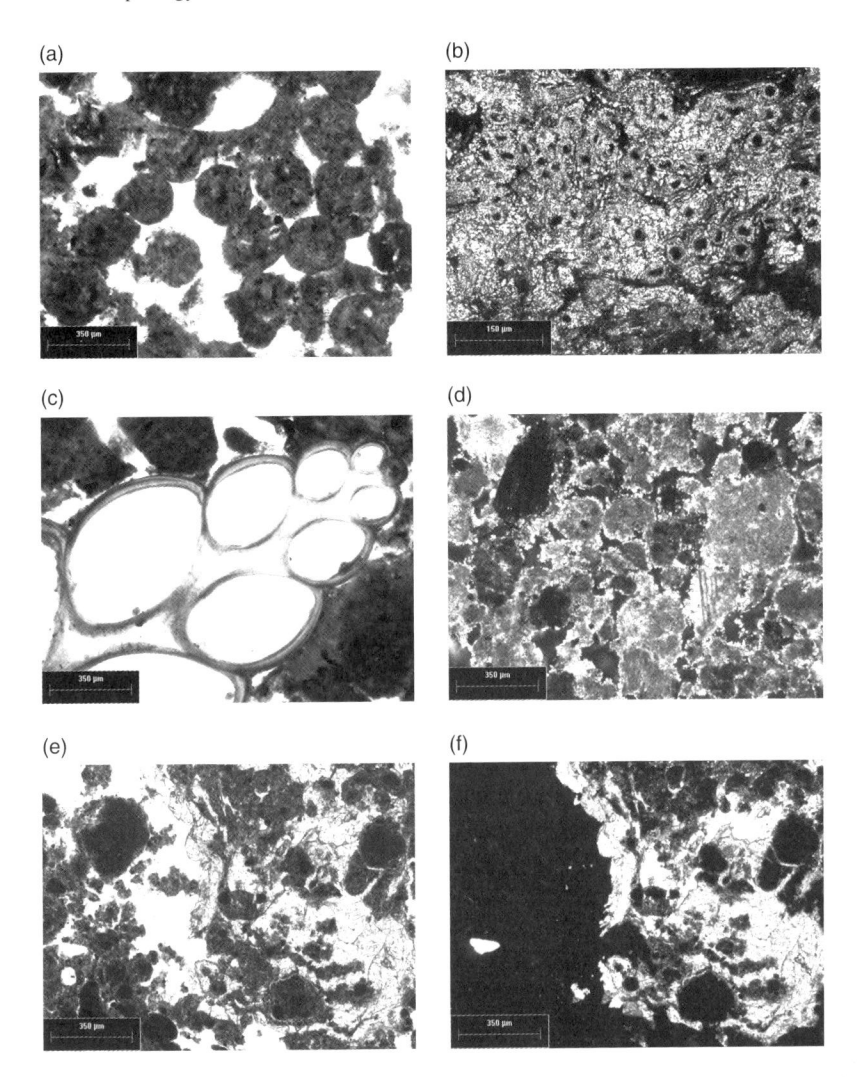

Fig. 3 Micromorphology of the wetland soils from El Eden toposequence. PPL – plane polarized light, XPL – crossed polarizers. (**a**) Micritic ooids, Bk horizon, Leptic Calcisol, PPL, (**b**) Channeled micritic aggregate, Bk horizon, Leptic Calcisol, XPL, (**c**) Mollusk shell, Bk horizon, Leptic Calcisol, PPL, (**d**) Granular structure, micritic groundmass with high birefringence in Bk horizon of upper Calcisol member of polygenetic profile, XPL, (**e**) Microarea cemented by coarse crystalline calcite (*center to right part*), Ah horizon of the lower Cambisol member of polygenetic profile, PPL, (**f**) Same as (**e**), XPL, note high birefrigence of calcitic cement

horizon is an analogue of laguna Calcisols; its material is dominated by neoformed biogenic calcite (3d). However, the underlying Ah and Bw horizons resemble the Phaeozems of elevated landsurfaces. Their groundmass is free of carbonates and contains abundant silicate clay. The Ah horizon has a granular structure and is enriched with organic material, although it also has some clusters of biogenic micrite that may be derived from the overlying carbonate horizon. The micromorphology of Bw is somewhat similar to the Bg horizon of the Phaeozem in karst depression, being rather compact, clayey, having blocky structure and containing frequent ferruginous nodules. A peculiar feature observed from both Ah and Bw horizons are microzones, cemented by large crystals of calcite, filling all pore spaces and "trapping" the soil aggregates (Fig.3e and 3f).

The upland soil, recently affected by fire, is heterogeneous: dark grey-brown humus-rich clods are mixed with red aggregates, enriched in iron-clay fine material; large black charcoal particles, often with the cell wall structure preserved, are very frequent (Fig. 4a). A peculiar feature of this material is the presence of loose concentrations of silt-size crystalline particles with high interference colors, and of irregular or rhombohedral shape in the pores and on the surfaces of the charcoal particles. Some of the rhomboids, being observed under crossed polarizers at higher magnifications, demonstrate non-uniform, undulating extintion patterns, indicating that they are not monocrystals but rather aggregates of minor crystals (Fig. 4b).

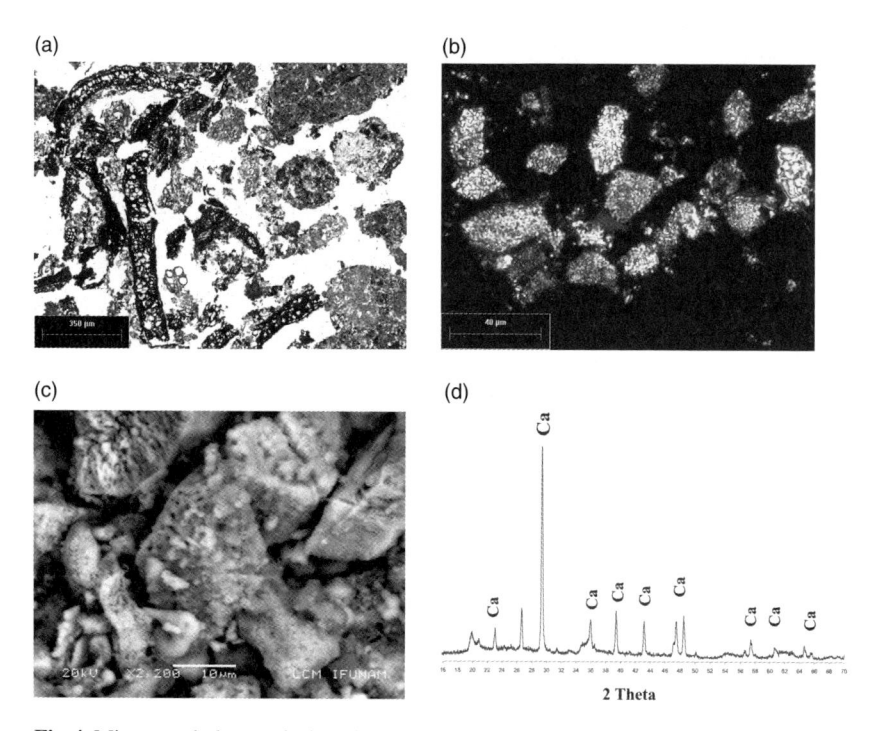

Fig. 4 Micromorphology and mineralogy of Rendzina, affected by forest fire. (**a**) Micromorphology of A horizon: clay and humus-clay aggregates (*right*), charcoal particles (*left*), PPL, (**b**) Neoformed carbonate grains in A horizon of irregular and rhombohedral shape. Note irregular extinction pattern, XPL, (**c**) Scanning electron microscopy of neoformed carbonates: note aggregative morphology of the rhombohedral grain, (**d**) Powder X-ray diffraction pattern of the soil material, enriched with neoformed grains; Maxima corresponding to calcite are marked with Ca

Under SEM we observed the rhombohedral grains which consist of tiny (about 1 μ) compactly packed particles that confirms the aggregative nature of at least part of the rhomboids (Fig. 4c), that could be micritic pseudomorphs after large whewellite (calcium oxalate) crystals (Canti 2003). About 10 particles of this shape that were subjected to EDX microprobe analysis showed the presence of Ca as the dominant element. An X-ray diffractogram from the material enriched with white powdery particles indicated calcite as the only crystalline non-silicate phase (Fig. 4d). This calcitic material is not derived from limestone fragmentation since its morphology is completely different from micritic or biomorph limestone carbonates.

4.3 Texture, Chemical Properties and Clay Mineralogy

All soils have pH values close to 7, somewhat lower in the upland Phaeozems and Cambisol, and a bit higher in the Leptosol and lowland Calcisols (Table 1). A horizons of the upland Rendzinas show pH values close to neutral, and with high values of humus (above 8% of organic carbon) and clay content (more than 70% in the Phaeozems about 50% in Leptosol). More than 15% of organic carbon (!) are found in the Leptic Gleyic Phaeozem in the small karst depression. The wetland, water-saturated with Calcisol, has much less humus than the upland Rendzinas (about 5%). This is a rare case of a soil toposequence in which well drained upland

Table 1 Selected properties of the studied soils

Horizon	Depth (cm)	Color dry	Color wet	OC* (%)	pH	P** ppm
Profile 1, Leptic Calcaric Phaeozem						
Ah	0–15	5YR 3/3	5YR 3/2	8.93	6.8	7.31
AC	15–27			6.10	6.8	3.50
Profile 2, Leptic Calcaric Phaeozem						
Ah	0–13	5YR 3/3	5YR 3/2	9.15	6.8	10.10
Profile 3, Rendzic Leptosol						
Ah	0–14	10YR 3/2	10YR 2/2	11.15	7.1	7.35
Profile 4, Leptic Gleyic Phaeozem						
Ah	0–11	10YR 3/2	10YR 2/2	15.3	6.3	7.31
Bg	11–24			0.97	6.9	7.31
Profile 13, Chromic Cambisol						
A	5–20	5YR 3/3	5YR 3/2	5.05	6.5	2.24
B	20–35			2.69	6.5	1.12
Profile 5, Polygenetic profile						
Bk	0–5	7.5YR 7/1	7.5YR 4/2	2.59	7.5	7.31
Ah	5–14	7.5YR 3/4	7.5YR 3/2	1.04	7.6	3.50
Bw	14–24	7.5YR 3/4	7.5YR 3/2	0.68	7.6	3.50
Profile 8, Leptic Calcisol						
Bk	1–3			5.20	7.5	3.01

*Organic carbon.

** Available phosphorus

members demonstrate higher accumulation of organic materials, than hydromorphic lowland members. Clay content of wetland soils is also lower.

X-ray diffraction patterns of all upland soils and lower subprofile of a two-phase profile in the forest-wetland transition are rather uniform. They show high and sharp 14 Å and 7 Å maxima, staying unchanged after glycolation; in heated specimens, 14 Å peak shifts to 12–13 Å peak with a shoulder to larger angles and 7 Å maximum disappears (Fig. 5). We suppose that this set of peaks is diagnostic of dioctahedral vermiculite. Incomplete contraction after heating (to 12–13 Å and not to 10 Å) points to formation of a fragmental additional Al-hydroxide interlayer between 3 layer

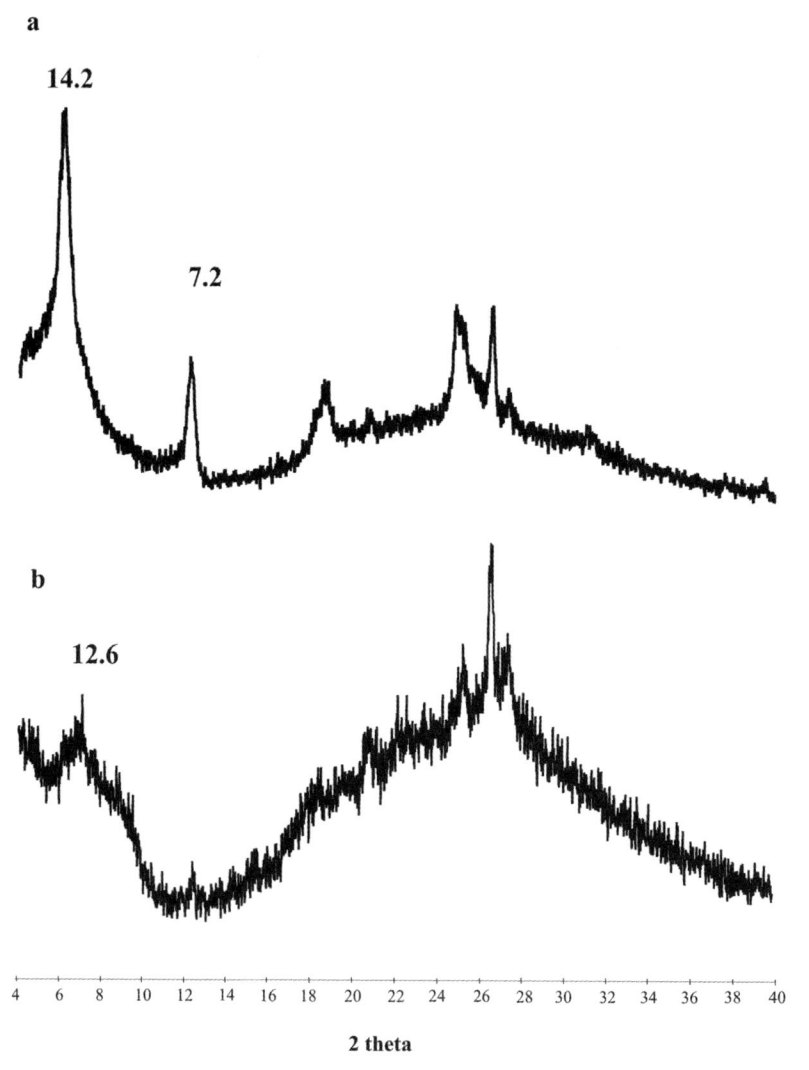

Fig. 5 X-ray diffractogram from the clay fraction of the Leptic Calcaric Phaeozem Ah horizon. Above: specimen saturated with ethylene glycol, below: specimen heated to 400°C

units (Dixon and Weed 1989). The 7 Å peak is likely a second order maximum of vermiculite, rather than a first order of kaolinite. The reason is that it disappears after heating to 400°C – temperature too low for destruction of kaolinite lattice, but enough for vermiculite contraction and transformation of its diffraction pattern.

No clay minerals were identified in the clay fraction of the Bk horizon of the wetland Leptic Calcisol. Only weak peaks of calcite are present, indicating, that this mineral, forming major part of the fine materialpartly remains after the HCl pretreatment.

5 Discussion

5.1 Interplay of Pedogenesis and Karst Erosion – the Reason of Pedodiversity in North-Eastern Yucatán

It is clear from first glance that soils formed on limestone in the northern Yucatán Peninsula differ greatly from the "central image" of humid tropical pedogenesis, perceived as thick, deeply weathered Ferralsols or soils with illuviation of kaolinitic clay (Acrisols-Lixisols). Leptic Phaeozems are shallow, discontinuous and often stony. Their pH values are neutral to slightly acid (Table 1), indicating incomplete leaching. They give an impression of very young, underdeveloped profiles. Their groundmasss shows the presence of primary carbonates and weatherable minerals and is dominated by clay and iron oxides, with few resistant quartz grains.

We still do not have enough data to detect the origin of the precursors of these components. At least, a volcanic source is unprobable. We found neither rests of primary volcanic minerals, nor allophanes or halloysite – typical weathering products of pyroclastic deposits. Further speculations are limited by the lack of knowledge about composition of non- carbonate components of limestone. The results of Aguilera-Herrera (1963) and Bautista-Zúñiga (2003) (both propped the residual genesis of the Yucatán soils) are few and contradictory. The former states that the limestones of Yucatán contain smectites, whereas the latter detected halloysite.

However, independent from the primary origin of the groundmass material (from limestone weathering or allochtonous sediment, as discussed above), properties clearly indicate the advanced weathering stage, which is not expected for a primitive soil. We further speculate that this property resulted from mineral transformation in situ and not from the deposition of pre-weathered sediment. Our reasoning is related to the geomorphological distribution pattern of weathering products. The most altered soil materials were found on elevated land surfaces which provide a leaching environment, and not in lower wetlands, which present better conditions for accumulation of any kind of sediment but less opportunities for advanced in situ mineral alteration, because of poor internal drainage.

Having rather high weathering status, the upland soils however do not reach the most advanced "ferrallitic" weathering stage. Their clay material is dominated by 2:1 mineral – Al-hydroxide interlayered vermiculite known to be an intermediate

product of weathering in humid environments, resulting from incongruent dissolution of illites and chlorites (Dixon and Weed 1989). They do not contain Al
hydroxides and kaolinite – the "end members" of the weathering products sequence,
characteristic for Ferralic diagnostic horizon. This differentiates the soil mantle of
the Yucatán from that of the neighbour Carribean islands (Jamaica, Cuba) where
Ferrallitic soils are frequently found on limestones.

Contrary to undisturbed upland soil, we found abundant neoformed calcite in the
Rendzina, recently affected by forest fire. Components of similar composition and
morphology are described in the burned plant materials (Canti 2003, Courty et al.
1989) This observation points to the discontinuity of the process of carbonate leaching in the upland soils. It is interrupted by the periods of carbonate addition with
plant ash produced by forest fires, the latter becoming more frequent due to present
and past human activities.

We arrived at what seems to be a contradiction between high weathering status
of soil materials from one side, and soil thinness, uneven distribution, and abrupt
contact with the limestone, from the other. The interaction of pedogenetic and geomorphological processes is supposed to be responsible for such a situation. In general, we believe that considerable soil redeposition occurs in geomorphologically
unstable karst environments of the northern Yucatán Peninsula. However, our studies of the soil catenary sequence at El Edén revealed no signs of lateral sheet erosion, natural or man-induced, which would transport fine material from the uplands
to the wetlands. In the latter we have not found thick pedosediments. Wetland soils
are often thinner than upland counterpart. The evidence, even more convincing, is
that the groundmass composition of the Calcisols in wetland is very specific and
cannot result from redeposition of materials, originating either from upland Phaeozems or limestone outcrops.

Calcisols lack the silicate clay, which could serve as a marker of Phaeozem-derived
pedosediments. Their groundmass consists mostly of carbonates; however, those are
not the products of limestone fragmentation, but rather the result of biochemical
calcite precipitation from dissolved carbonates, mostly by algae. We relate this process directly to the functioning of algal community developed in wetlands during the
rainy season. This conclusion agrees with results reported by Leyden et al. (1996).
These authors also found low erosion rates and absence of anthropogenic lateral soil
transport enhancement. The erosional dynamics in the northern Maya Lowlands
seems to differ remarkably from that in the southern Maya Lowlands, where various
authors report the strongest evidences of intensive lateral soil transport corresponding
to ancient Maya occupation periods. Evidence for erosion in the southern lowlands
includes thick layers of redeposited soil on lower land surfaces (Beach 1998a; Beach
et al. 2003, Dunning et al. 2002) as well as enhanced sediment accumulation in lakes
(Rice 1996). Both phenomena are lacking in our study area. We agree with Leyden
et al. (1996) that this difference isrelated to the character of relief.

However, although we did not find evidence of long distance lateral soil
redeposition we assume that short-distance, mostly vertical soil transport is quite
intensive towards cracks, hollows, and small but steep depressions of karst origin.

We found a lot of such forms, empty as well as partly filled with fine pedosediments in all parts of our toposequence. Adjacent to these karstic forms, limestone outcrops are often free of soil. The Gleyic Phaeozem in the small karst depression has enhanced thickness, due to addition of redeposited soil from above, as indicated by the presence of charcoal fragments, even in its lowest Bg horizon. The higher soil thickness is accompanied by the manifestation of redoximorphic processes (frequent ferruginous nodules) indicating that these locations receive both additional fine material and moisture. Only here we found profiles with vertic properties, which are reported to be much more common in depressions of the southern Maya Lowlands (Beach et al. 2003) and the signs of the incipient clay illuviation.

Since many karst sinkholes are connected to a vast underground fissure and solution system, part of the soil material is probably transported deeper, contributing to cave and cenote sediments. In fact, each profile on the land surface is the result of a balance between pedogenesis and soil loss due to vertical transportation towards karst sinkholes. Rate differences between these two processes cause the variety of in situ formed soils, from the thick red Chromic Cambisols to the thinnest Rendzic Leptosols, and finally bare rock. It should be noted that the groundmass of Rendzic Leptosols, although dominated by plant detritus and primary carbonates, still contains some clay-rich aggregates – most probably the last remnants of an earlier continuously weathered horizon, such as those in the neighboring Phaeozems. We think that these Leptosols are not young, but rather a result of degradation of a previous well developed profile, due to unbalanced soil loss caused by enhanced karst erosion. High intensity of karstification processes in Yucatán could be also be responsible for the absence of deeply weathered ferrallitic soils in this region. Similar "balance" models for soil cover development in karst regions were proposed for other natural zones such as boreal forests (Goryachkin and Shavrina 1997) and Mediterranean environments (Atalay 1997).

5.2 Two-Storeyed Soil in the Wetland Periphery: An Indicator of Recent Environmental Change

A soil profile formed in the marginal part of the wetland, close to the upland forest has signs of the overlapping of two different pedogenetic events. The upper unit, similar to wetland Calcisol, is superimposed on a Cambisol, which is typical for upland soil formation. We interpret this as a result of the extension of wetland area, affected by the regular floods towards the territories earlier occupied by the upland forest ecosystem. We also attribute partial cementation of the Cambisol groundmass (originally carbonate-free) by large crystals of secondary calcite to the same landscape dynamics, which caused periodical saturation of a soil with surface water, from which calcium carbonate could precipitate. The timing of this environmental change is unknown and we have no instrumental or archaeological dating for either of the two soil units. The driving force of this dynamic is also uncertain and could imply climatic change as well as groundwater table rising, related to sea level fluctuations.

The data on the stratigraphy of paleosol-sedimentary sequences of the southern Maya Lowlands give some hints for the preliminary correlation and interpretation of the Calcisol-Cambisol. In a number of papers where such data are presented (Pohl and Bloom 1996, Pope and Pohl 1996, Voorhies 1996), the following stratification regularity is described: well developed humus-rich paleosols are overlain by pale "gray clay," or "gray silt," often with carbonates and gypsum, sometimes accompanied by layers of freshwater mollusk shells.

The lower paleosols indicate a phase of relative geomorphological stability, while the overlying layers "are the result of natural deposition representing both alluviation and colluviation in floodplains and associated backswamps, accentuated by deposition of calcium carbonate and gypsum from ground and surface water" (Pohl and Bloom 1996). We suppose that the Cambisol unit (profile 5) could be associated to organic-rich paleosols and the Calcisol, the overlying carbonate-rich sediments. If this association is correct, then the two-storeyed profile of the wetland periphery reflects landscape processes which took place within the ancient Maya occupation period.

5.3 Pedogenetic Properties in Relation to Ancient Land Use

Fray Diego de Landa, a Franciscan missionary who arrived to Yucatán soon after the Spanish occupation and left informative descriptions about various aspects of Maya life wrote: "Yucatán is the country with the least earth that I have seen... it is marvelous that the fertility of this land is so greaton top of and between the stones" (Tozzer 1941, p. 186; original 1566). In this passage he refers to the extraordinary fertility of thin calcareous soils of Yucatán (Leptic Phaeozems and Rendzic Leptosols) which, from the contemporary standpoint, seem to be so unsuited to agriculture. In fact, these soils lack the majority of disadvantages common to the usual, thick but extremely leached, humid, tropical soils. The characteristics of Yucatán soil are developed by the unique combination of weathered groundmass, rich in silicate clay of 2:1 type and iron oxides, and limestone located quite near the surface. The permanent dissolution of the latter (evidenced by various corrosion forms) gives sufficient dissolved calcium carbonate to provide pH values close to neutral and to avoid the problems of typical tropical soils, related to excessive acidity and Al toxicity. It also provides the stabilization and accumulation of dark disperse humus.

Already Kubiëna (1970) in his classic study of soil development on limestone in the Vienna Woods underlined that formation of the Mull Rendzina with abundant dark colloidal humus is conditioned by accumulation of sufficient quantities of clay components in the topsoil above calcarious rock. In the studied soils, humus, vermiculitic clay, and iron oxides are the agents of stable aggregate formation which provide favorable physical properties, regarding soil moisture availability, aeration, and root penetration. There is no tendency to compaction and hardening, that is usual for cultivated tropical Acrisols and Lixisols (Van Wambeke 1992). These pedogenetic features could also be responsible for relatively rapid recovery of calcareous soils in humid tropics of Mexico during fallow period, recently reported by Mendoza-Vega and Messing (2005).

We agree with Beach (1998b) that the major soil limitations for agriculture are its thinness and uneven surface distribution, as well as very low available P contents (Table 1), showing deficiency in this nutrient. An important way to match cultivation to such soil conditions was planting in natural karst features filled with soil material (Kepecs and Boucher 1996) and fertilization of the upland soils, for which purpose P-rich periphyton evidently was used, being collected and transported from the wetlands (Fedick et al. 2000, Palacios et al. 2003).

Acknowledgements This research has been funded by UC MEXUS – CONACYT collaborative grant program 2002 ("Reconstructing ancient Maya sustainable intensive agriculture: a multidisciplinary evaluation of the periphyton fertilizer hypothesis"), DGAPA, project IX121204 "Suelos de paisajes cársticos de Yucatán como la base de la agricultura antígua de los Mayas" and CONACYT project 58948 "Cubierta edáfica de Yucatán: génesis, memoria y funcionamiento en los geosistemas kársticos". We thank Juan Castillo-Rivero and James Rotenberg for their valuable guidance during field research in El Edén, as well as J. Gama-Castro, T. Méndez and E. Jiménez for technical assistance.

References

Aguilera-Herrera N (1963) Los recursos naturales del Sureste y su aprovechamiento: Suelos. Chapingo. Revista de la Escuela Nacional de Agricultura. Epoca II, 3(11–12): 1–54.

Ahmad N, Jones RL (1969) Genesis, chemical properties and mineralogy of limestone-derived soils, Barbados, West Indies. Tropical Agriculture, 46: 1–15

Atalay I (1997) Red Mediterranean soils in some karstic regions of Taurus mountains, Turkey. Catena, 28: 247–260.

Bautista-Zúñiga F, Jiménez-Osornio J, Navarro-Alberto J, Manu A, Lozano R (2003) Microrelieve y color del suelo como propiedades de diagnóstico en Leptosoles cársticos. Terra, 21(1): 1–11.

Bautista-Zuñiga F, Estrada-Medina H, Jiménez-Osornio J, González-Iturbe JA (2004) Relación entre el relieve y unidades de suelo en zonas cársticas de Yucatán. Terra Latinoamericana, 22(3): 243–254.

Beach T (1998a) Soil catenas, tropical deforestation, and ancient and contemporary soil erosion in the Petén, Guatemala. Physical Geography, 19: 378–405.

Beach T (1998b) Soil constraints on Northwest Yucatán, México: Pedoarchaeology and Maya subsistence at Chunchucmil. Geoarchaeology, 13: 759–791.

Beach T, Dunning N, Luzzadder-Beach S, Scarborough V (2003) Depression soils in the lowland tropics of Northwestern Belize: Anthropogenic and natural origins. In: Gómez-Pompa A, Allen M., Fedick S, Jiménez-Orsonio J. (eds) The lowland Maya Area: Three Millennia at the Human-Wildland Interface. Haworth Press, Binghampton, NY.

Borg LE, Banner JL (1996) Neodymium and strontium isotopic constraints on soil sources in Barbados, West Indies. Geochimica et Cosmochimica Acta, 60(21): 4193–4206.

Bruce JG (1983) Patterns and classification by Soil Taxonomy of the soils of the Southern Cook Islands. Geoderma, 31(4): 301–323.

Brückner H, Schnütgen A (1995) Soils on coral reef tracts – the example of Huon Peninsula, New Guinea. Z. Geomorph. NF, Suppl.-Bd.99, Berlin, Stuttgart, pp. 1–15.

Bullock P, Fedoroff N, Jongerius A, Stoops G, Tursina T, Babel U (1985) Handbook for Soil Thin Section Description. Waine Research Publications, Wolverhampton, U.K.

Canti MG (2003) Aspects of the chemical and microscopic characteristics of plant ashes found in archaeological soils. Catena, 54: 339–361.

Courty MA, Goldberg P, Macphail RI (1989) Soils and micromorphology in Archaeology. Cambridge University Press, Cambridge.

Culbert TP, Rice DS (eds) (1990) Prehistoric Population History in the Maya Lowlands. University of New Mexico Press, Albuquerque.

Dixon JB, Weed SB (eds) (1989) Minerals in soil environments. Madison: SSSA.

Dunning NP, Luzzadder-Beach S, Beach T, Jones JG, Scarborough V, Culbert TP (2002) Arising from the Bajos: The evolution of a neotropical landscape and the rise of Maya civilization. Annals of the Association of American Geographers, 92: 267–283.

Fedick SL (ed) (1996) The Managed Mosaic: Ancient Maya Agriculture and Resource Use. University of Utah Press, Salt Lake City.

Fedick S., Morrison BL, Andersen BJ, Boucher S, Ceja Acosta J, Mathews JP (2000) Wetland manipulation in the Yalahau region of the northern Maya lowlands. Journal of Field Archaeology, 27: 131–152.

Flannery KV (ed) (1982) Ancient Maya Subsistence: Studies in Memory of Dennis E. Puleston. Academic Press, New York.

García E (1988) Modificaciones al sistema de clasificación climática de Köpen (special edition made by author).

Goryachkin SV, Shavrina EV (1997) Evolution and dynamics of soil-geomorphic systems in karst landscapes of the European North. Eurasian Soil Science, 30: 1045–1055.

Harrison PD, Turner II BL (eds) (1978) Pre-Hispanic Maya Agriculture. University of New Mexico Press, Albuquerque.

Hernández XE, Bello BE, Levy TS (1985) La roza-tumba-quema en Yucatán. In: La milpa en Yucatán: un sistema de producción agrícola tradicional. Colegio de Postgraduados, Montecillo, México.

Kepecs S, Boucher S (1996) The pre-Hispanic cultivation of rejolladas and stone-lands: New evidence from Northeast Yucatán. In: Fedick SL (ed) The Managed Mosaic. Ancient Maya Agriculture and Resource Use. University of Utah Press, Salt Lake City.

Kubiëna WL (1970) Micromorphological Features of Soil Geography. Rutgers University Press, New Brunswick, New Jersey.

Leyden BW, Brenner M, Whitmore T, Curtis JH, Piperno DR, Dahling BH (1996) A record of long- and short-term climatic variation from Northwest Yucatán: Cenote San José Chulchacá. In: Fedick, SL (ed) The Managed Mosaic: Ancient Maya Agriculture and Resource Use. University of Utah Press, Salt Lake City.

López-Ramos A (1975) Geología General y de México. 1ª. Edición, México, tomo III.

Lugo-Hubp J, Aceves-Quezada J, Espinasa-Pereña R (1992) Rasgos geomorfológicos mayores de la península de Yucatán. Revista del Instituto de Geografía, UNAM, 10: 143–150.

Mendoza-Vega J, Messing I (2005) The influence of land use and fallow period on the properties of two calcareous soils in the humid tropics of southern Mexico. Catena, 60: 279–292.

Miranda F (1959) Estudios acerca de la vegetación. In: Beltrán E. (ed) Los recursos naturales del sureste y su aprovechamiento, 2, Instituto Mexicano de Recursos Naturales Renovables, México, pp. 251–271.

Muhs DR (2001) Evolution of Soils on Quaternary Reef Terraces of Barbados, West Indies. Quaternary Research, 56: 66–78.

Muhs DR, Bush CA, Stewart KC, Rowland TR, Crittenden RC (1990) Geochemical evidence of Saharan dust parent material for soils developed on Quaternary limestones of Caribbean and western Atlantic islands. Quaternary Research, 33(2): 157–177.

Munsell Soil Color Charts (1975) Macbeth Division of Kollmorgen Corporation, Baltimore, Maryland.

Ortega Sastriques F (1984) El humus de los suelos de Cuba. II. Suelos automórficos sobra calizas duras. Academia de Ciencias de Cuba, Ciencias de la Agricultura, 21: 91–103.

Palacios S, Anaya AL, González-Velásquez E, Huerta-Arcos L, Gómez-Pompa A (2003) Periphyton as potencial biofertilizer in intensive agriculture of the ancient Maya. In: Gómez-Pompa A, Allen M, Fedick SL, Jiménez-Orsonio JJ (eds) The Lowland Maya Area: Three Millennia at the Human-Wildland Interface. Binghampton, NY, Haworth Press.

Pohl M (ed) (1985) Prehistoric Lowland Maya Environment and Subsistence Economy. Papers of the Peabody Museum of Archaeology and Ethnology, Vol. 77. Harvard University, Cambridge.

Pohl M, Bloom P (1996) Prehistoric Maya farming in the wetlands of Northern Belize: More data from Albion Island and beyond. In: Fedick SL (ed) The Managed Mosaic: Ancient Maya Agriculture and Resource Use. University of Utah Press, Salt Lake City.

Pool L (1980) El estudio de los suelos calcimórficos con relación a la producción maicera. In: Hernández XE, Padilla PO (eds) Seminario sobre producción agrícola en Yucatán. Gob. Del Estado de Yucatán,-SARH-SPP-CP Mérida, Yucatán, México.

Pope KO, Pohl MD (1996) Formation of ancient Maya wetland fields: natural and anthropogenic processes. In: Fedick SL (ed) The Managed Mosaic: Ancient Maya Agriculture and Resource Use. University of Utah Press, Salt Lake City.

Rice DS (1996) Paleolimnological analysis in the Central Petén, Guatemala. In: Fedick SL (ed) The Managed Mosaic: Ancient Maya Agriculture and Resource Use. University of Utah Press, Salt Lake City.

Scholten JJ, Andriesse W (1986) Morphology, genesis and classification of three soils over limestone, Jamaica. Geoderma, 39(1): 1–40.

Sharer RJ (1994) The Ancient Maya. Fifth edition. Stanford University Press, Stanford.

Singer A (1988) Properties and genesis of some soils of Guanxi Province, China. Geoderma, 43 (2–3): 117–130.

Tozzer AM (1941) Landa's Relación de las Cosas de Yucatán: A Translation. Papers of the Peabody Museum of American Archaeology and Ethnology, Vol. XVIII. Cambridge, Harvard University.

USDA (1993) The Soil Survey Manual. Handbook No. 18. United States Department of Agriculture, Washington.

Van Wambeke A (1992) Soils of the Tropics: Properties and Appraisal. McGraw-Hill, New York.

Vogt T (1987) Quelques microstructures de croutes calcaires quaternaires d'Afrique du Nord. In: Fedoroff N, Bresson ML, Courty MA (eds), Micromorphologie des sols.

Voorhies B (1996) The transformation from foraging to farming in Lowland Mesoamerica. In: Fedick SL (ed) The Managed Mosaic: Ancient Maya Agriculture and Resource Use. University of Utah Press, Salt Lake City.

Wetzel RG (ed) (1983) Periphyton of Freshwater Ecosystems. Dr. W. Junk Publishers, The Hague.

White CD (ed) (1999). Reconstructing Ancient Maya Diet. University of Utah Press, Salt Lake City.

WRB (1998) World Reference Base for Soil Resources. World Soil Resources Reports 84. Food and Agricultural Organization of the United Nations, Rome.

Soil Evolution Along a Toposequence on Glacial and Periglacial Materials in the Pyrenees Range

Jaume Boixadera, Montserrat Antúnez, and Rosa Maria Poch

Abstract Data from eight selected soils located in the La Cerdanya basin (Catalan Pyrenees, Spain) are presented in this study in order to elucidate soil forming processes throughout the Quaternary on glacial and periglacial deposits, at altitudes between 1130 and 2390 m. The soils are classified as Umbrisols, Podzols, Arenosols, Regosols, and Luvisols (FAO/ISRIC/IUSS 2006) and Dystrocryepts, Dystrudepts, Fragiorthods, Udalfs (Pale-, Fragi- and Haplo-) and Ustorthents (Soil Survey Staff 1999). The present moisture and temperature regimes are ustic and mesic in the low altitude soils and udic and frigid/cryic in the high altitude soils. The parent materials of the soils are silicate rock end deposits and consist of lateral moraine tills, slope deposits affected by present-day periglacial processes and fluvio-glacial materials at the bottom of the sequence. The soil forming processes in the area relate to the climate changes during the Pleistocene and Holocene, lithology, altitude, and local hydrological conditions. They consist of in-situ deep weathering of granites, organic matter accumulation, fragipan and clay formations, illuviation, and podzolization. The soils have morphological evidences of the past permafrost of the glacial periods, such as the silt and fine sand cappings, lamellar Bt horizons, vesicular structures, and abundant redoximorphic features, which are inherited from moister conditions, probably during the wet interglacial periods. The clay minerals of the soils are mainly illitic, with some kaolinite and smectite in the valley bottom. Chlorite is the dominant

Jaume Boixadera
Department of Environment and Soil Sciences, University of Lleida Av. Rovira Roure 191, 25198 Lleida, Catalonia; Secció d'Avaluació de Recursos Agraris, DARP, Generalitat de Catalunya

Montserrat Antúnez
Department of Environment and Soil Sciences, University of Lleida Av. Rovira Roure 191, 25198 Lleida, Catalonia

Rosa Maria Poch
Department of Environment and Soil Sciences, University of Lleida Av. Rovira Roure 191, 25198 Lleida, Catalonia, e-mail: rosa.poch@macs.udl.cat

S. Kapur et al. (eds.), *New Trends in Soil Micromorphology*,
© Springer-Verlag Berlin Heidelberg 2008

clay mineral at higher altitudes (due to mica weathering), whereas hydromica, kaolinite, and halloysite were found in the lower altitudes, due to feldspar weathering. The high mineral weathering and clay illuviation in soils of the lower altitudes indicate that the degree of erosion of the upslope soils have probably been very intense. On upslope soils, the accumulation of organic matter, with clay minerals formation and podzolisation have lead to the development of umbric, cambic, and spodic horizons, whose formation are probably still active at present.

Keywords Soil formation · podzolisation · clay illuviation · permafrost · soil micromorphology · mountain soils · pyrenees

1 Introduction

Pedogenesis in mountain environments has received little attention compared with that of soils from agricultural regions. The identification of the soil forming processes during the Quaternary could help to assess the present soil quality and the use and management. Some studies deal with soil formation in relation to organic matter and biological activity (Cassagne et al. 2000, 2004), podzolisation (Bech et al. 1981, Remaury et al. 1999), soil morphology and evolution at regional scale (Martí and Badia 1995, Badia and Martí 1999). The types and distribution of vegetation, the evolution of soil climates during the Quaternary, and the changes in land use makes it difficult to distinguish the past and present soil forming processes. Our objective was to identify soil forming processes occurring along a 4 km transect on the northern side of La Cerdanya Valley, in the Catalan Pyrenees and classify the soils using micromorphology.

2 Physiographic Setting

The Cerdanya valley (40 km long, 10 km wide) is a graben formed during the Upper Miocene running parallel to the Pyrenean range (Fig. 1).

The valley bottom is at an altitude of 1000 m and is crossed by the Segre River. The mean altitudes of the southern and northern flanks of the valley are between 2500 and 2800 m. The mean annual rainfall at the valley bottom is between 600 and 700 mm, but increases with the altitude, reaching more than 1200 mm in the highest points. The mean annual temperature is about 10°C at 1000 m, decreasing with altitude at a rate of −0.9°C/100 m. The estimated soil water and temperature regimes using Soil Survey Staff (1999) according to these gradients are presented in Table 1 for the profiles studied, the udic moisture regime occurs above 1500 m, and frigid/cryic above 1700 m.

The geological formations along the flanks are metamorphic Paleozoic rocks with granitic intrusions in the north, and metamorphic and sedimentary (limestone) in the

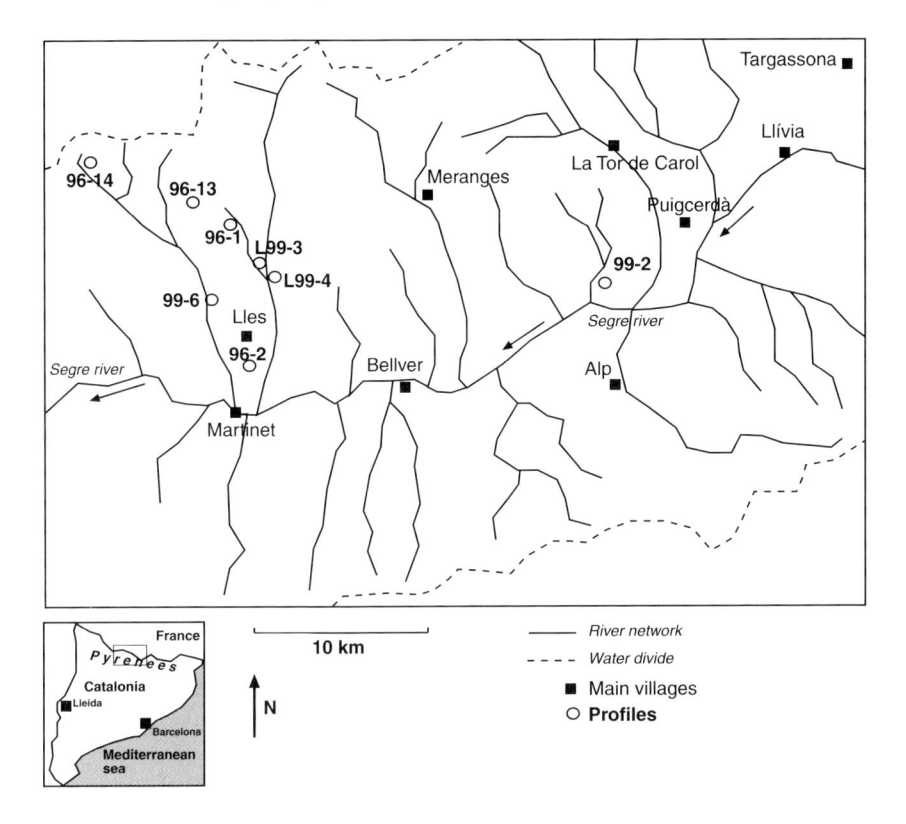

Fig. 1 Location of the studied area

south. The valley bottom is filled with a thick layer of sedimentary rocks deposited during the Upper Miocene and Pliocene (Roca 1986). During the Quaternary, particularly the recent Pleistocene, glacial, periglacial, and fluvial processes have created a mosaic of tills, slope formations and fluvial terraces covering older materials (Gómez-Ortiz 1987). Glacial lakes appear at altitudes above 2100 m. Glacial tills are found along lateral or bottom moraines, covering platforms or forming frontal moraines near the valley bottom. There is a great variety of superficial slope formations, such as the coarse clast-supported materials and *grèzes litées* in higher altitudes, as well as thick colluvia of granitic sands, which may have been inherited from the milder periods of the Miocene. These warmer and moister climates induced the weathering of granites that were later transported and deposited at the valley bottom. Some of the deposits which were not eroded by ice have remained at mid altitudes along the Northern flank of the valley (Balasch 2003).

The vegetation is distributed according to three altitudinal stages, namely the alpine, subalpine and montane. The alpine stage, above 2200 m, corresponds to the alpine meadows. The vegetation comprises the *Festucion airoidis* communities on siliceous soils, or the *Festucion gautieri* (=*Festucion scopariae*) on neutral to basic soils. The subalpine stage is found between 1600 (2000) and 2200 m, and consists of black pine

Table 1 The environment of the studied profiles

Profile	Location	Altitude (m)	Aspect	Parent material	Landform	Moisture and temperature regimes (Soil Survey Staff, 1999)	Vegetation/ Land use
99-2	X: 410250 Y: 469700	1130	S	Fluvio-glacial deposits	Pleistocene (early) terrace	ustic, mesic	Old agricultural field
96-2	X: 039195 Y:469310	1330	SSW	In situ weathered granites	Terraced slope (10–20%)	ustic, mesic	Pasture with scattered shrubs
L99-4	X: 039350 Y: 469680	1500	W	Till, mainly schists, gneiss and granite, covered by 1 m-thick slope deposits.	Lateral moraine	udic, mesic	Forest (*Pinus sylvestris, Buxus sp*)
L99-3	X: 039305 Y: 469840	1550	N	Granitic till with granite boulders covered by slope deposits.	Lateral moraine	udic, mesic	Forest (*Pinus sylvestris, Buxus sp*)
99-6	X: 038870 Y: 469765	1840	N	Granitic till covered by slope colluvium	Lateral moraine	udic, frigid	Forest (*Pinus sylvestris, Vaccinium myrtillus, Genista purgans*)
96-1	X: 39155 Y: 469912	2050	NE	Quartzitic till	Slope (10%)	udic, frigid	Forest (*Pinus uncinata*) and heath vegetation
96-13	X: 38700 Y: 469989	2130	SSW	Till deposits in the slope over in situ weathered granite	Slope (30%)	udic, frigid	Forest (*Pinus uncinata, Vaccinium myrtillus, Genista purgans*)
96-14	X: 38472 Y: 474900	2390	NW	Glacial till	Glacial cirque	udic, cryic	Pasture alpine meadows

forests, either on non-calcareous (*Rhododendro-Pinetum uncinatae*) or on calcareous soils (*Pulsatillo-Pinetum uncinatae*). The most important shrub communities consist of heather vegetation in cold and humid areas (*Saxifrago-Rhododendretum*) and *Genisto-Arctostaphyletum* along the sun-facing slopes. Some areas are occupied by secondary pastures of the *Nardion* when developed on acidic soils, and of *Festucion gautieri, Mesobromion* and *Primulion* associations on basic or neutral soils.

The montane stage, under 1600 (2000) m, is occupied by oak (*Buxo-Quercetum pubescentis*) or pine (*Pinus sylvestris*) forests. Shrub communities are found on the steeper slopes, and belong to the *Mesobromion, Xerobromion, Aphyllanthion* and *Ononidion* associations, depending on pH and soil moisture regimes, whereas the intensive pasturelands (*Arrhenatherion*) occupy flatter areas. On valley bottoms and on soils with high watertables some remains of the riparian forests are found, namely as the *Saponario-Salicetum, Alnetum catalaunicae* and *Brachypodio-Fraxinetum* (Pedrol 2003).

Although the natural timber line is at 2200 m, meadows and pastures are noticeable at 1700 m altitudes, due to forest clearing for cattle-grazing, since the ancient times. Anthropic influence is more noticeable on soils and landforms towards lower altitudes (below 1500 m), due to cultivation. Slope modifications by terracing on steep slopes and charcoal holes are the remains of an intense human pressure of the past along with forest clearing. Most of these terraces are abandoned and recolonized by shrubs today, while few are still used as irrigated pastures. The anthropic pressure which was the major cause of soil erosion was more effective between 1000 and 1500 m altitudes, manifested by the predominant bare soils and rock outcrops of today. The valley bottom is occupied by numerous dairy farms together with the cultivation of forage crops, cereals and some fruit trees (apples and pears).

3 Materials and Methods

Eight soil profiles developed on glacial and periglacial deposits at altitudes between 1130 and 2390 m were selected for morphological, physical, chemical, mineralogical, and micromorphological analyses. The soils are found on siliceous and granitic materials at a sequence of lateral moraine tills, and slope deposits affected by present-day periglacial processes and fluvio-glacial materials (Fig. 2) (Table 1).

Profiles were described and soil samples were collected for physical and chemical analyses according to the Manual of Soils of the Spanish Ministry of Agriculture – SINEDARES- (CBDSA 1983). Undisturbed samples were also collected at selected profiles and horizons for micromorphological and mineralogical analyses.

Vertical thin sections, of 13 cm length and 5.5 cm width, were made from air-dried undisturbed soil blocks according to Murphy (1986) and studied following the guidelines of Stoops (2003). The physical and chemical analyses of the soils were conducted according to Porta et al. (1986), MAPA (1993) and Page et al. (1982). Particle-size distribution was determined by the pipette method, after the removal of organic matter using H_2O_2 and dispersed with Na-hexametaphosphate. Cation-exchange capacity

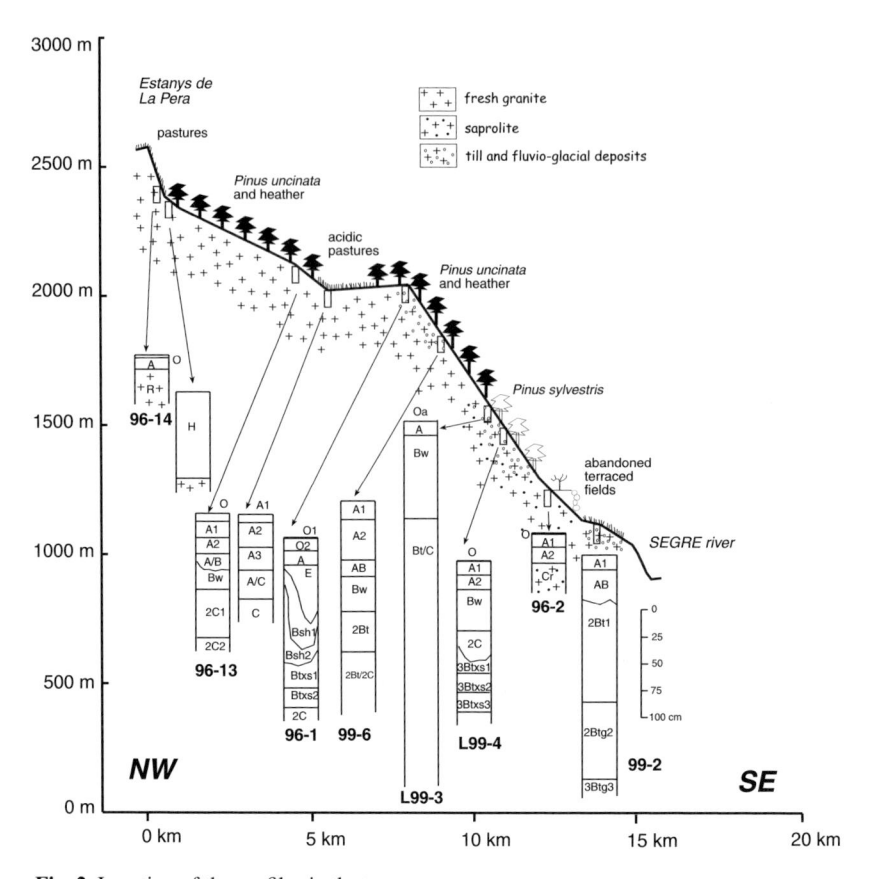

Fig. 2 Location of the profiles in the toposequence

and exchangeable cations were determined by the NaOAc – NH$_4$OAc (pH 7) and 1M KCl extractions. Organic carbon, total nitrogen and extractable phosphorous were determined following Walkley-Black, Kjeldahl and the Olsen methods.

The mineralogical identification of the clay fraction was made by X-ray diffraction using a Philips X'Pert diffractometer (graphite monochromated CuKα radiation). XRD patterns were obtained using: (a) random powder, and the following oriented suspended clay slides: (b) air dried, (c) ethylene glycol solvated, (d) heated to 300°C for 3 h, and (e) heated to 500°C for 3 h.

4 Results and Discussion

4.1 Soil Forming Processes

Several processes have been identified in the studied soils, according to the data shown in Tables 2 and 3.

Table 2 Morphological properties of the soils studied

Profile	Horizon	Depth (cm)	Munsell colour (moist)	Structure	Consistence	Mottling	Coarse fragments (%)	Bulk density Mg/m^3
99-2	A$_1$	0–10	10R6/4	1 f gr	–	abs.	16–35	–
	AB	10–34	10R5/4	1 m sbk	–	abs.	35–70	–
	2B$_t$	34–97	–	–	–	abs.	70–90	–
	2B$_{tg1}$	97–160	–	–	–	>50 %; 2.5 YR 2/1	70–90	–
	3B$_{tg2}$	160–220	–	–	–	>50 %; 2.5 YR 2/1	70–90	–
96-2	A$_1$	0–10	10YR3/3	2 m gr	C.	abs.	5–15	–
	A$_2$	10–23	10YR3/4	1 vf sbk	C.	abs.	5–15	–
	C$_{rt}$	23–300	10YR5/6	–	C.	abs.	–	–
L99-4	O$_a$	0–0.5	–	–	–	–	–	–
	O$_e$	0.5–2	10YR2/3	–	–	–	–	–
	A$_1$	2–10	10YR3/4	1 f gr	Low C	abs.	<15	–
	A$_2$	10–36	10YR4/4	2 f gr	Low C	abs.	<15	–
	B$_w$	36–79	10YR5/4	2 m sbk	Low C	abs.	<15	–
	2C	79–88/106	10YR5/6	abs.	Low C	abs.	35–50	–
	3B$_{tx1}$	88/106–120	2.5YR6/4	abs.	High C	Fe, Mn in cracks	15–35	–
	3B$_{tx2}$	120–146	5YR4/6 2.5Y6/4	abs.	Very high C	Fe, Mn in cracks, 7.5YR5/8	15–35	–
	3B$_{tx3}$	146–166	2.5Y6/6	abs.	Very high C	Fe, Mn in cracks	15–35	–
L99-3	O$_a$	0–1	–	–	–	–	–	–
	A$_1$	1–5	10YR2/2	1 f gr	Low C	abs.	<15	–
	A$_2$	5–15	10YR4/3	1 f gr	Low C	abs.	<15	–
	B$_w$	15–74	10YR5/6	1 m gr	Low C	abs.	–	–
	B$_t$/C$_1$*	74–130	Bt (7.5YR5/4)	–	Low C	abs.	–	–
	B$_t$/C$_2$*	130–350	C(2.5Y6/6)	–	Low C	abs.	–	–
99-6	A$_1$	0–12	10YR3/4	1 f gr	Very low C	abs.	<15	–
	A$_2$	12–52	10YR4/4	1 f gr	Very low C	abs.	<15	–
	AB	52–71	10YR4/6	2 m sbk	C	abs.	15–35	–
	B$_w$	71–100	7.5YR5/6	1 m sbk	C	abs.	35–70	–

Table 2 (Continued)

Profile	Horizon	Depth (cm)	Munsell colour (moist)	Structure	Consistence	Mottling	Coarse fragments (%)	Bulk density Mg/m³
	$2B_{tx}$	100–145	10YR5/6	massive	High C, slakes in water	abs.	>70	1.92
	$2B_t/2C^*$	145–160	–	–	–		–	(2.05)
96-1	O_a	0–10	–	–	Low C	abs.	–	–
	A	10–20	7.5YR5/2	abs.	–	abs.	>70	–
	E	20/22–62	7.5YR5/3	abs.	–	abs.	36–70	–
	B_{sh1}	22/62–33/77	7.5YR4/5	abs.	C	2–20%	16–35	–
	B_{sh2}	33/77–78/88	7.5YR5/6	abs.	C	1–2%	16–35	1.76
	B_{tsx}	78/88–105	10YR6/6	2 m pl	High C, B, Slakes in water.	<1%	16–35	2.11
	B_x	105–120	10YR 6/6	massive	C, slakes, partly, in water	<1%	35–70	–
96-13	O	0–5	7.5YR2/2	2 m cr	Low C	abs.	1–5	–
	A_1	5–17	7.5YR2/2	1 vf cr	Low C	abs.	1–5	–
	A_2	17–28	7.5YR3/2	1 vf cr	Low C	abs.	1–5	–
	AB	28–32/37	10YR3/2	2 f sbk	Low C	–	1–5	–
	B_w	32/37–51	10YR5/6	1 f sbk	Low C	–	1–5	–
	$2C_1$	51–86	10YR6/5	abs.	–	–	15–35	–
	$2C_2$	>86	10YR5/4	abs.	–	–	>70	–
96-14	O	0–1	–	–	–	–	–	–
	A_1	1–26	5YR2/3	1 f gr	–	abs.	35–70	–
	A_2	26–55	5YR2/3	2 m gr	–	abs.	35–70	–
	R	>55	–	–	–	–	–	–

1=weak, 2=mod., 3=strong; vf=very fine, f=fine, m=medium; gr=granular, abk=angular blocky, sbk=subangular blocky, pl=platy, cr=crumb; C=compactness; B=brittleness
*Lamellar clay illuviation horizon

Table 3 Properties of the soils studied

Profile	Horizon	Depth (cm)	pH (H$_2$O 1:2.5)	OM (%)	Texture (USDA)	Sand (%)	Clay (%)	V (%)	pH (H$_2$O 1/1)	C/N
99-2	A$_1$	0–10	5.7	3.9	SL	69	6.9	–	–	–
	AB	10–34	5.5	0.7	SL	65.8	13.5	–	–	–
	2B$_t$	34–97	5.8	0.5	SL	67.2	17.7	100	–	–
	2B$_{tg1}$	97–160	6.6	0.1	SCL	68.9	21.8	100	–	–
	3B$_{tg2}$	160–220	6.7	0.3	SCL	66.5	20.9	100	–	–
96-2	A$_1$	0–10	7.0	2.8	–	–	–	61.6	–	–
	A$_2$	10–23	7.0	1.4	SL	76.2	10.4	70.6	–	–
	C$_{rt}$	23–300	7.4	–	Si	72.0	11.6	–	–	–
L99-4	O$_e$	0.5–2	4.9	73.5	–	–	–	–	–	19.4
	A$_1$	2–10	5.6	10.2	LS	79.1	3.3	71.0	–	–
	A$_2$	10–36	6.8	1.8	LS	80.3	6.1	100	–	–
	B$_w$	36–79	7	1.4	LS	76.7	7.0	100	–	–
	2C	79–88/106	7.6	0.9	LS	77.0	4.0	100	–	–
	3B$_{tx1}$	88/106–120	7.7	0.3	LS	74.9	5.3	100	–	–
	3B$_{tx2}$	120–146	7.4	0.3	LS	71.9	6.3	100	–	–
	3B$_{tx3}$	146–166	7.6	0.3	LS	73.2	7.1	100	–	–
L 99-3	O$_a$	0–1	6.0	54.2	–	–	–	–	–	20.5
	A$_1$	1–5	6.3	11.1	SL	73.2	8.7	89.0	–	–
	A$_2$	5–15	6.5	3.4	LS	83.8	6.7	–	–	–
	B$_w$	15–74	7.1	1.1	LS	77.1	5.9	–	–	–
	B$_t$/C	74–130	–	–	–	–	–	–	–	–
	B$_t$	130–350	8.3	0.2	S	88.5	6.0	–	–	–
	C	130–350	–	–	–	–	–	–	–	–
99-6	A$_1$	0–12	5.5	1.7	LS	78.2	6.6	48.1	–	–
	A$_2$	12–52	5.1	1.3	SL	72.9	10.2	25.3	–	–

Table 3 (Continued)

Profile	Horizon	Depth (cm)	pH (H$_2$O 1:2.5)	OM (%)	Texture (USDA)	Sand (%)	Clay (%)	V (%)	pH (H$_2$O 1/1)	C/N
	AB	52–71	5.2	1.2	LS	77.9	6.5	18.5	–	–
	B$_w$	71–100	5.6	0.4	S	96.3	1.1	34.8	–	–
	2B$_{tx}$	100–145	5.7	0.3	SL	67.3	4.6	56.1	–	–
	2B$_t$/2C	145–160	6.8	0.2	LS	83.4	5.8	100.0	–	–
96-1	O$_a$	0–10	3.7	32.7	LS	–	–	–	3.9	–
	A	10–20	4.1	3.1	SL	80.8	7.3	31.7	4.5	–
	E	20–22/62	4.1	0.4	SL	75.2	2.7	25.7	4.6	–
	B$_{sh1}$	22/62–33/77	4.5	5.1	SL	65.0	18.1	7.7	3.7	–
	B$_{sh2}$	33/77–78/88	4.8	7.7	SL	65.2	10.8	6.1	5.6	–
	B$_{tsx}$	78/88–105	5.3	0.1	SL	56.0	10.3	4.5	5.5	–
	B$_x$	105–120	–	–	–	–	–	–	–	–
96-13	O$_a$	0–5	4.7	19.6	SL	74.3	17.6	7.6	4.0	27.2
	A$_1$	5–17	5.0	8.7	SL	75.1	11.9	4.4	4.3	26.8
	A$_2$	17–28	5.2	5.9	SL	69.0	16.8	22.8	4.9	19.0
	AB	28–32/37	5.4	3.0	SL	67.1	13.5	–	–	–
	B$_w$	32/37–51	5.5	1.9	SL	75.3	9.1	–	5.3	–
	2C$_1$	51–86	5.9	0.5	S	87.4	4.4	–	5.4	–
	2C$_2$	>86	6.1	–	S	87.6	2.4	–	–	–
96-14	O	0–1	4.6	22.3	–	–	–	10.9	–	19.7
	A$_1$	1–26	4.5	9.6	SL	65.2	18.7	5.8	–	13.3
	A$_2$	26–55	4.7	8.7	SL	69.2	14.6	1.3	–	16.8

4.1.1 Organic Matter Accumulation and Formation of Umbric Horizons

Organic matter accumulation causes the formation of an umbric horizon above the altitudinal limits where water moisture regime becomes udic. This process is highly prominent in profiles 96-13 and 96-14, where organic matter contents of the mineral horizons varies between 1.7 and 20% at acidic (pH 4.2–5.5) conditions with base saturations varying between 4.4 and 48%. The C/N ratios of these surface horizons range from 9 to 27 under forest vegetation and acidophyllic pastures (Table 3 and data not presented).

The acidic character and the high C/N ratios qualify for the humus type of the umbric horizons as mor, while in the lower positions is eu- or dysmoder, or even agrimull at the bottom of the valley (Alcañiz 2003). The accumulation of organic matter is due to the imbalance between mineralisation and organic matter supply by the vegetation under a frigid/cryic soil temperature regime. The organic matter supply by tree residues in La Cerdanya forests has been estimated as 7–15 Mg ha^{-1} year^{-1}, with C/N ratios varying between 60 and 100 (Gracia et al. 2001). In the pastures there are less organic matter accumulation (between 3 and 5 Mg ha^{-1} year^{-1}), due to cattle grazing (Alcañiz 2003). The accumulation of organic matter gives rise to Histosols (Saprists) in the glacier-excavated platforms with impeded drainage (Boixadera and Poch 2003).

4.1.2 Formation of Cambic Horizons

The structure and colour of the Bw horizons of profiles 96-13 and 99-6 qualify for a cambic horizon (FAO/ISRIC/IUSS 2006), which is associated to the silicate clay mineral formation and free Fe formation. Weatherable minerals are determined to be abundant in thin sections (Fig. 3). Abundant chlorite-like clay minerals and mica were determined in profile 96-13 by XRD (Table 4). This could probably be a Fe-hydroxy interlayered vermiculite, following the normal evolution of mica to vermiculite (Ghabru et al. 1990). Nevertheless, the presence of some pedogenic chlorite should also be considered, since soil acidity and strong leaching tend to destroy the expanding 2:1

Fig. 3 Biotite alteration to hydroxil layered vermiculite. Horizon B$_w$, Profile 99-6. Frame length 2.1 mm. PPL, XPL

Table 4 Mineralogical and chemical attributes of some selected horizons
Table 4a Clay mineralogy

Profile	Horizon	Depth (cm)	%				I/Sm
			Chlorite	Kaolinite	Illite (I)	Smectite (Sm)	
99-2	A_1	0–10	1	6	76	1	16
	AB	10–34	1	4	83	1	12
	$2B_t$	34–95	–	4	80	1	15
	$3B_{tg2}$	160–220	1	5	75	3	15
96-13	A_1	5–17	68	–	32	–	–
	AB	28–37	71	–	29	–	–
	B_w	37–51	78	–	21	–	–
	$2C_2$	51–96	tr.	tr.	tr.	–	–
96-1	B_{sh1}	22/62–33/77	Qualitative: Quartz, mica, hydromica, feldspars, halloysite				
	B_{sh2}	33/77–78/88					

Table 4b Iron and aluminum analyses

Profile	Horizon	Cp g/kg	Fe_0 g/kg	Fe_d g/kg	Fe_o/Fe_d	Al_0 g/kg	$Al_0+1/2$ Fe_0 (%)	ODOE
96-1	O_a	–	–	–	–	–	–	–
	A	6.1	0.5	2.3	0.22	0.6	0.085	0.019
	E	1.5	0.2	9.7	0.02	0.7	0.080	0.003
	B_{sh1}	12.2	10.6	16.6	0.64	5.9	1.12	0.244
	B_{sh2}	8.3	3.7	4.2	0.88	5.1	0.70	0.215
96-13	O_a	27.3	3.7	4.2	1.09	3.2	0.5	–
	A_1	13.7	2.7	7.0	0.39	3.1	0.25	–
	A_2	6.6	2.3	5.8	0.43	3.6	0.48	–
	B_w	3.1	1.8	4.9	0.37	3.1	0.40	–
	$2C_1$	1.3	1.2	5.2	0.23	2.9	0.35	–

Cp: Carbon extracted by pyrophosphate, Fe_o, Al_o: extracted by oxalate, Fe_d: Iron extracted by dithionite-citrate, ODOE: Optical density of oxalate extract.

clays or develop hydroxyl interlayers, and convert vermiculites to pedogenic chlorite (Fanning and Keramidas 1977).

A further evolution to kaolinite (Rebertus et al. 1986) does not take place here, indicating the prevailing moderate weathering conditions, which probably are due to the low temperatures. The chemical analysis (Table 4) clearly points to the formation of free iron, which is a significantly important process, despite the absence of a spodic horizon. This clearly points out to the dominant soil forming process taking place on the rather stable granitic colluvia above ca. 1700 m, where at lower altitudes this process would be masked by soil erosion due to the long standing forest clearing and cultivation.

4.1.3 Mineral Weathering and Clay Migration

The soil located at the bottom of the geomorphic sequence (99-2) is formed on very old fluvio-glacial materials, possibly of Pleistocene age, which are affected by very intense in situ weathering of granites, schists, and strong clay illuviation. The clay minerals in the soil are mainly illitic, derived from feldspar alteration to sericite as seen in thin sections together with rare kaolinite and smectite. The clay size fractions in the B_t horizons have most probably been transported during the very wet conditions of the Pleistocene forming thick illuvial coatings on surfaces or infillings, occasionally stained by Fe and Mn oxides (Fig. 4). The restricted leaching indicated by strongly pronounced mottling, could be the reason for the formation of the 2:1 instead of the 1:1 clay minerals. This manifests the past moister and warmer conditions that would favor the deep alteration.

Profile 96-2 is also at a low altitude (1300 m), and is developed on a thick (>5 m), highly weathered, granitic *in situ* saprolite. Such intensely weathered materials are common at certain altitudes (from the bottom of the valley up to more than 1600 m, and present in some places above 2000 m) at the northern flank of La Cerdanya, which have given rise to extensive boulder formations such as the Targassona Chaos. Although the *in situ* granite weathering may be Tertiary, the alteration in profile 99-2 must be younger, dating from a wet and warm interglacial period, prevailing before the incision of the valley and the buildup of the present river terraces, which were not degraded by the subsequent less intense glaciers of the Pleistocene (Gómez-Ortiz 1987). Clay migration has been the dominant process in such materials when surfaces were stable (data not presented), but human-induced erosion has been so strong that only a few spots with this soil remain.

4.1.4 Formation of Lamellar Argic Horizons

Two of the profiles on till deposits (L99-3 and 99-6) show banded clay illuviation (3 cm thick), which increase in thickness and spacing to a depth of 3 m. Similar

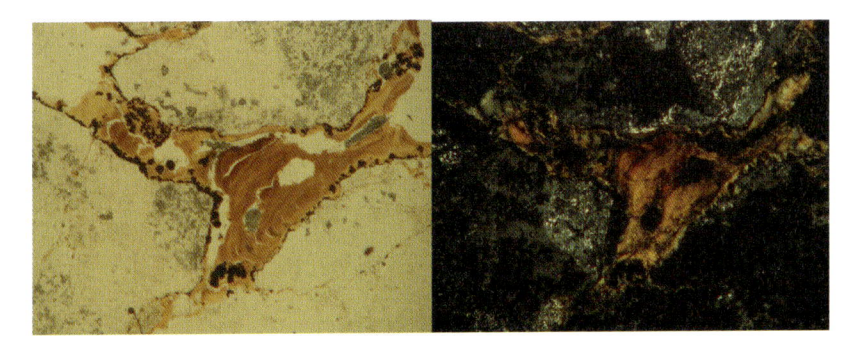

Fig. 4 Clay coatings and manganese nodules on the saprolite till of horizon B_{tg2}, Profile 99-2. Frame length 4.2 mm. PPL, XPL

Table 5 Micromorphological description of the soils

Profile	Horizon	Porosity	Coarse elements	Micromass	Organic material	Pedofeatures
99-2	A_1	60%, complex packing pores	Fine and medium quartz sand, few plagioclases, schists and mica flakes (fresh)	Undifferentiated b-fabric	Root and leaf remains, excrements and organic pigment	None
	$2B_t$	50%, planar voids, complex packing pores	Unso rted sand of quartz, quartzites and schists (fractured)	Crystallitic b-fabric	Few tissue remains and organic pigment	Clay coatings and infillings, microlaminated, in planar voids, around coarse fragments and in fissures of schist fragments, with Fe hypocoatings.
	$2B_{tg1}$	45%, complex packing voids	Unsorted sand of quartz, fragmented and rubefacted schists, feldspars (cross-linear and dotted alteration to sericite), fragmented mica flakes and alterites.	2:1 clay, micro-laminated.	Few tissue remains and organic pigment.	Virtually all micromass is as coatings or infillings, microlaminated, bridging grains, and related to the present packing pores. Some of the coatings have lost their orientation.
	$3B_{tg2}$	45%, complex packing voids	Unsorted sand of quartz, fragmented and rubefacted schists, feldspars (cross-linear and dotted alteration to sericite), fragmented mica flakes and alterites.	2:1 clay, micro-laminated.	Few tissue remains and organic pigment	Virtually all micromass is as coatings or infillings, microlaminated. They are related to the present packing pores. Some of the coatings have lost their orientation. Impregnative nodules and punctuations of Fe-oxyhydroxides
L99-4	$3B_{tx1}$	25%, vesicles and vughs, few planar voids and biopores	Quartzite and quartz (fresh), plagioclases (strong mottled alteration to sericite), biotite (linear parallel alteration to chlorite)	Mostly silicic silt, speckled b-fabric	Root remains	Limpid 2:1 clay coatings and infillings, microlaminated, around and in pores and around some coarse fragments. Silt cappings on some biopores and coarse fragments, up to 2 mm thick, with impregnative Fe-hypocoatings in the cappings.

Table 5 (Continued)

Profile	Horizon	Porosity	Coarse elements	Micromass	Organic material	Pedofeatures
	$3B_{tx2}$	10%, vesicles and vughs	Quartzite and quartz (fresh), plagioclases (mottled alteration to sericite), biotite (linear parallel alteration to chlorite), schists	Mostly silicic silt, crystallitic b-fabric	None	Limpid 2:1 clay coatings and infillings, microlaminated, around and in pores Silt and fine sand capping on quartzite gravel, with a vesicular microstructure. Vesicular microstructures above coarse fragments. Impregnative subhorizontal intercalations and nodules of Fe-oxyhydroxides.
	$3B_{tx3}$	25%, vesicles and vughs, packing voids	Quartzite and quartz (fresh), plagioclases (mottled alteration to sericite), biotite (linear parallel alteration to chlorite)	Mostly silicic silt, speckled b-fabric	Root remains	Abundant silt cappings, 0.5 to 2 mm thick, on coarse fragments Limpid clay coatings, microlaminated, around pores and below some coarse fragments Silt intercalations Hypocoatings and impregnative nodules of Fe-oxyhydroxides
L99-3	B_t/C^* 100 cm	50%, planar and simple packing voids, vesicles and vughs	Quartz (fresh), biotite (parallel linear alteration), plagioclases and orthoclases (moderate mottled alteration to sericite)	Siliceous silt and clay, crystallitic b-fabric	None	Frequent coatings of 2:1 clay, mottled, microlaminated, along 2 cm horizontal Bt bands (lamellar argillic) Few silt and fine sand cappings on quartzite grains in the Bt and C bands.

Table 5 (Continued)

Profile	Horizon	Porosity	Coarse elements	Micromass	Organic material	Pedofeatures
	B$_t$/C* 350 cm	40%, simple packing voids	Quartz (fresh), plagioclases and orthoclases (moderate mottled alteration to sericite), biotite (mostly fresh)	Siliceous silt and clay, crystallitic b-fabric	None	Very frequent limpid clay coatings, up to 0.5 mm thick, around 70% of packing voids in Bt bands.
99-6	B$_w$	50%, complex packing pores, planar and biopores	Very coarse to medium sand of fresh quartz and quartzite, some granitoid fragments, feldspars altered to sericite and mica flakes, some altered to chlorite. Oriented at random.	Fine clay and silt, undifferentiated b-fabric	Some root remains, sclerotia, phlobaphenized organic material	Few silt cappings on coarse quartzites Abundant silt coatings on pores. Limpid clay coatings and infillings in the lower half of the section, microlaminated, up to 0.5 mm wide, some of them broken.
	2B$_t$/2C*	25%, vesicles, few complex packing pores and planar voids	Quartzite gravels, broken, without orientation,; very coarse to medium sand of quartz, few feldspar and micas. Feldspars altered to sericite and some micas to chlorite.	Fine clay and silt, undifferentiated b-fabric	None	Silt cappings on every gravel, 2mm wide. Vughs and planar voids under the gravels, 1 mm wide, some of them infilled with loose quartz sand, and some of them with silt coatings. Clay coatings, microlaminated, concentrated in bands, up to 0.25 mm thick, often under the gravels, some of them broken and mixed with the micromass.

Table 5 (Continued)

Profile	Horizon	Porosity	Coarse elements	Micromass	Organic material	Pedofeatures
96-1	B$_{sh1}$	30%, complex packing pores between welded rounded aggregates	Quartz sand, muscovite flakes	Brownish mixture of clay, silt and organic matter, undifferentiated b-fabric	Frequent black root remains,	None
	B$_{txs}$80 cm	10%, vesicles and vughs	Mostly quartz sand, muscovite flakes, fresh	Quartz silt and coarse kaolinitic clay, crystallitic b-fabric	None	Frequent coatings and infillings of clay and fine silt around pores, microlaminated Silt cappings, 1 to 2 mm thick Few silica pendents
	B$_{txs}$ 100 cm	25%, vesicles, vughs and planar voids	Mostly quartz and quartzite sand, sericite after plagioclase grains.	Quartz silt and coarse kaolinitic and illitic clay, crystallitic b-fabric	None	Impregnative nodules and punctuations of Fe-oxyhydroxides. Frequent coatings and intercalations of fine sand, silt and clay, well sorted. Intercalations of kaolinite. Coatings of mottled clay around pores, microlaminated. Few silica pendents

Table 5 (Continued)

Profile	Horizon	Porosity	Coarse elements	Micromass	Organic material	Pedofeatures
	B_x	20%, simple packing pores, vughs and vesicles.	Quartz and quartzite sand, sericite after plagioclase grains	Quartz silt and coarse kaolinitic and illitic clay, crystallitic b-fabric (fine quartz)	None	Impregnative nodules and punctuations of Fe-oxyhydroxides. Frequent coatings and intercalations of fine silt and clay, mottled. Intercalations of quartzitic silt. Coatings of mottled clay along rock fissures. Few silica pendents
96-13	$2C_1-2C_2$	25%, vesicles and vughs	Sand of quartz and mica flakes, fresh	Fine sand and silt	None	Absent

horizons with thicknesses up to 5 m are also present on till deposits of the tributaries of the Segre (*Vall de la Llosa*), dissected at present by the river network. The original material has about 80% sand and is highly porous, with simple packing pores and some vesicles in the B_t bands. Frequent clay coatings and infillings, limpid and microlaminated, are identified in the clayey lamellae. Silt cappings on coarse fragments are found both in the clayey and non-clayey bands. These morphologies suggest the formation during cold periods, when ice lensing would develop in the original porous till. In finer sediments this process gives rise to a layered structure indicative of permafrost (Van Vliet-Lanoë 1985).

The layered microstructure is absent due to the coarse texture of the till, but the macromorphology of the B_t bands show that thickness and spacing increase with depth. This may reflect the original morphology of the ice lenses in the till, where the few aggregates and clods containing micromass are strongly compacted and coated by microlaminated clay, indicating that the clay was illuviated after the compaction of the material, probably due to frost action.

To explain the formation of illuvial-clay lamellae in wind-blown deposits, Kemp and McIntosh (1989) propose a mechanism concurrent with the deposition of the parent material, which cannot be the case in tills, since their formation is not gradual. Bond (1986) studied the development of illuvial clay bands in laboratory columns experiment of sand and concluded that the lamellae formed simultaneously through downward flow, by flocculation of the clay transported from the overlying layer, as soon as the soil solution attained a maximum concentration of suspended clay. The triggering mechanism could be enhanced by developing certain layering in the original columns, where the packing of the sands would create some bands with smaller pores where clay could start to accumulate by a sieving effect.

In our study, macro and micromorphological evidence revealed a pre-existing layering developed by frost lenses and a later melting of water, which was forced to stagnate and percolate horizontally in near-saturated conditions, being responsible of the banded clay illuviation. The absence of redoximorphic features indicates that the availability of water was limited at that time. The silt coatings would correspond to a later process of freezing and thawing in the active layer when the permafrost was disappearing.

4.1.5 Formation of Fragic Horizons

Compact, horizons of low porosity were found in L99-4, 99-6 and 96-1 soils, which have been classified as soils with fragic horizons (Soil Survey Staff 1999). They are characterized by porosities below 25%, dominated by vesicles and vughs, frequent silt cappings on coarse fragments up to 2 mm thick (Fig. 5), limpid and microlaminated clay coatings and impregnative nodules and hypocoatings of Fe-oxyhydroxides. Vertical fissures are not very common, probably due to the abundance of coarse fragments. Few silica pendents under coarse fragments are observed in soil 96-1 (Fig. 6).

Fig. 5 Silt capping under coarse fragment. Note the compactness at the top of the fragment due to heave. Horizon 2B$_{tx}$, Profile 99-6. Frame length 4.2 mm. PPL, XPL

Fig. 6 Silica pendent. Horizon B$_{tsx}$, Profile 96-1. Frame length 6.5 mm. PPL, OIL, XPL, OIL+XPL

Some of the coatings and infillings are sorted from clay at the bottom to sand at the top (Fig. 7), indicating flow regimes from low to high energy during the illuviation. This would correspond to the flows due to melting of ice, reaching higher discharges at the end of the process. Fragipan formation is explained in glacial conditions by the compaction of the material below the permafrost table, by desiccation due to ice growth during cold periods, and melting during warmer periods when silt cappings and vesicular pores would be formed (Courty et al. 1989). Other processes leading to fragipan formation are silica cementation, pore clogging by

Fig. 7 Clay, silt, and sand coating on the fragipan of Profile 96-1, showing an inverse particle sorting. Frame length 5.5 mm. PPL, XPL

clay illuviation (Smeck et al. 1989), or degradation of former illuvial argic horizons (Bockheim 2003).

Studying several fragipans on Tertiary deposits of NW Spain, Macías and Guitián Ojea (1976) point to clay illuviation as one of the main reasons for fragipan formation, together with wetting/desiccation cycles and periglacial processes. They recognise several cycles of clay accumulation since the Tertiary, in which the vesicular porosity characteristic of saturated flow do not exist (Macías et al. 1976). In our case, only one of the profiles (96-1) has some silica accumulation that cannot be responsible for the compaction, although it shows strong clay and silt illuviation that could bind particles and be the reason for the high consistency and brittleness.

The groundmasses of the three horizons ($3B_{tx}$ of L99-4, $2B_t/2C$ of 99-6 and B_{tsx}, B_x of 96-1) present porphyric c/f related distribution pattern, highly packed fabrics and low intrapedal packing porosity which suggest a compaction process by physical tightening, due to desiccation (Bryant 1989). In profile 96-1 the coatings are on these preexisting pores, and therefore their ability to bind particles is dubious. The morphology of some sericite masses, originally pseudomorphs after feldspar grains, confined and between loose grains, suggests that the horizon has undergone strong pressure that could be the reason for the brittleness and hard consistency. Although clay illuviation is not believed to be the reason of fragipan formation in profile 96-1, it could also be related to the fragipan in the sense that the sodium released by albite weathering could enhance clay and fine silt dispersion (Karathanasis 1989).

Free silica would also be released during silicate weathering at low pH, forming the opaline pendents. The plagioclase particles in profile 96-1 are highly weathered to sodium-free micaceous clays, i.e., the neo-formed micas, which increase the SAR of the soil solution and cause the formation of the illuvial microlaminated clay and silt coatings, which display a perfect preferred orientation along pore walls. Other field evidences, as boulder vertical orientation, location of silt cappings in the profile and podzolisation -particularly in soils L99-4 and 96-1, can be explained by permafrost in a similar way proposed by FitzPatrick (1976) for Scottish fragipans.

4.1.6 Podzolisation

Profile 96-1 is a Podzol formed on an ablation till, consisting mainly of quartzitic sand. The land occurs on a gentle slope with evidences of saturated subsurface flow and seepage that gives rise locally to the development of some Histosols. The podzolisation process has developed on top of the low permeability fragic horizon with the following chemical properties: (i) extreme acidity (pH 3.7 to 4.8) and (ii) high aluminium and Fe contents throughout the profile that increase with depth to a maximum in the B_{sh2} horizon. This is congruent with the fact that in Podzols, Al and Fe are accumulated independently, due to differences in solubilities of Al- and Fe-organic complexes. The high molar ratios of Al/Si detected in the soils studied is most probably due to the presence of minerals rich in readily soluble aluminium (Camps Arbestain et al. 2002), such as halloysite (kandite) instead of gibbsite in this context (Table 4).

The micromorphology of the underlying horizon shows that mica and hydrous mica present in the clay could partly derive from a very intense feldspar weathering to sericite. The B_{hs} horizons are alumunium-rich spodic horizons, with a well developed microstructure and polymorphic organic matter (De Coninck 1980), with recognisable root remains, indicating that the origin of the organic matter is partly due to in-situ decomposition of the organic tissues (Phillips and FitzPatrick 1999). Coatings are observed, and the presence of well-preserved plant tissues suggests impeded drainage conditions that inhibit the aerobic decomposition of the organic matter (Buurman et al. 2005).

Bockheim (2003), studying Fragiorthods in the Upper Great Lakes Region (USA) proposed a formation model where podzolisation could be the initial process, contemporary or not with argilluviation, followed by a subsequent fragipan formation through degradation of the B_t horizon. This author does not suggest the formation of fragipans to frost action, but to wetting/drying cycles causing cracking, and to water saturation, due to spring snowmelt responsible for the vesicular porosity and structural collapse. In our case, there are several evidences of podzolisation being a present-day process. In the field both albic and spodic horizons show tonguing prograding into the fragic horizon, thus indicating the age of pedogenesis.

The formation of the fragic horizon is a relict feature, as explained above. We have also observed that mobilization of Fe and organic matter is a general process on granitic materials in the udic/cryic section of the transect, which in some places are even cemented giving rise to the development of "ortstein" horizons (Boixadera and Poch 2003). In profile 96-1 podzolisation is fully expressed with the development of an albic horizon, due to more acidic nature of the substratum and the local hydrological conditions creating media for limited clay formation in the soil. In similar conditions in the Alps, smectite is the end product of the sequence (Egli et al. 2001), indicating to the recent development of podzolisation in the area studied. In the Alps the clay content of podzolic soils does not exceeds 10%.

Podzolisation is an active present day process and has been observed in another geomorphic position above 2000 m (data not presented), in slopes without the formation of an albic horizon and in some shallow soils over granite bedrock, where lateral flow takes place and a placic horizon develops.

4.2 Soil Development Along the Toposequence

Higher positions (>1700 m) have been strongly affected by glaciers during the Quaternary. These glaciers removed almost all the regolith derived from granites, in such a way that soils develop either on fresh granite or on coarse rocky slope deposits. Below 1500 m altitudes deeply weathered granites are dominant, and along some valleys, glacial tills in lateral moraines are present at 1200 m altitudes. The lowest parts of the toposequence are partly occupied by fluvio-glacial deposits of the early Quaternary age. These deposits are granite and schist gravels and boulders, reworked by the actual river system with rock outcrops occupying the rest of the footslopes.

Fragipans are only found in tills, normally on rather stable positions of slight to moderate slopes, although they are not dominant. After their formation, other processes, as banded clay illuviation or podzolisation, took place on these fragipans, depending on the mineralogy of the parent material and local hydrological conditions.

Above 2000 m organic matter accumulation and strong leaching lead to the formation of umbric and histic epipedons. Iron mobilization and accumulation form cambic horizons, even in steep slopes, whereas, below 2000 m, in more stable positions, cambic horizons form by clay formation and Fe release from mineral weathering.

Soils occurring between 1500 m and the valley bottom are normally on steep slopes, with a southern aspect and have undergone an intense erosion. The representative soils are very shallow, developed on granitic regolith without evidences of any pedogenetic process. Only on very few stable landforms small remnants of old argic horizons have been preserved.

Clay illuviation proceeds along the toposequence in a wide range of altitudes and conditions. Besides the aforementioned banded argic horizons on tills, the fluvio-glacial deposits of the lowest position of the toposequence have undergone deep weathering. The characteristic soil consists of an ochric overlying an argic horizon formed on granitic sand, where all clay is illuviated as pore coatings and infillings, juxtaposed to Fe and Mn oxides that indicate past redoximorphic conditions. Some of these surfaces are irrigated, but most of them are abandoned at present.

4.3 Soil Classification

The studied soils are classified according to Soil Taxonomy (Soil Survey Staff 1999) and WRB (FAO/ISRIC/IUSS 2006) (Table 6). For Soil Taxonomy, a temperature and a moisture regime has to be given to each soil. According to the available data, 1500 m is the altitudinal limit from ustic (below) to udic moisture regimes. Above 1700 m we consider the temperature regime to be frigid or cryic, and only cryic

Table 6 Classification of the soils studied

Profile	Soil Taxonomy (Soil Survey Staff,1999)	WRB (FAO, ISRIC, IUSS, 2006)
99-2	Typic Paleustalf	Cutanic Luvisol (Hypereutric, Profondic, Skeletic)
96-2	Typic Ustorthent	Haplic Regosol (Eutric)
L99-4	Typic Fragiudalf	Haplic Cambisol (Fragic, Hypereutric)
L99-3	Lamellic Hapludalf	Lamellic Arenosol (Hypereutric)
99-6	Typic Fragiudalf	Haplic Cambisol (Fragic, Hyperdystric)
96-1	Typic Fragiorthod	Haplic Podzol (Fragic)
96-13	Humic Dystrudept (Spodic Humic Dystrudept)	Cambic Umbrisol (Humic, Hyperdystric, Greyic, Endoarenic)
96-14	Humic Dystrocryept	Endoleptic Umbrisol (Humic, Hyperdystric)

above 2200 m. Identification of a fragipan allows to classify profiles L 99-4 and 99-6 as Typic Fragiudalfs and profile 96-1 as Typic Fragorthod. The latter profile has also an albic and a spodic horizon.

Pedon 96-13, having an umbric epipedon and an udic moisture regime, is classified as Dystrudept. Pedon 96-14, located at an altitude of 2340 m (above the present timber line below in an Alpine meadow), has a cryic temperature regime and an umbric epipedon, therefore, can be classified as a Dystrochrept. Profile 96-13 presents both a cambic horizon, and a high dithionite-citrate extractable iron and aluminium content. So, we propose to classify it as Spodic Humic Dystrudept. This is in agreement with the dominant pedogenic processes such as organic matter accumulation and leaching of iron and aluminium.

The soil of valley bottom is classified as Paleustalf on the basis of clay distribution along the profile, as well as the redox features and strong clay illuviation. Lithological discontinuities are present in this profile (data not given) but they are considered to be present in the original parent material. Other profiles of the area, not included in this study, show an abrupt textural change, indicating a better expression of the process.

Using criteria of the World Reference Base (FAO/ISRIC/IUSS 2006) Profiles 96-13 and 96-14 are classified as Umbrisols having high organic carbon contents. Profile 96-1 is a Haplic Podzol (Fragic) due to the presence of fragic and spodic horizons besides an umbric horizon.

An argic horizon was not determined in profiles L 99-4, L 99-3, and 99-6, due to the clay contents of these horizons, which was less than 8%. The soils having only a fragipan key out in the cambisols, which was not possible with the previous versions of the WRB. The present classification system allows more information to be given in the soil name and reflects better the processes acting in the soils.

5 Conclusions

Soils have been studied from landscape to the microscope level. Soil formation processes have been mainly identified through micromorphological techniques, that allowed the scaling down from soil-landscape relationships to horizon morphology.

The profiles studied are dominant along the examined toposequence and present a reliable data to understand the soil forming processes that are active in the area. The soil formation in the sequence is related to the climatic change during the Pleistocene and Holocene, lithology, the altitude, and local hydrological conditions. It consists of the in-situ deep weathering of granites, organic matter accumulation, fragipan formation, clay formation and illuviation, and podzolization. There are morphological evidences of past permafrost during glacial periods. Silt and fine sand cappings, lamellar B_t horizons, vesicular structures and many redoximorphic features are inherited from moister conditions, probably prevailing during wet interglacial periods. The very strong expression of mineral weathering and clay illuviation in the soil located in the valley bottom, indicates that the degree of erosion of upslope soils has probably been very intense, which likely inhibited the occurrence of any other process than organic matter accumulation, clay formation, and podzolization, which are probably still active at present. Therefore, only umbric, cambic, and spodic horizons, could develop.

Acknowledgments We are grateful to A.R.Mermut and an anonymous referee for their constructive observations and reviews.

References

Alcañiz JM (2003) Dinàmica de la matèria orgànica. In: Boixadera J and Poch RM (eds) Estudi de Camp dels Sòls de La Cerdanya. Course notes. Universitat d'Estiu de la UdL, La Seu d'Urgell. Unpublished

Bech J, Vallejo VR, Josa R, Fransi A and Fleck I (1981) Study of the podzolic character in acid soils on the high mountain of Andorra Spain, soil morphology and profiles. Anales de Edafología y Agrobiología, 40(1/2):119–132

Badia D and Martí C (1999) Suelos del Pirineo Central: Fragen. INIA, UZ, CPNA, IEA, Huesca (Spain)

Balasch JC (2003) Geologia de La Cerdanya. In: Boixadera J and Poch RM (eds) Estudi de Camp dels Sòls de La Cerdanya. Course notes. Universitat d'Estiu de la UdL, La Seu d'Urgell. Unpublished

Bockheim JG (2003) Genesis of bisequal soils on acidic drift in the Upper Great Lakes Region, USA. Soil Sci. Soc. Am. J. 67:612–619

Boixadera J and Poch RM (2003) Excursion guide. In: Boixadera J and Poch RM (eds) Estudi de Camp dels Sòls de La Cerdanya. Course notes. Universitat d'Estiu de la UdL, La Seu d'Urgell. Unpublished

Bond WJ (1986) Illuvial band formation in a laboratory column of sand. Soil Sci. Soc. Am. J. 50:265–267

Bryant RB (1989) Physical processes of fragipan formation. In: Smeck NE and Ciolkosz EJ (eds) Fragipan, their occurrence, classification and genesis. SSA Sp Pub 24. SSSA Madison: 141–150

Buurman P, Van Bergen PF, Jongmans AG, Meijer, EL, Duran B and Van Lagen B (2005) Spatial and temporal variation in podzol organic matter studied by pyrolysis-gas chromatography/mass spectrometry and micromorphology. Eur. J. Soil Sci. 56:253–270

Camps Arbestain M, Barreal ME and Macías F (2002) Phosphate and sulfate sorption in Spodosols with Albic horizon from Northern Spain. Soil Sci. Soc. Am. J. 66:464–473

Cassagne N, Remaury M, Gauquelin T and Fabre A (2000) Forms and profile distribution of soil phosphorus in alpine Inceptisols and Spodosols (Pyrenees, France). Geoderma 95(1–2):61–72

Cassagne N, Bal-Serin MC, Gers C and Gauquelin T (2004) Changes in humus properties and collembolan communities following the replanting of beech forests with spruce. Pedobiologia 48(3):267–276

CBDSA (1983) SINEDARES, Manual para la descripción codificada de suelos en el campo. Ministerio de Agricultura, Pesca y Alimentación de España. Madrid (Spain)

Courty MA, Goldberg P and Macphail R (1989) Soils and micromorphology in archaeology. Cambridge University Press. Cambridge

De Coninck F (1980) Major mechanisms in the formation of spodic horizons. Geoderma 24: 101–128

Egli M, Mirabella A and Fitze P (2001) Clay mineral formation in soils of two different chronosequences in the Swiss Alps. Geoderma 104:145–175

Fanning DS and Keramidas VZ (1977) Micas. In: Dixon JB and Weed SB (eds) Minerals in soil environments. SSSA. Madison, Wisconsin. pp. 195–258.

FAO/ISRIC/IUSS (2006) World Reference Base for Soil Resources 2006. 2nd edn. World Soil Resources Reports 103. FAO, Rome

FitzPatrick EA (1976) Cryons and Isons. Proceedings of the North of England Soils Discussion Group (1974) 11:31–43

Ghabru SK, Mermut AR and St Arnaud RJ (1990) Isolation and characterization of an iron-rich chlorite-like mineral from soil clays. Soil Sci. Soc. Am. J. 54:281–287

Gómez-Ortiz A (1987) Contribució geomorfològica a l'estudi dels espais supraforestals pirenencs. Gènesi, organització i dinàmica dels modelats glacials i periglacials de la Cerdanya i l'Alt Urgell. Institut Cartogràfic de Catalunya. Barcelona (Spain)

Gracia C, Burriel JA, Ibáñez JJ, Mata T and Vayreda J (2001) Inventari ecològic i forestal de Catalunya – Regió forestal I i II. CREAF, Bellaterra

Karathanasis AD (1989) Solution chemistry of fragipans. Thermodynamic approach to understanding fragipan formation. In: Smeck NE and Ciolkosz EJ (eds) Fragipan, their occurrence, classification and genesis. SSA Sp Pub 24. SSSA. Madison, p. 113–139

Kemp RA and McIntosh PD (1989) Genesis of a texturally banded soil in Southland, New Zealand. Geoderma 45:65–81

Macías F, García Paz C and Guitián Ojea F (1976) Suelos de la zona húmeda Española. VIII Suelos con Fragipan. 3. Micromorfología. Anales de Edafología y Agrobiología 35:837–861

Macías F. and Guitián Ojea F (1976) Suelos de la zona húmeda Española. VIII Suelos con Fragipan. 4. Génesis y sistemática. Anales de Edafología y Agrobiología 35:863–876

MAPA (1993) Métodos Oficiales de Análisis. Ministerio de Agricultura, Pesca y Alimentación de España. Madrid, Spain

Martí C and Badia D (1995) Characterization and classification of soils along two altitudinal transects in the Eastern Pyrenees, Spain. Arid Soil Res. Rehabil. 9:367–383

Murphy CP (1986) Thin Section Preparation of Soils and Sediments. AB Academic Publishers. Berkhamsted, UK

Page AL, Miller RH and Keeney DR (eds) (1982) Methods of Soil Analysis II. Chemical and microbiological properties. 2nd edn. ASA-SSSA. Madison, USA

Pedrol J (2003) Vegetació. In: Boixadera J and Poch RM (eds) Estudi de Camp dels Sòls de La Cerdanya. Course notes. Universitat d'Estiu de la UdL, La Seu d'Urgell. Unpublished

Phillips DH and FitzPatrick EA (1999) Biological influences on the morphology and micromorphology of selected Podzols (Spodosols) and Cambisols (Inceptisols) from the eastern United States and north-east Scotland. Geoderma 90:327–364

Porta J, López-Acevedo M and Rodríguez R (1986) Técnicas y Experimentos en Edafología (Vol 1). Col·legi Oficial d'Enginyers Agrònoms de Catalunya. Barcelona

Rebertus RA, Weed SB and Buol SW (1986) Transformations of biotite to kaolinite during saprolite-soil weathering. Soil Sci. Soc. Am. J. 50:810–819

Remaury M, Benmouffok A, Dagnac J and Gauquelin Th (1999) Pedogenesis and distribution of humic substances in Pyrenean soils, France. Analusis 27(5):402–404

Roca E (1986) Estudi geològic de la fossa de la Cerdanya. MSc. Thesis. Faculty of Geology. Univ. de Barcelona. Unpublished

Smeck NE, Thompson ML, Norton LD and Shipitalo MJ (1989) Weathering discontinuities: a key to fragipan formation. In: Smeck NE and EJ Ciolkosz (eds) Fragipan, their occurrence, classification and genesis. SSA Sp Pub 24. SSSA. Madison, p. 99–112

Soil Survey Staff (1999) Soil Taxonomy. A basic system of soil classification for making and interpreting soil surveys. 2nd edn. Agriculture Handbook 436. NRCS-USDA. Washington, USA

Stoops G (2003) Guidelines for Analysis and Description of Soil and Regolith Thin Sections. Soil Sci. Soc. Am., Madison, WI

Van Vliet-Lanoë B (1985) Frost effects in soils. In: Boardman J (ed) Soils and Quaternary Landscape Evolution. Wiley p. 117–158

A Micromorphological Study of Andosol Genesis in Iceland

G. Stoops, M. Gérard, and O. Arnalds

Abstract Icelandic Andosols form in climatic and geological conditions that are different from those common for Andosol formation, namely a frigid oceanic climate and the steady aeolian addition of fresh and reworked tephra material. Three representative profiles were sampled in different ecological zones of Iceland for micromorphological studies. They are classified as Orthidystri-Vitric Andosol, Dystri-Vitric Andosol and Thaptohistic-Vitric Andosol.

The parent material consists entirely of relatively young volcanic ashes, and organic layers. Multiple lithological discontinuities are recognised. Microstratification is observed in thin sections, both in organic and inorganic horizons. The coarse mineral fraction is characterised by variable proportions of pumice, different types of glass, hypo- and holocrystalline pyroclasts and euhedral and subhedral crystals of feldspar and augite. Locally, green-grey volcanic glass shows pellicular alteration to an orange component, considered as palagonite.

Some more clayey horizons (described as B horizons in the field) consist of soil aggregates or nodules of various compositions, mixed with pyroclasts covered by a coating of micromass. They may represent locally transported material of older, more evolved soils. The microstratification of both organic and mineral material and the (sub)-horizontal orientation of organ and tissue residues points to a gradual deposition of material, with short interruptions, not disturbed by pedoturbation, especially bioturbation. It means that every sub-layer corresponds to a former soil surface in a temporarily relatively stable environment. A weakly developed lenticular or iso-band microstructure, pointing to freeze-thawing phenomena, is observed in several

G. Stoops
Laboratorium voor Mineralogie, Petrologie en Micropedologie, Universiteit Gent, Belgium

M. Gérard
UR GEOTROPE, Institut de Recherche pour le Développement, Bondy, France

O. Arnalds
Agricultural University of Iceland, Reykjavik, Iceland

horizons. Capping, often with reverse sorting, observed between 100–230 cm depth on organ residues in one of the profiles are signs of former frost activity. It is perhaps the first time that such cappings are reported on organic components.

The relative pureness of most mineral layers, the freshness of the volcanic components and their sorting indicate in many cases direct deposition during eruptions, rather than a water or air transport over land.

Keywords Weathering · cryogenic microfabrics · isoband fabric · organic deposits · tephra

1 Introduction

Icelandic soils form in different climatic and geologic environments than are common for Andosols, characterised by frigid oceanic climate and steady aeolian additions of tephra materials due to intensive wind erosion on Icelandic desert areas and during volcanic eruptions. The soils were recently reviewed by Arnalds (2004), including their description, chemical and physical properties. However, micromorphological studies of Icelandic soils are scarce. The most detailed chapter was published by Romans et al. (1980) dealing with young soils on till exposed by the retreating Breidamerkur glacier. Excellent micromorphological descriptions of soil materials formed on aeolian tephra were given by Simpson et al. (1999) and by Milek (2006) as part of a detailed archaeological research near Hofstaðir, North Iceland. Other accounts include those presented in Ph.D. publications by Gudmundsson (1978) and Arnalds (1990).

A EU funded "COST Action-622" titled "Soil Resources of European Volcanic Systems" was established in 1998 and concluded in 2004 (Oskarsson and Arnalds 2004). A strategic sampling of reference pedons was undertaken in several European countries in relation to this program and some results have been published (Bartoli et al. 2003, Arnalds and Stahr 2004). A systematic microscopic description of the soils sampled in Iceland revealed interesting features that prompted a deeper study of their micromorphology.

The purpose of this chapter is to present a micromorphological description of three Andosols in Iceland. Other properties of the soils have been studied in detail and are published elsewhere (Arnalds et al. 2007).

2 Physical Environment

Iceland is situated between 63° and 66° latitudes in the North Atlantic Ocean. The island is mountainous with lowland areas and river plains along the coastline. The climate is humid cold temperate to low arctic, with cool summers and relative mild winters. Rainfall varies between 500 and 2000 mm yr^{-1} in lowland areas. Glaciers cover about 10 000 km^2. Glacial rivers carry large quantities of vitric sediments that are deposited at the margins of the glaciers and on floodplains.

Volcanic activity cuts across the country from the Southwest and Central South to Northeast. Volcanic eruptions are frequent and more than once every 5 yrs on average. Tephra deposition is common during eruptions. The volcanic materials are predominately basaltic and andesitic. Some of the most active volcanoes such as Katla and Grimsvötn are covered with glaciers, which results in explosive volcanism and tephra deposition, commonly associated with catastrophic floods that are important contributors of sand to unstable desert surfaces (see Arnalds et al. 2001). Some volcanoes, such as Hekla and Askja periodically produce rhyolitic tephra that, being light coloured, are excellent markers for dating soil profiles. A two coloured tephra layer (black ash above light coloured rhyolite) marks the settlement of Iceland, shortly before 900 AD.

Icelandic soils were recently described by Arnalds (2004). In this chapter we will use WRB (FAO 1988), Soil Taxonomy (Soil Survey Staff 1998) and Icelandic classification of soils (Arnalds 2004). Most of Icelandic soils form in tephra materials deposited during eruptions or as glacio-fluvial sediments, or redistributed by aeolian processes. Deserts are widespread with soils that classify as Andisols according to Soil Taxonomy (Arnalds and Kimble 2001), but Vitrisols according to the Icelandic scheme. These surfaces are unstable and cause large-scale redistribution of vitric materials. Soils under vegetation are also Andisols (Arnalds et al. 1995). The age of the parent materials becomes gradually younger from about 9000 yrs at the bottom to <500 yrs or even <100 yrs in the surface horizons. Underneath these andic materials are commonly Quaternary glacial deposits or lavas (Tertiary, Quaternary or Holocene). Permeability of the active volcanic belt is high, with water tables usually well below the described soils, while the permeability of older Tertiary rocks is slower, resulting in a water table commonly rising up to the surface, creating wetlands. Drainage and distance from volcanoes/aeolian sources determine what kind of Andosols will form, with a gradient of increased organic content and lower pH with increased distance from these sources. Typical Andosols (vitric and allophanic Brown and Gleyic Andosols, Icelandic scheme) are found on the volcanic active belt, while allophanic and aluandic Gleyic and Histic Andosols are typical soils at distance from the volcanic belt.

3 Material and Methods

Three pedons were described and sampled in August 1999 by A.G. Jongmans, F. van Oort and O. Arnalds. The location of the pedons is shown in Fig. 1 and profile descriptions are summarised in Tables 1 and 2. Physical and chemical data for the pedons will be published elsewhere (see Arnalds et al. 2007), but following is a summary of some important characteristics. All the sampled pedons meet criteria for Andosols. The first profile (EUR-7) is typical of somewhat poorly drained Histic Andosol (Icelandic system, Arnalds 2004) with >12% C down to 65 cm depth with only 2–5% allophane but 2–10% ferrihydrite. It is classified as Orthidystri-Vitric Andosol (Quantin and Spaargaren 2007) according to WRB (Driessen et al. 2001).

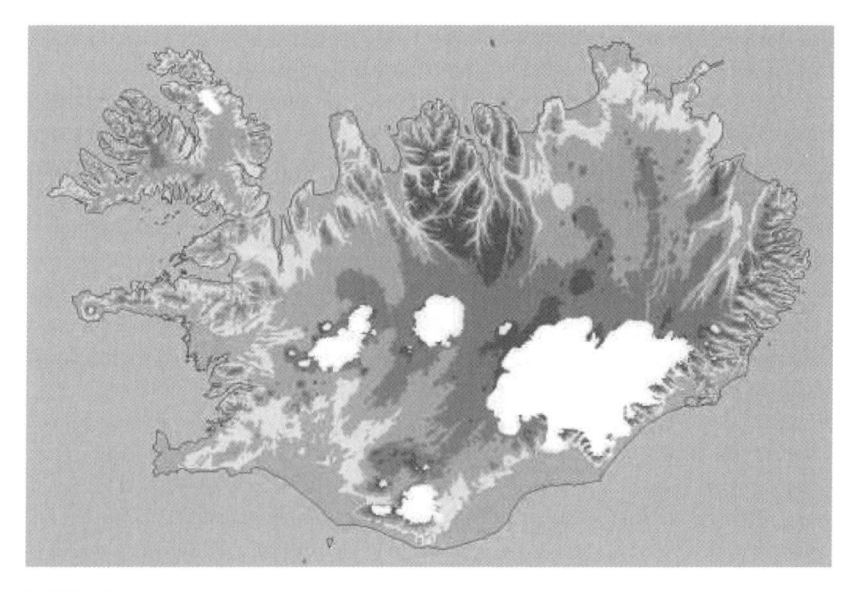

EUR07 : Hvammstangi
EUR08 : Auðkúluheiði
EUR09 : Hella

Fig. 1 Location of the reference profiles

Table 1 Profile location and environmental conditions

	Profile EUR-7	Profile EUR-8	Profile EUR-9
Location	NW Iceland, intersection route N1 and Hvammstang iroad; 40 m altitude	Central North Iceland, Audukuluheidi, 400 m altitude	South Iceland, Hella, 50 m altitude
Soil Climate	Cryic/Frigid, 550 mm rainfall, 2.5°C MAT, mild winter	Cryic, 800 mm rainfall, 1.0°C MAT, 6–9 months frost	Cryic/Frigid, 1150 mm rainfall, 4.5°C MAT, mild winter
Landform	glaciated upland	upland glacial plain	glacial plain
Land element	west facing slope	bottom of slope	middle slope
Microtopography	hummock	hummock	hummock
Vegetation	mostly grass	heath and moss	grass
Parent rock	aeolian and tephra deposits	aeolian and volcanic ash over glacial moraine	aeolian tephra and organic remains
Erosion	little or no	wind and water	none
Drainage	somewhat poorly drained	well drained	somewhat poorly drained
Classification (USDA)	Ashy, amorphic Eutric Pachic Fulvicryand	Ashy, amorphic Typic Vitricryand	Ashy, amorphic Eutric Pachic Fulvicryand
Classification (WRB)	Orthidystri-Vitric Andosol	Dystric-Vitric Andosol	Thaptohisti- Vitric Andosol (Umbric and Pachic)

Table 2 Profile descriptions

Horizon depth cm	Colour	Texture	Structure/ consistency	Boundary	Remarks
Profile EUR-7					
O 0–5	5YR 3/2	organic	very weak fine granular	clear and smooth	–
Ah 5–17	5YR 3/3	loam	weak, fine subangular blocky/ granular	abrupt and wavy	–
AC 17–35/50	2.5YR 3/4	organic clay loam	weak, very fine subangular blocky	abrupt and wavy	few fine basalt gravel, altered basaltic tephra
2BC 35/50–65	5YR 3/2 and 5 YR 2.5/2	organic clay loam	moderately fine platy	smooth	stratified, iron mottles
3BC 65–73	10YR 6/6 and 10 YR 5/4	organic loam	none	smooth	distinct iron mottles
3CB 73–82	10YR 5/4 and 5 YR 2.5/1	loam	moderately fine platy	abrupt and smooth	stratified, fine and medium iron mottles
4Bw 82/90–100	5YR 3/3	organic clay loam	fine subangular blocky	abrupt and wavy	few partly altered basaltic gravel; few distinct iron mottles
4B/Cg >90/100	5Y 4/1	clay	none	–	many coarse gravel, prominent iron mottles
Profile EUR-8					
O 0–3	3	organic		abrupt and smooth	–
Ah1 3–11/19	7.5XR 3/	sandy loam	moderate fine to medium platy	abrupt and wavy	irregular bright spots
Ah2 11/19–21/27	5YR 3/3	sandy loam	none	abrupt and tonguing	–
Bw1 21/27–26/34	7.5YR 4/4	sandy loam	none	abrupt and tonguing	–
2Bw2 26/34–37/42	10YR 5/5	sandy loam	moderately fine platy	abrupt and wavy	bright discontinuous tephra (Hekla 2800 yrs BP); cryoturbation features
3Bw3/4C 37/42–59/62	7.5YR 3/5	loam	none	abrupt and smooth	–
4C >59/62	10YR 4/4	sandy clay loam	none	–	glacial till, cryoturbation

Table 2 (Continued)

Horizon depth cm	Colour	Texture	Structure/ consistency	Boundary	Remarks
Profile EUR-9					
Ah 0–55	5YR 3/3	loam	–	–	disturbed stratification, cryoturbation, collemboli
2C 55–60	2.5YR 2.5/0	loamy sand	–	clear and wavy	basaltic ash layer
3H 60–95	7.5YR 3/3	organic?	–	clear and wavy	stratified plant remains and basaltic tephra
3C 95–100	10YR 3/2	–	none	abrupt and wavy	"Settlement tephra" layer (870 AD)
4H 100–230	5 YR 2.5/1.5	organic	–	–	stratified organic and aeolian material

The second profile (EUR-8) is located closer to the volcanic active belt and is a typical freely drained Brown Andosol (Icelandic system) with 6–12% allophane and 1–3% ferrihydrite. The organic content is 2–7%. It is classified as Dystri-Vitric Andosol (WRB).

The last profile (EUR-9) is a Gleyic Andosol (Icelandic system) and was sampled in a recent drainage ditch in wetland near Hekla volcano in South Iceland. The area commonly receives tephra from Hekla (andesitic) and Katla, (basaltic) and large aeolian additions from desert areas in the vicinity, a total of >0.5 mm yr^{-1} over past 1000 yrs. The soil is 230 cm deep and has both histic and vitric characteristics. The surface 60 cm have about 10% carbon, but this increases to about 20% below in the soil that formed during a period of lower aeolian activity. Clay content is low with 2–6% allophane. Thick tephra layers are sandy and vitric. It is classified as Thaptohistic-Vitric Andosol (WRB).

Undisturbed and oriented samples were taken for micromorphological studies. Prior to impregnation, water was removed by the acetone replacement method (Murphy 1986) for Pedon EUR-9, but by air-drying for Pedons EUR-7 and EUR-8 in the Laboratory of Soil Science at the Wageningen University. Polished thin sections of 6 by 7 cm were prepared in the laboratory of UR GEOTROPE of the "Institut de Recherche pour le Développement" (IRD, former ORSTOM), Bondy (France). Additional thin sections were prepared in the laboratory of the "Centre de Pédologie Biologique", CNRS, Vandoeuvre-lez-Nancy (France) of air-dried samples of the Ah1 and Ah2 or Bw horizons. Thin sections were described systematically according to the concepts and terminology proposed by Stoops (2003). After separation of the sand from the silt and clay fraction by wet sieving, sand and silt were subdivided into 3 grain size fractions (2000–500 μm, 500–250 μm and 250–50 μm respectively) by dry sieving. The fractions were treated with H_2O_2 to remove organic material and with dithionite-sodium citrate to remove free iron. Grain mounts were prepared for study with a polarising microscope. After identification of the minerals, line counting was used for quantification. Results of the 500–250 μm and the 250–50 μm fraction are presented, transparent grains totaling

100%, whereas the opaques are presented as an extra value (Edelman method). Results were published by Stoops and Van Driessche (2002 and 2007).

Selected thin sections were carbon-coated and examined with a scanning electron microscope (SEM) (Cambridge Stereoscan 200) coupled to an energy dispersion spectra microprobe (EDS) (AN10000 Link System), calibrated with a Co sample. Quantitative chemical microanalyses used ZAF correction. Scanning electron microscopy (SEM) images on sample fragments were obtained using gold-coated samples.

4 Results

4.1 Pedon EUR-7

The study of the sand fraction (Table 3) of this pedon shows a clear stratification confirmed by micromorphological analysis (Table 4). Glass shards, hypocrystalline pyroclasts, pumice fragments and grains of feldspar and augite, all fresh, are recorded in the different layers in various proportions. Elongated, often serrated phytoliths are abundant throughout the profile, besides diatoms with bilateral symmetry even as sponge spicules in horizons rich in organic matter.

Only in the surface horizons (Fig. 2a) a pedal granular microstructure is found, in the deeper layers it occurs always as an intrapedal microstructure to which later fine platy and/or lenticular structures are superimposed (Fig. 2d). The top 2 cm, practically devoid of mineral material, clearly resembles a moder (Kubiëna 1953), overlying a peaty layer where the yellowish to reddish brown organic residues are generally well preserved, often showing interference colours. In the 2CB and 3BC horizons, large isotropic orange organ residues (partly leafs) show a subhorizontal orientation (Fig. 2b,c).

The 3BC horizon consists of rounded soil nodules, of various compositions (different colours, different limpidity) together with rounded pumice fragments covered by thin coatings and hypocoatings of fine material, suggesting a reworked older, more developed soil material. Starting with the 2CB horizon, reddish brown hypocoatings surround root channels, which are sometimes infilled by limpid

Table 3 Mineralogical composition of sand fractions

A. Medium sand fraction: grains between 250 and 500 µm

Sample	Feldspar	Augite	Others	Rock fragments	Glass	Opaque
EUR-7						
O	1	–	1	43	55	7
3BC		–	2		98	5
3CB	5	–	5	67	23	19
4Bw		1	–	2	97	3
4B/Cg	3	–	6	91	–	4
EUR-8						
Bw1	–	–	–	7	93	5
2 Bw	–	–	–	3	97	4

Table 3 (Continued)

A. Medium sand fraction: grains between 250 and 500 μm

Sample	Feldspar	Augite	Others	Rock fragments	Glass	Opaque
3 Bw3/4C	2	–	1	82	15	31
4C	1	–	1	91	7	13
EUR-9						
3H	1	–	–	–	99	4
3C	2	–	–	14	84	2

B. Fine sand fraction: grains between 50 and 250 μm

Sample	Feldsp.	Augite	Others	Rock fragments	Glass	Opaque
EUR-7						
O	2	–	–	20	78	14
3BC	–	–	–	–	100	7
3CB	5	–	5	67	23	19
4Bw	3	–	–	4	93	3
4B/Cg	22	–	–	74	4	3
EUR-8						
Bw1	1	–	–	12	87	4
2 Bw	3	–	–	7	90	3
3 Bw3/4C	–	–	–	80	20	15
4C	3	1	–	89	7	16
EUR-9						
3H	2	–	–	1	97	3
3C	2	–	–	4	94	2

Table 4 Summary of the micromorphological description of the three profiles

Horizon	c/f*	Microstructure pores	Coarse mineral material **	Organic material	Remarks
EUR-7					
O 0–2	1/100	fine granular	few glass and hypocrystalline pyroclasts, few phytoliths and diatoms	fresh organ and tissue residues; fungal hyphae	–
O 2–5	1/10	fine granular	hypocrystalline pyroclasts and pumices, brown glass, F, A, phytoliths	fresh organ and tissue residues; few roots	–
Ah 5–17	1/10	angular blocky with moderate granular intrapedal	idem	as above	–
AC 17–35/50	1/100	as above	rare F and hypocrystalline pyroclasts	rare organ and tissue residues; few roots	strongly pedoturbated

Table 4 (Continued)

Horizon	c/f*	Microstructure pores	Coarse mineral material **	Organic material	Remarks
2CB 35/50–65	1/2	platy with weak granular intrapedal; isoband fabric	brown green glass, F, common phytoliths and diatoms	elongated tissue and organ residues 20% with subhorizontal orientation	reddish brown channel hypo-coatings and reddish isotropic limpid infillings
3BC 65–73	1/1–1/2	massive, platy with weak granular intrapedal	microlayering of angular F and A, rounded pumice, hypocrystalline pyroclasts, phytoliths	isotropic organ residues with subhorizontal orientation; fresh roots	reddish brown channel hypocoatings; internal hypocoatings of fine material in pumice; up to 80% soil nodules of different composition
3CB 73–82 **light**	3/1	pellicular grain to vughy	rounded pumice, few phytoliths	<5% tissue fragments	thin hypocoatings of fine material on pumices
3CB 73–82 **dark**	variable	subangular blocky and granular with weak tendency to fine platy; weak isoband fabric	as above, with few diatoms and very rare sponge spicules	few tissue and organ residues, rarely anisotropic; sklerotia	subhorizontal orientation of elongated organ residues.
4Bw 82–99/100	1/2	granular with tendency to fine platy or lenticular	A, F, green glass, feldspathic pyroclasts, black scoria; chlorite aggregates; phytoliths	isotropic organ residues, subhorizontal	rare thin coatings on pumices reddish brown channel hypocoatings
4B/Cg >90/100	20/1	packing of lithoclasts with interstitial soil material with channels	rounded feldspathic pyroclasts, A, F, opaques	rare tissue residues; few roots	yellowish and reddish brown hypocoatings on root channel
EUR-8					
Ah1 0–5	2/1	weakly blocky and platy, granular intrapedal, weakly developed isoband fabric.	rounded pumices in upper 5 cm; rare F, A and pyroclasts	subhorizontal isotropic tissue residues (50–10%); roots	diffuse coatings of fine material around pumice

Table 4 (Continued)

Horizon	c/f*	Microstructure pores	Coarse mineral material **	Organic material	Remarks
Ah1 5–11/19	2/1	idem, enaulic,	green glass	idem	as above
Ah2 11/19–21/29	5/1	granular, enaulic, blocky	pumices, green glass, rare phytoliths	isotropic organ residues (2%)	internal hypocating and coatings of fine material around pumice
Bw1 21/27–26/34	10/1	granular, enaulic	pyroclasts, pumices; layers with and without green glass; phytoliths	isotropic organ and tissue residues	clear thick (100 μm) coatings of fine material on pyroclasts; nodules of similar composition
2Bw2 26/34–37/42	1/2	massive; packing pores	basaltic pyroclasts and green glass, rare pumice embedded in colourless fine glass shards	none	rounded typic soil nodules of denser fine material and similar coatings on pyroclasts;
3Bw3/4C 37/42–59/62	10/1	granular, close fine enaulic	pyroclasts, green glass with orange alteration internal hypocoatings	none	unoriented cappings of reddish brown fine material on pyroclasts
4C >59/62		massive	unweathered tuff with rounded basalt fragments	–	–
EUR-9					
Ah 0–55		channels and vughs	pumice, A, phytoliths, diatoms	95%. anisotropic elongated organ residues, tissue residues and cells; sklerotia	parallel, but not subhorizontal layering of organ residues; pedoturbated
3H 60–95		fine platy; packing pores	vesicular green glass, few F, phytoliths and diatoms	80% anisotropic organ residues, few tissue and cell fragments	alternation of organic and inorganic layers; subhorizontal orientation of leafs

Table 4 (Continued)

Horizon	c/f*	Microstructure pores	Coarse mineral material **	Organic material	Remarks
3C 95–100	variable	–	(a) colourless pumice and glass shards; (b) green glass; rare diatoms and phytoliths	large organ residues (20%), cells (75%); black particles (5%)	alternation of organic and inorganic layers
4H 100–230 mineral		–	(a) green brown glass, rare pyroclasts; (b) large F and pumice, rare euhedral A; brown glass	5% large orange, weakly anisotropic organ residues	cappings with reverse sorting on leaf residues
4H 100–230 organic		–	colourless vesicular glass, phytoliths and diatoms	95% weakly anisotropic leaf residues, cell residues, fungal hyphae	subhorizontal orientation of leafs; rare excrements

* c/f ratio
** A: augite; F: feldspar

reddish isotropic material, rich in iron (Table 5, sample 1) pointing to active oxido-reduction processes. Inside some voids surrounded by hypocoatings, ferruginised plant remains (Fig. 3a and Table 5, sample 2) and typic fine coatings of fibrous goethite are observed. Similar ferruginised plant fragments were observed also by Milek (2004). The coatings consist of several parallel layers, composed of thin goethite needles with a fanlike orientation pattern (Fig. 3b). The material is practically isotropic, pointing to a stacking of very fine, weakly crystalline goethite particles (Stoops 1983a). The high amount of Si and Al detected by microprobe in this material (Table 5, samples 2 and 3) supports the conclusion that it is an early stage of transformation of ferrihydrite to goethite. According to mineralogical analyses, the ferrihydrite content in this Pedon varies between 2 and 10% (Arnalds et al. 2007).

4.2 Pedon EUR-8

This pedon is situated in the northern highlands that experience 6–9 months of frost annually (Table 1), and is therefore the coldest of the profiles studied. The three main sources of parent materials are: (1) deposition by aeolian processes from various desert surfaces, mainly consisting of vitric materials (Vitrisols, Arnalds 2004), (2) aeolian redistribution of Brown Andosols, and (3) by direct deposition of tephra, especially from Mt. Hekla volcano, which is at about 150 km distance to the south.

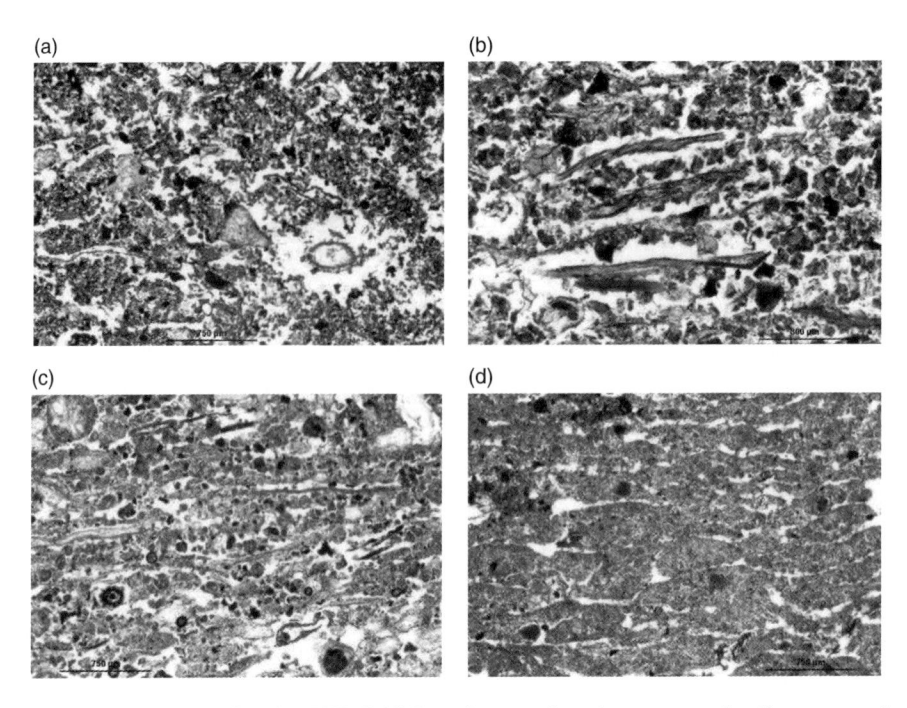

Fig. 2 Micrographs of Pedon EUR-7 (**a**) Irregular granular microstructure, locally aggregated to crumbs, with root section. Ah horizon, PPL. (**b**) Parallel arrangement of organ residues and granular microstructure. 2CB horizon, PPL. (**c**) Parallel arrangement of organ residues. Note differences in colour and limpidity of granules. 3BC horizon, PPL. (**d**) Moderately expressed lenticular microstructure in 3CB horizon. PPL

The last such event took place in 1980 (faint basaltic layer at about 3 cm depth not described, but noted by O. Arnalds).

In the 3Bw3/4C horizon the green glass shows a remarkable pellicular altera-tion to an orange isotropic substance, considered as palagonite (Fig. 4d). Optical observations at high magnification show distinct layering in the orange weathering product. Results from SEM and EDS analyses point to the existence of two types of weathering. In type 1 (Fig. 5 a,b) the following sequence is observed: intact glass (Table 5, sample 4), a first internal layered hypocoating (Table 5, sample 5), followed by a second, non-laminated internal hypocoating (Table 5, sample 6) con-taining feldspar microliths. In type 2 (Fig. 5c) the glass shows a complex layered orange hypocoating (Table 5, sample 7). The chemical analyses show that the beige-green glass has a basaltic composition. The first internal hypocoating shows marked depletion of silica and earth alkali elements, accumulation of iron, and hydration. In the outer hypocoating, the depletion and hydration of the material between the microliths is less pronounced, but accumulation of iron is evident. In the case of the layered coating (type 2), silica is slightly depleted, and alkali and earth alkaline elements are practically absent, but a relative strong accumulation of iron and alu-minium is noticed.

Table 5 Results of EDS analyses

	1	2	3	4	5	6	7	8	9
SiO_2	16.4	12.7	10.3	48.96	22.64	41.86	36.6	72.7	49.9
TiO	0.6	0.6	nd	2.90	5.55	2.51	7.4	0.3	2.1
Al_2O_3	3.6	2.9	0.6	13.78	16.40	18.97	20.8	14.6	13.4
Fe_2O_3	59.6	52.3	40.5	14.78	22.28	19.41	20.4	2.7	14.3
MnO	nd	nd	nd	0.25	nd	nd	nd	nd	0.2
MgO	nd	nd	nd	6.58	nd	4.64	nd	nd	6.3
CaO	0.7	0.6	nd	10.35	1.31	3.46	1.5	0.8	11.4
Na_2O	nd	0.5	nd	2.46	nd	nd	nd	3.5	2.1
K_2O	0.1	nd	nd	0.30	nd	0.20	nd	4.4	0.2
P_2O_5	nd	nd	nd						
SO_3	3.4	2.9	1.2						
Total	84.4	72.5	52.6	100.38	74.41	91.05	86.8	99.0	99.9

1: EUR-7, 2CB: iron hypocoating
2: EUR-7, 2CB: ferruginated organic material
3: EUR-7, 2CB: fanlike goethite coating
4: EUR-8, 3BwC: unaltered glass
5: EUR-8, 3BwC: alteration type 1, first internal hypocoating
6: EUR-8, 3BwC: alteration type 1, second internal hypocoating
7: EUR-8, 3BwC: alteration type 2, internal hypocoating
8: EUR-9, C: pumice
9: EUR-9, C: beige glass shards
nd: not detected

(a) (b)

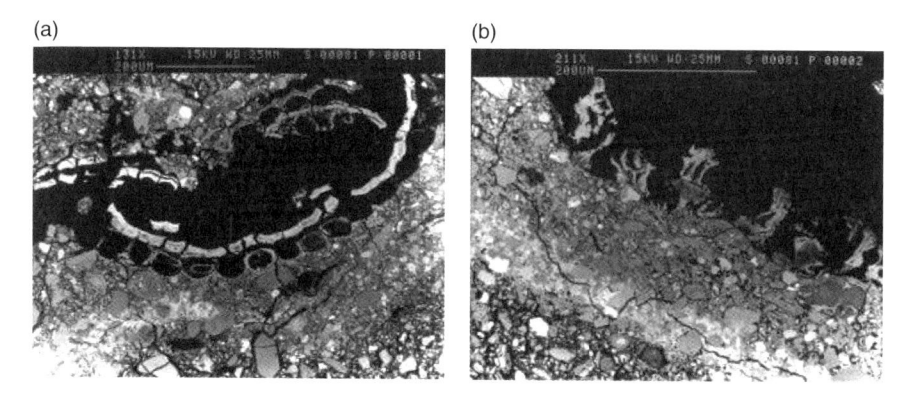

Fig. 3 SEM images of ferruginised material in horizon 2CB of Pedon EUR-7. (**a**) Ferruginised organ residue in channel surrounded by hypocoating of ferrihydrite. (corresponding to EDS analyses Table 5, 1 and 2). (**b**) Coating of fibrous goethite in channel (corresponding to EDS analysis Table 5, 3)

Layering is less prominent in the field than for Pedon 9 (Table 2). The analysis of the sand fraction (Table 3 and 4) shows that vesicular or fibrous colourless glass (Fig. 4a) and dark green aggregates are dominant components in the upper part, whereas hypocrystalline pyroclasts dominate the lower horizons. In the upper 5 cm, pumice grains are clearly dominant, pointing to a recent addition of rhyolitic

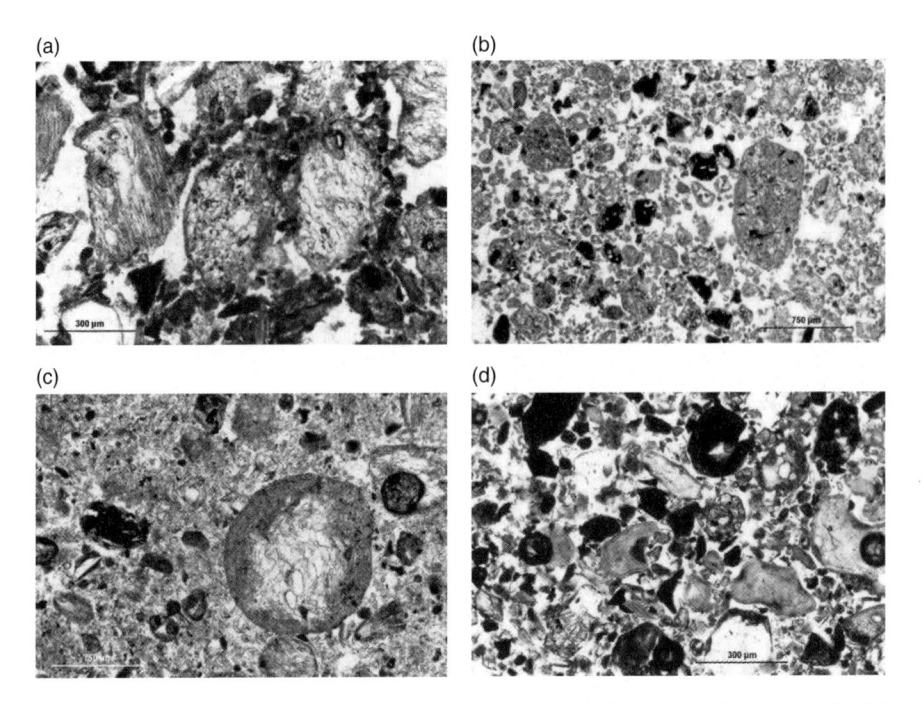

Fig. 4 Micrographs of Pedon EUR-8 (**a**) Pumice fragments with coatings of fine material. Ah2 horizon. PPL. (**b**) Granules and subangular aggregates of different limpidity. Bw1 horizon. PPL. (**c**) Pumice fragment with coating of fine glass (so called armoured pyroclast) formed during eruption. Horizon 2Bw2. PPL. (**d**) Beige-green volcanic glass with orange alteration hypocoating determined as palagonite. 3Bw/4C horizon. PPL

material redistributed by wind erosion of Brown Andosols in the vicinity of the profile. Pumice and pyroclasts are rounded throughout the profile also indicating reworking and/or abrasion by wind erosion. Locally the pumice preserved a coating of fine glass (armoured pyroclast) formed during the eruption (Fig. 4c).

The microstructure throughout the profile is fine granular (Fig. 4b) to enaulic in the more sandy layers, except for the 2Bw2 horizon which is more massive. It is a very heterogeneous material consisting for a large part of rounded soil nodules of different colour and limpidity pointing to a transported material, probably by saltation or wind. O. Arnalds identified the horizon to be 2800 yrs old rhyolitic tephra from Mt. Hekla volcano.

Dense, reddish brown clayey cappings on pyroclasts in the 3Bw3/4C horizon have no specific orientation, suggesting a pedoturbation of the original soil material e.g. by cryoturbation. Romans et al. (1980) described cappings in young soils on till in southern Iceland. In the 4C the cappings are better expressed, related to former permafrost.

Subhorizontal oriented tissue residues are observed throughout the Ah1 horizon, but are missing in the Ah2. Compared to the other two pedons (Pedons 7 and 9), the

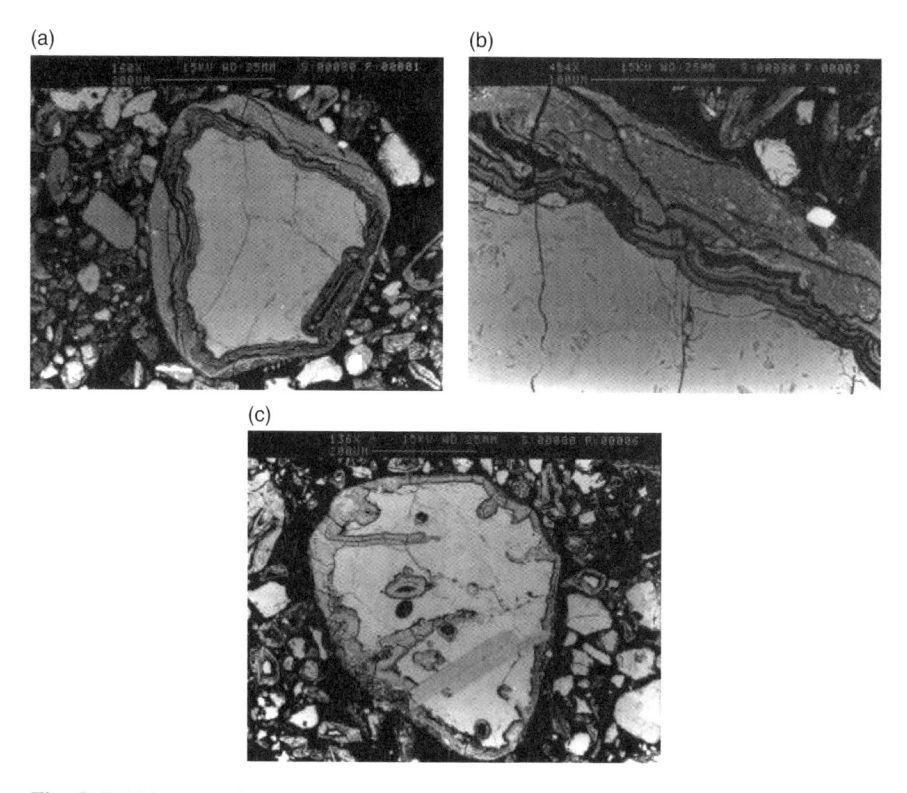

Fig. 5 SEM images of orange alteration hypocoatings on volcanic glass in horizon 3Bw/4C of Pedon EUR-8. (**a**) Orange layered alteration hypocoating on glass grain (**b**) idem, detail (**c**) Layered orange alteration coating. Note the presence of unaltered lath shaped minerals

presence of organic material is much less, most likely due to more freely drained landscape position.

Starting from the 2Bw2 horizon, a different type of material appears, probably locally transported soil sediment, overlying an in situ weathered, but pedoturbated glaciated deposit, as shown by the random arrangement of the cappings.

4.3 Pedon EUR-9

This pedon is situated in an area with milder climate compared to the other two pedons studied, with mild winters and cool summers (Table 1). It is located in the wetland lowlands of South Iceland (Fig. 1). Catastrophic wind erosion in the vicinity of the area has resulted in large additions of aeolian materials, mainly andesitic glass and rhyolitic tephra originating from Mt. Hekla, while the tephra is both andesitic (Mt. Hekla) and basaltic (Katla and other volcanoes in the south).

The EDS results (Table 5, samples 8 and 9) clearly point to the presence of both basaltic and rhyolitic components in this profile. Settlement of the area, about AD 870, caused an increased amount of aeolian deposition, which results in a clear boundary between the 3C (Settlement tephra, identified by O. Arnalds) and the 4H horizons. The parent materials of the lower half were deposited over about 8000 yrs, from the end of Quaternary to the start of human settlement).

According to the field description (Table 2), stratification is present throughout the pedon, although somewhat disturbed by cryoturbation in the top layers where also a marked collemboli activity is noticed. Alternative layers of organic and mineral materials are observed throughout the pedon (Table 3; Fig. 6a). Within the organic layers, sub-layers of less than 1 mm can often been distinguished, consisting of alternations of larger organ residues (mainly leafs) and laminae composed mainly of mashed, isotropic cells and cell residues mixed with amorphous fine material.

The leafs and other larger organ residues show distinct parallel orientation, clearly sub-horizontal in the lower part of the pedon (Fig. 6b), but often tilted in the top layers. Even in the horizon designated as C, the alternation of organic and mineral material is evident where the mineral layers of a few mm, generally with

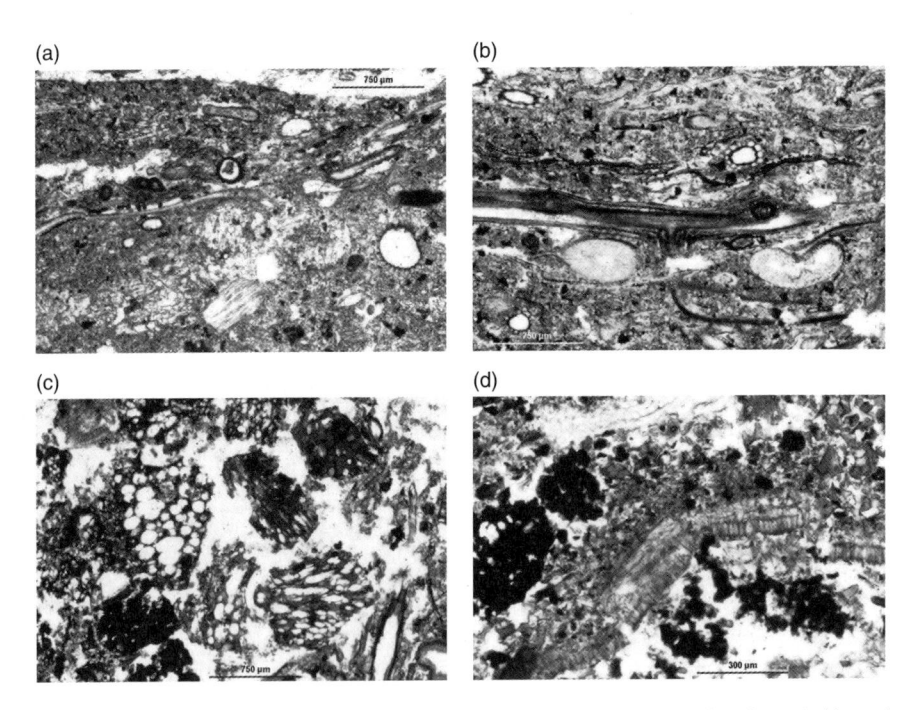

(a)   (b)
(c)   (d)

Fig. 6 Micrographs of Pedon EUR-9 (**a**) Sharp contact between mineral micro-layer (with sand sized pumice grains) and overlying micro-layer of organ residues. 3C horizon, PPL. (**b**) Parallel orientation of well preserved organ residues mixed with fine, relatively fresh, organic fine material. 3C horizon, PPL. (**c**) Vesicular scoria of basaltic composition and organ residues. 4H horizon, PPL. (**d**) Capping of fine material on top of organ residue. 4H horizon, PPL

a coarse monic c/f related distribution, are either composed of sand sized, fresh colourless pumice in a matrix of fine colourless glass shards, or of generally well sorted greenish, fresh volcanic glass shards with concave faces and sharp edges (Fig. 6c).

The intensity of the interference colours in the organic matter decreases with depth as a result of increased degradation. The mashed organic material has a reddish brown colour, reflective of peat deposits, indicating very slow humification. High amounts of elongated phytoliths, even as some diatoms with bilateral symmetry, are found in the layers with mashed tissue residues, reflecting the wet surface conditions. Fungal hyphae are omnipresent in the organic layers.

The platy microstructure in 3H is considered to be related to the parallel layering of the leafs, rather than to freeze – thaw phenomena. However, cappings (Fig. 6d) often with reverse sorting, observed between 100–230 cm on organ residues are signs of former frost activity. It is, as far as we know, the first time that such cappings are reported on organic components. The presence of large birch fragments (up to 20–30 cm in diameter) in these horizons indicates a dryer and perhaps warmer climate during their formation, which is well established in Icelandic wetlands (about 9000–7000 and 5000–2500 BP; Einarsson 1994). The thin sections made of air-dried samples (Ah1 and Ah2) are comparable to those described above, except for the presence of larger shrinkage cracks, and contraction of areas dominated by amorphous constituents.

5 Discussion

5.1 Depositional Features

Two striking features distinguish the parent material of these three pedons from that of most other Andosols: a clear microstratification and a (sub)horizontal, parallel orientation of the elongated plant residues in the organic layers.

The microstratification is caused by an alternation of fine (sometimes only a few millimetre thick) layers composed of tephra of variable composition, e.g. basaltic glass alternating with rhyolitic components. Sometimes this feature is very striking as for instance in Pedon EUR-9, where thin layers with pumice and with green basaltic glass alternate. Many layers are composed of angular, well-sorted pyroclasts, and devoid of organic material, pointing to a rapid tephra fallout as also suggested by Simpson et al. (1999). This does, however, not exclude aeolian origin, as the wind abrasion often attacks rather uniform, large tephra deposits. In other layers a mixed composition of the material is observed, most probably related to aeolian transported sediments. The microstratification observed in thin sections in these profiles illustrates clearly the lithological discontinuities, better than those based on analyses of bulk samples (e.g. Buurman et al. 2004 and optical mineralogical analyses of grain mounts).

The sub-horizontally oriented orange organ (mainly leaf) and elongated tissue residues, even in deeper parts of the profile, were also observed by Simpson et al. (1999) and by Milek (2004) in the natural soils of Hofstaðir. A similar observation was made also by Romans et al. (1980) who explained it as the result of a "glacial outwash or a glacier burst". Such explanation is improbable in the studied soils in tephra deposits in the area studied. It can rather be explained by a slow deposition and a rapid burial of a litter layer by a mineral layer, before humification takes place.

The sub-horizontal orientation of the organic residues, the microstratification of both the organic and mineral layers and the alternating composition of the latter, clearly reflect the event driven deposition of the soil parent material and the limited pedoturbation (bioturbation and cryoturbation) that would homogenise the soil, both with respect to composition and arrangement. The relative pureness of the mineral layers, the freshness of the volcanic components and their sorting indicate direct deposition during eruptions, rather than a water or air transport over land. It means that every sub-layer corresponds to a former soil surface in a temporarily relatively stable environment.

5.2 Pedoturbation

At present, bioturbation seems limited as shown by the preservation of micro–stratification and the concentration of pumice in the upper 5 cm of Pedon 8. In Indonesia similar recent pumice deposit of the Krakatau eruption in 1883 is already much more incorporated into the topsoil (Fauzi and Stoops 2004). Milek (2004), however, describes a considerable biological activity in the soils, pre-dating the human occupation of the Hofstaðir site, formed in the Settlement tephra sequences (Landnám in Icelandic). This may reflect warmer climatic conditions during these earlier times, as confirmed by the organic matter in the deeper part of Pedon 9.

5.3 The Coarse Mineral Fraction

Data of the optical analysis of the sand fraction in grain mounts are confirmed by thin section studies. The latter are not quantitative, but are qualitatively of a much higher standard. Grain mounts are for practical reasons restricted to relative limited size fractions, with a lower limit at 50 µm and an upper limit at about 500 µm. Smaller grains are difficult to identify with normal optical techniques, whereas larger grains become too thick to allow identification in transmitted light. Larger and compound grains (rock fragments, pyroclasts) can generally not be identified in mounts, but form no problem in thin section studies. Therefore thin section studies yield much more useful information. It is also important to recall that soil analyses, made on bulk samples, represent an average composition of each horizon, whereas micromorphology clearly shows sub-layers of a few millimetres with different mineralogical and/or petrographic composition, allowing the distinction of specific mineral associations or para-geneses.

The presence of internal orange alteration coatings is only found in one horizon of Pedon 8 and therefore considered as being part of the parent material, namely as the result of a palagonitisation process.

5.4 Organic Matter

The organic constituents of Pedons EUR-7 and EUR-9 display characteristics of materials deposited in stagnant water, a conclusion supported by the presence of diatom skeletons with bilateral symmetry and sponge spicules. The organic material has peat-like appearance. Neither pyrite nor other iron minerals were detected, in contrary to the findings of Gudmundsson and FitzPatrick (2004) for Icelandic Histosol. The observed alternation of reddish brown, well preserved organ residues overlying reddish brown fine (mainly amorphous) organic material, characteristic for some layers suggests limited humification before burial. Similar observations were made by Shoba et al. (2004) in Kamchatka soils.

5.5 Soil Sediments and Micromass Coatings

The presence of heterogeneous material composed of soil aggregates or nodules in some horizons (Pedon EUR-7, 3BC, Pedon EUR-8, 2Bw2) with different colour and limpidity suggest that the deposition of soil sediments derived from older, pedo-genic more evolved soils. In these layers pyroclasts are always surrounded by coatings of fine material, comparable to that of the nodules. The transport process is most likely soil creep or saltation.

A typic feature observed in Pedons EUR-7 and EUR-8, but common in many other COST 622 reference profiles, is the presence of a coating of fine material (micromass) surrounding the pyroclasts. In the case of porous pumice and scoria, an internal hypocoating is formed consisting of the same material filling the vacu-oles at the rim of the particle. In the case of the Bw horizons, these coatings and hypocoatings have the same variable composition as the above-mentioned nodules or aggregates.

Thick coatings of fine material enveloping coarse grains or aggregates have been described by Rose et al. (2000) in Devensian sediments in South England. The authors explain the genesis of these "snowballs" structures by aeolian transport, probably with accretion on a damp surface. As the feature is quite common in young volcanic materials we attribute this to a pedogenic process. Hypocoatings and coatings of fine material on pumice grains have been observed in several young volcanic ash soils of Europe, e.g. Santorini, Massif Central, and southern Italy (Stoops and Gérard 2004, Stoops and Gérard 2007).

5.6 Microstructure and Freeze-Thaw Fabric

The microstructures observed in the mineral horizons of the pedons grade from monic over enaulic to granular. When more fine material is present, a blocky or lenticular microstructure appears, however always with an intrapedal granular one. The origin of the omnipresent granular microstructure, either as a pedal microstructure or as an intrapedal one, is most likely not due to intense activity of the mesofauna, as often considered, because all other micromorphological features point to a very low biological activity. As a granular microstructure is universally observed in Andosols (Stoops 1983b, Stoops and Gérard 2004), it seems to be rather linked to the composition of the colloidal fraction. Its high stability is illustrated by the fact that it survives to freeze-thaw processes, and persists inside lens shaped peds.

The fine platy or lenticular microstructures as observed in Pedon EUR-7 and EUR-9 were also reported by Romans et al. (1980) for young soils they studied in southern Iceland. It corresponds to the isoband fabric described first by Dumanski and St. Arnaud (1966) in eluvial horizons of soils of Saskatchewan. Other authors (e.g. Van Vliet-Lanoë 1976, Mermut and St. Arnaud 1981, Van Vliet 1985) have described similar fine platy structures in deeper horizons, and explained its origin as a freeze-thaw fabric. The observed platy peds are somewhat different from those described in literature. Van Vliet-Lanoë and Coutard (1984) mention that the platy peds are "separated by smooth fissures, gently undulating and with unconformable walls, as a main contrast with desiccation cracks". In the present study, the surfaces of the peds are less smooth, due to the fine granular intrapedal microstructure. The weaker expression of an isoband fabric in Pedon 8, where climatic conditions are considered suitable for the formation of this feature, can be explained by the relative coarse texture of the parent material.

In the 3Bw3/4C horizon of Pedon 8, reddish brown clayey cappings occur on pyroclasts without specific orientation, suggesting pedoturbation of the original soil material. Van Vliet (1985) described similar features and considered them as an indication of gelifluction. Other specific characteristics of freeze-thawing, such as uplifted stones, vertical orientation of gravels and vesicles described by the same author were not observed the pedons studied. The latter can be explained by the relative coarse texture of the material near the surface. In the interpretation of cryogenic features, much attention is often given to striated b-fabrics (Huijzer 1993). Due to the isotropic nature of the micromass in the volcanic ash soils studied, this criterion could not be used.

6 Conclusions

According to the results of our study and additional data from literature, the micro–morphology of the volcanic ash soils of Iceland differs from those of the temperate regions:

1. absence of weathering at microscopic scale;
2. restricted humification of the organic matter and its accumulation as peat-like material;
3. dominance of monic and enaulic c/f related distributions, grading to a granular microstructure;
4. presence of stratification and microstratification and (sub)horizontal, parallel oriented organ residues, both pointing to an event driven deposition of the parent material, either as tephra or as aeolian sediment, and the absence of pedoturbation, especially bioturbation;
5. presence of, sometimes weakly expressed, platy and/or lenticular microstructures, corresponding to the so-called isoband fabric, if sufficient fine material is present;
6. Bw horizons composed of transported, more developed soil material.

All these characteristics can be related to the special physical environment of Iceland, namely its cryic/frigid climate and the ongoing volcanic and aeolian activity.

Many other micromorphological characteristics correspond to those typical for Andosols in temperate regions and young Andosols in tropical regions, namely:

- a dominance of granular microstructures throughout the profile, either on pedal or intrapedal level;
- the coarse material of the groundmass being formed of volcanic rock fragments and volcanic minerals, especially subhedral feldspar and augite;
- an undifferentiated b-fabric in the micromass;
- the occurrence of coatings and internal hypocoatings of micromass on larger coarse constituents.

Acknowledgements The authors thank Dr. F. Bartoli for providing thin sections of air-dried samples of the A and B horizons of the three pedons. Mrs. K. Milek is thanked for providing unpublished information on micromorphology of the Hofstaðir excavations.

References

Arnalds O (1990) Characterization and erosion of Andisols in Iceland. Ph.D. dissertation, Texas A&M University, College Station, Texas

Arnalds O (2004) Volcanic soils of Iceland. Catena 56: 3–20

Arnalds O, Kimble J (2001) Andisol of deserts in Iceland. Soil Science Society of America Journal 65: 1778–1786

Arnalds O, Stahr K (eds.) (2004) Volcanic Soil Resources: Occurrence, Development and Properties. Catena Special Issue, Vol. 56

Arnalds O, Hallmark CT, Wilding LP (1995) Andisols from four different regions of Iceland. Soil Science Society of America Journal 59: 161–169

Arnalds O, Gisladottir FO, Siburjonsson H (2001) Sandy deserts of Iceland. Journal of Arid Environments 47: 359–371

Arnalds O, Bartoli F, Buurman P, Garcia-Rodeja E, Oskarsson H, Stoops G (eds.) (2007) Soils of Volcanic Regions of Europe. Springer Verlag, Berlin, Heidelberg, New York

Bartoli F, Buurman P, Delvaux B, Madeira M (eds.) (2003) Volcanic soils: Properties and processes as a function of soil genesis and land use. Geoderma Special Issue, Vol. 117

Buurman P, Garcia Rodeja E, Martinez Cortizas A, van Doesburg JDJ (2004) Stratification of parent material in European volcanic and related soils studied by laser-diffraction grain-sizing and chemical analysis. Catena 56: 127–144

Driessen P, Deckers J, Spaargaren O, Nachtergaele F (eds.) (2001) Lecture Notes on the Major Soils of the World. World Soil Resources Report 94, FAO, Rome

Dumanski J, St. Arnaud RJ (1966) A micropedological study of eluvial soil horizons. Canadian Journal of Soil Science 46: 287–292

Einarsson Th (1994) Myndun og mótun lands. Jardfraedi. (Geology). Mal og Menning, Reykjavik, Iceland

FAO (1988) World Reference Base for Soil Resources. World Soil Resources Report 84. FAO, Rome

Fauzi AI, Stoops G (2004) Reconstruction of a toposequence on volcanic material in the Honje Mountains, Ujung Kulon Peninsula, West Java. Catena 56: 45–66

Gudmundsson Th (1978) Pedological Studies of Icelandic Peat Soils. PhD thesis, University of Aberdeen

Gudmundsson Th, FitzPatrick EA (2004) Micromorphology of an Icelandic Histosol. In: Oskarsson H, Arnalds O (Eds.) Volcanic Soil Resources in Europe. Cost Action 622 final meeting, Abstracts. Rala Report no. 214, Agricultural Research Institute, Reykjavik, Iceland, 79–80

Huijzer AS (1993) Cryogenic microfabrics and macrostructures: Interrelations, processes, and paleoenvironmental significance. Ph.D. thesis Universiteit Amsterdam

Kubiëna WL (1953) The Soils of Europe, Thomas Murby & Co, London

Mermut AR, St Arnaud RJ (1981) Microband fabric in seasonally frozen soils. Soil Science Society of America Journal 45: 578–586

Milek KB (2006) Aðalstraeti, Reykjavik, 2001: Geoarchaeological report on the deposits within the house of the soils immediately pre- and post-dating its occupation. In: Roberts HM (ed.), Excavations at Aðalstraeti 2003. Reykjavik: Fornleifastofnun Islands, 73–114

Murphy CP (1986) Thin Section Preparation of Soils and Sediments. A B Academic Publishers, Berkhamsted

Oskarsson H, Arnalds O (2004) Volcanic Soil Resources in Europe. COST Action 622 final meeting. Abstracts. Rala Report no. 214. Agricultural Research Institute, Reykjavik, Iceland

Quantin P, Spaargaren O (2007) Classification of the Reference Pedons: World Reference Base for Soil Resources and Soil Taxonomy. In: Arnalds O, Bartoli F, Buurman P, Garcia-Rodeja , Oskarsson H, Stoops G (eds.), Soils of Volcanic Regions of Europe. Springer Verlag, 231–249

Romans JCC, Robertson L, Dent DL (1980) The Micromorphology of Young Soils from South-East Iceland. Geografiska Analer A62: 93–103

Rose J, Lee AJ, Kemp RA, Harding PA (2000) Palaeoclimate, sedimentation and soil development during the Last Glacial Stage (Devensian), Heathrow Airport, London, UK. Quaternary Science Review 19: 827–847

Shoba S, Targulian V, Sedov S, Sakharov A, Zacharichina L (2004) Pedogenesis and weathering on tephra: Climate and the dependency. 12th Intern. Meeting on Soil Micromorphology, Adana, Extended Abstracts 37–39

Simpson IA, Milek KB, Gudmundsson G (1999) A reinterpretation of the Great Pit at Hofstaðir, Iceland using Sediment Thin Section Micromorphology. Geoachaeology 14: 511–530

Soil Survey Staff (1998) Keys to Soil Taxonomy, 8th ed. USDA-NRCS, Washington, DC

Stoops G (1983a) SEM and light microscopic observations of minerals in bog-ores of the Belgian Campine. Geoderma 30: 179–186

Stoops G (1983b) Mineralogy and micromorphology of some andisols of Rwanda. In: Beinroth FH, Neel H, Eswaran H, (eds.), Proc Fourth Intern Soil Class Workshop, Rwanda 2–12 June 1981. Part 1: Papers. ABOS-AGCD, Agricultural Editions 4, 150–164

Stoops G (2003) Guidelines for the Analysis and Description of Soil and Regolith Thin Sections. SSSA. Madison, WI.

Stoops G, Van Driessche A (2002) Mineralogical composition of the sand fraction of some European volcanic ash soils. Preliminary data. Mainzer naturwiss Archiv 40: 31–32

Stoops G, Gérard M, (2004) Micromorphology of the volcanic ash soils of the Cost-622 reference profiles. In: Oskarsson H, Arnalds O (eds.), Volcanic Soil Resources in Europe. Cost Action 622 final meeting, Abstracts. Rala Report no. 214. Agricultural Research Institute, Reykjavik, Iceland, 12–13

Stoops G, Gérard M (2007) Micromorphology. In: Arnalds O, Bartoli F, Buurman P, Garcia-Rodeja E, Oskarsson H, Stoops G (eds.), Soils of Volcanic Regions of Europe. Springer Verlag, 129–140

Stoops G, Van Driessche A (2007) Mineralogy of the sand fraction – results and problems. In: Arnalds O, Bartoli F, Buurman P, Garcia-Rodeja E, Oskarsson H, Stoops G (eds.), Soils of Volcanic Regions of Europe. Springer Verlag, 141–153

Van Vliet-Lanoë B (1976) Traces de ségrégation de glace en lentilles associées aux sols et phénomènes périglaciaires fossiles. Biuletyn Peryglacjalny 26: 41–55

Van Vliet B (1985) Frost Effects in Soils. In: Boardman, J. (ed.), Soil and Quaternay Landscape Evolution, John Wiley & Sons, 117–158

Van Vliet-Lanoë B, Coutard J-P (1984) Structures caused by repeated freezing and thawing in various loamy sediments: A comparision of active, fossil and experimental data. Earth Surface Processes and Landforms 9: 553–565

Improved Paleopedological Reconstruction of Vertic Paleosols at Novaya Etuliya, Moldova Via Integration of Soil Micromorphology and Environmental Magnetism

A. Tsatskin, T.S. Gendler, and F. Heller

Abstract The integration of micromorphology, environmental magnetism and Mössbauer spectroscopy along with identification of pedogeomorphic classes of paleosols may improve the quality of Pleistocene paleopedological reconstructions and their applicability to loess stratigraphy. Our case study from the SW Black Sea area demonstrates that microfabrics, mineralogy and magnetism of paleosols older than 0.5 Ma are comparable with certain types of Mediterranean soils from Israel and Turkey. The diversity of paleosols in terms of pedogeomorphic, genetic, and diagenetic typology is prominent. Two welded accretionary paleosols, i.e. pedocomplexes PK4 (~0.5 Ma) and PK8 (~0.9–1.0 Ma), have a distinct rubefied A (AB) horizon with micromorphological features of high biological activity, quasi-isotropic humus-clayey groundmass and compacted excremental fabric. Vertic micromorphological features, more intense in the older PK8, are superposed upon bio-related microstructures. Both paleosols show magnetic enhancement in a topsoil due to increased concentration of pedogenic <20 nm superparamagnetic (SP) and stable single domain magnetite which is partly oxidized. In these PKs the concentration of ferrimagnetics is correlated with antiferromagnetic SP hematite/goethite and paramagnetic Fe(III)-clay content. In contrast, other >0.5 Ma old paleosols which are related to a transformational welded type, show a small magnetic susceptibility signal, a high concentration of paramagnetic clays and SP Fe oxyhydroxides. Fe/Mn impregnations are juxtaposed in the microfabric with abundant soft masses of micritic calcite and aggregates of stress coatings. The latter may have originated from decay of clay coatings under intermittent waterlogged conditions. In the time

A. Tsatskin
Zinman Institute of Archaeology, University of Haifa, Haifa 31905, Israel,
e-mail: tsatskin@research.haifa.ac.il

T.S. Gendler
Institute of Physics of the Earth, Russian Academy of Sciences, Bolshaya Gruzinskaya 10,
Moscow 123810, Russia, e-mail: gendler06@mail.ru

F. Heller
Institut für Geophysik, ETH Hönggerberg, CH-8093 Zürich, Switzerland,
e-mail: heller@mag.ig.erdw.ethz.ch

S. Kapur et al. (eds.), *New Trends in Soil Micromorphology*,
© Springer-Verlag Berlin Heidelberg 2008

interval ~0.5–~1.5 Ma the rubefied and magnetically enhanced pedocomplexes PK4 and PK8 are proposed as key stratigraphic markers in South Eastern Europe.

Keywords Paleopedology · pleistocene · vertic soils · micromorphology · magnetism · mössbauer effect

1 Introduction

Soil micromorphology along with macro- and mesomorphological investigations of soil profile organization are essential tools in paleopedology (Courty et al. 1989, Retallack 1990, Bronger 2003, Holliday 2004). In Pleistocene loess/paleosol sequences, the best terrestrial records of long-term global change, micromorphology aids in identifying specific soil-forming and sedimentary processes, and is increasingly complemented with other climatic proxies, e.g. magnetic susceptibility (Kemp 1998 and references therein). The last few decades have witnessed the substantial growth of rock-magnetic and paleomagnetic studies in loess/paleosol sequences in the world (Evans and Heller 2003). Pioneering work on the Chinese Loess Plateau (Heller and Liu 1982, 1986) enabled the establishment of geomagnetic polarity boundaries, including the Matuyama/Brunhes boundary (MBB), and showed that magnetically expressed loess/soil cycles above the MBB correspond well to the global marine oxygen isotope stages (Heller and Evans 1995). Since then, much effort has been put into understanding the nature of magnetic parameter fluctuations in loess/paleosol sequences in China, Europe, Siberia, North and South America, etc. A current prevailing view regarding China, central East Europe, and Middle Asia is that magnetic enhancement is due to pedogenic production and post-burial preservation of ultrafine-grained magnetite in topsoils (Maher 1998, Evans and Heller 2003).

In the East European plain, paleosols were shown to vary taxonomically with age. They were proposed as stratigraphic markers for different Pleistocene interglacials/-stadials taking into account zonal soil changes, i.e. those along the climatic gradient from humid to arid areas (Velichko 1973, Veklitch 1982, Sirenko and Turlo 1986, Morozova 1995). Catt (1988) questioned the validity of this approach while arguing that the concrete results of two leading Soviet teams appeared controversial. However, during recent interdisciplinary work in the western Black Sea area, we found that paleosols are indeed recognizable and change morphologically along the sequence from paleo-Chernozems towards reddish-brown Mediterranean soil types at the stratigraphic level around and below the MBB (Tsatskin et al. 1998, 2001, Gendler et al. 2006).

This chapter provides more detailed information and new insights into vertic reddish brown paleosols in a loess/soil section at Novaya Etuliya (45.5°N, 28.5°E) in the SW Black Sea area, as well as into Terra Rossa in the Lower Galilee, Israel, obtained by soil micromorphology, rock magnetism and Mössbauer spectroscopy. A comparison is made with the surface soils in Mediterranean climate with ~500–600 mm annual rainfall (Kapur et al. 1997, Yaalon 1997, Mermut et al. 2004). According to available paleomagnetic data, the reddish-brown paleosols at Novaya Etuliya are in the time range of ca. 0.5 Ma through ca. 1.5 Ma. Using a broader perspective, we will show that the integration of micromorphology with environmental magnetism techniques may improve the quality of Pleistocene paleopedological reconstructions and their applicability to stratigraphy.

2 Materials and Methods

Paleosols at Novaya Etuliya were described in the field according to Munsell color, texture, structure, consistence, porosity, secondary neo-formations (Catt 1990). Soil classification was based on the Russian and WRB systems (Classification 1986, Klassifikatsiya 2004, World Reference Base 1998). Petrographic 25×40 mm thin sections were prepared in the Institute of Geography, Russian Academy of Sciences (Moscow). Because of their small size, several thin sections were prepared from adjoining samples within the A horizons of major paleosols. Micromorphological descriptions follow the guidelines developed by Bullock et al. (1985) and Stoops (2003). Specific magnetic low field susceptibility χ_{LF} was measured on samples taken at 5 cm intervals along the section using a Bartington susceptibility meter at ETH Zürich (Spassov 1998). Magnetization (J_i) induced in a 0.45T magnetic field and saturation remanent magnetization (J_{rs}) acquired in a 0.85T magnetic field were measured in the Institute of Physics of the Earth, Russian Academy of Sciences on both bulk samples and separated clay fractions. In addition, J_i and J_{rs} were measured in a coercivity spectrometer, constructed by Pavel Yasonov (see Evans and Heller 2003), up to 0.3 T at ETH, Zurich. Temperature-dependence of $J_i(T)$ and $J_{rs}(T)$ in the temperature interval 20–700°C was measured for selected samples on a vibrating sample magnetometer and thermomagnetometer (produced in the Geophysical observatory, Borok). To identify paramagnetic Fe compounds, Mössbauer spectra were collected at room and liquid-nitrogen (ca. 80 K) temperature in the Physics Department of Moscow State University using a Mössbauer spectrometer with ^{57}Co(Rh) source in constant acceleration mode. The spectrometer was calibrated with a standard α-Fe absorber in a velocity range ±8.5 mm/s and $Na_2[Fe(CH)_5NO] \cdot 2H_2O$ in a velocity range ±3 mm/s. The spectra were computed using a least-squares fitting program (Gendler et al. 2006). To increase Mössbauer spectra intensity and resolution, clay fractions separated from paleosols and loess at Novaya Etuliya were measured. Only bulk samples of Terra Rossa were measured by magnetic and Mössbauer techniques.

2.1 Morphology, Pedogeomorphic Development and Magnetic Susceptibility of Paleosols at Novaya Etuliya

The loess/paleosol sequence at Novaya Etuliya is exposed in a gully cut into the Danube terrace of Pliocene age. The site is one of the best-known Quaternary type sections in the East European plain due to its rich faunal remains in the ancient alluvium and established paleomagnetic dating (Azarolli et al. 1997, Faustov et al. 1986, Nikiforova 1982, 1997, Velichko 1990, Virina et al. 2000a). More data on the site's geographical location and geomorphology are given in Tsatskin et al. (2001). The loess/soil sequence is up to ca. 30 m thick and contains 12 buried paleosols below a surface soil, designated PK1 (Fig. 1) with diagnostic features of a Chernozem (Klassifikatsiya 2004). The left column in Fig. 1 shows coded major morphologic features of the paleosols from PK1 through PK13, while only well preserved loesses L2, L3, L4 and L7 are indicated. Note that the L7 loess is the oldest typical yellowish loess in the section, which is related to the Matuyama chron (right column in Fig. 1). Field-determined Munsell colors of the A (AB) horizons of paleosols are shown next to the soil-stratigraphic column. The next column provides field assessment of the pedogeomorphic development in terms of incipience, welding, truncation, intensity of post-burial preservation or diagenesis (Yaalon 1971). All major paleosols show a Chernozem type profile, i.e. $A_{(k)}$, strongly bioturbated, with krotovinas B_k, and C_k horizons. However, the paleosols are consistently more complex than surface Chernozem due to over-thickness, a strong prismatic structure in the $A_{(k)}$ horizons and co-existence of various types of calcareous concretions in topsoils. Due to regional correlation among Novaya Etuliya and the previously studied Roxolany and Kolkotova Balka (Tsatskin et al. 1998) here we will refer to the paleosols in question as pedocomplexes (PKs).

The last column in Fig. 1 displays magnetic susceptibility variations and placement of the Matuyama/Brunhes boundary (MBB) in the upper part of PK7 and of the Jaramillo paleomagnetic subchron (ca. 0.9–1 Ma) in PK8 (Spassov 1998). Hence, paleomagnetism allows us to assess and approximate the chronological interval of the sequence. Pronounced magnetic susceptibility enhancement of paleosols vs. the intervening loess layers (Fig. 1), is characteristic of other Black Sea sections as well (Virina et al. 2000a, Tsatskin et al. 1998, 2001, Dodonov et al. 2006). A strongly developed, reddish-brown paleosol with vertic properties PK4 appears at ca.7 m depth. The PK4 shows a thick, double A horizon, with weak slickensides, and a maximal magnetic susceptibility up to $140 \times 10^{-8} m^3 kg^{-1}$ (Fig. 1). It correlates with the welded PK4 at Roxolany (Tsatskin et al. 1998) as well as with the Vorona pedocomplex of the East European plain, whose age is assessed at ca. 0.5 Ma (Velichko 1990, Velichko et al. 1999). Dodonov et al. (2006) question this correlation in the Roxolany type section, although not providing evidence for alternative interpretation. Below PK4, lies the sequence of PK5 through PK7 where the MBB (ca. 0.8 Ma) was found, and as a result, the assessed age of PK4 is supported. The PK5-PK7 show brown, strongly calcareous, firm A_k horizons and soft calcrete in the subsoil. The A horizons are less strongly developed than in the younger PK4, presumably because of truncation or post-burial diagenesis. The pedogeomorphic, or pedosedimentary (Kemp 2001), classification of these paleosols could not solely be established in the field. As will be shown later, micromorphology assisted in the resolution of this question.

At ca. 16 m depth, below the L7 loess, lies the strongly developed and welded PK8 pedocomplex, where the Jaramillo subchron was found, and hence it should be dated to the interval 0.9–1 Ma. The PK8 consists of two weakly calcareous paleosols

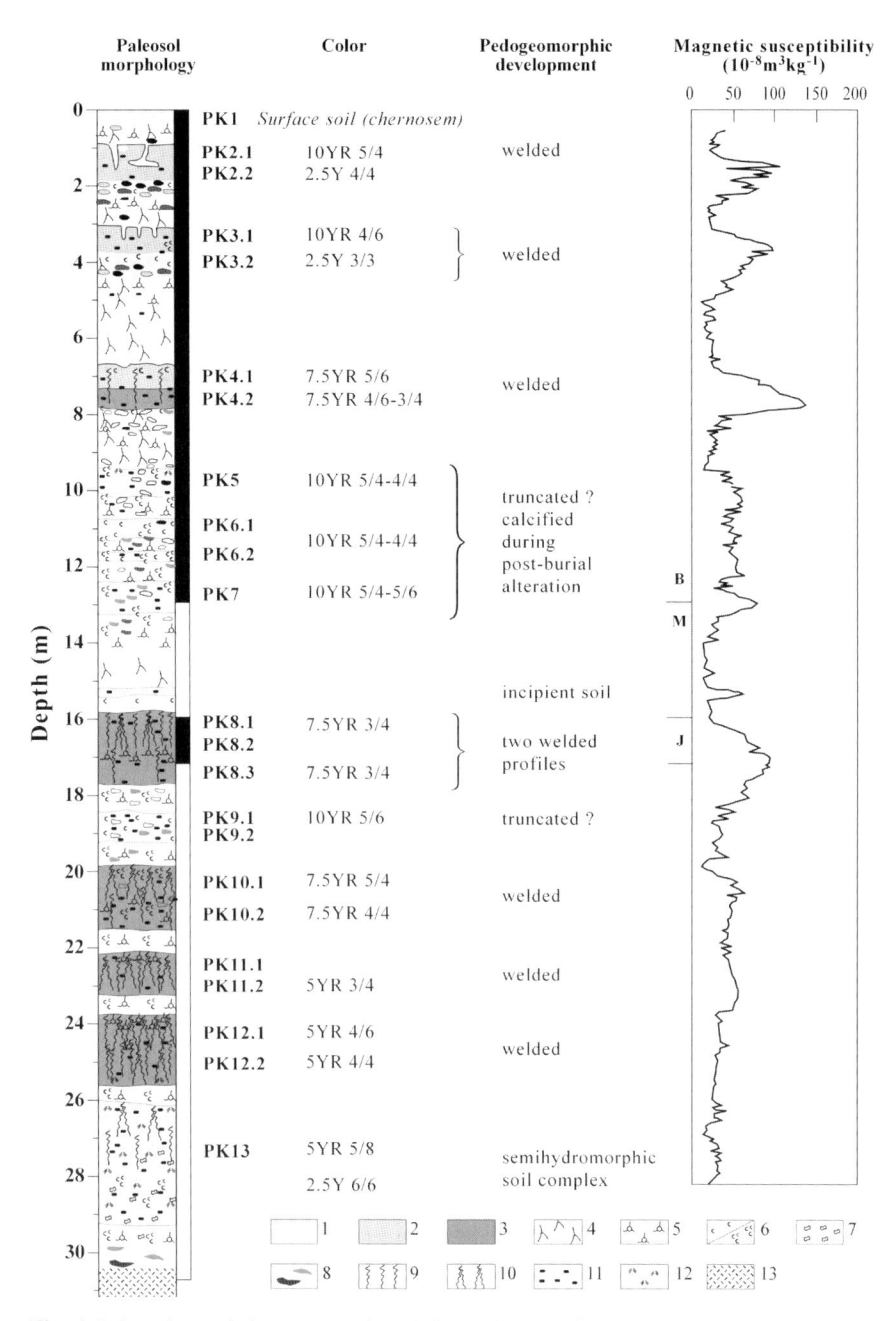

Fig. 1 Paleosol morphology, magnetic polarity and magnetic susceptibility depth functions at Novaya Etuliya. PK = pedocomplex; Magnetic polarity column: black – normal polarity; white – reversed polarity; B – Brunhes chron; M – Matuyama chron; J – Jaramillo subchron. 1 – weak A (AB$_k$) horizon; 2 – A(AB$_k$) horizon; 3 – strong A (AB$_k$) horizon; 4 – loess; 5 – calcareous nodules (beloglazka); 6 – soft powdery carbonates few/abundant; 7 – loess dolls (incipient calcrete); 8 – krotovinas (burrows) filled with light- or dark-colored material; 9 – weak slickensides and prismatic structure; 10 – strong oblique slickensides with shiny faces; 11 – Fe-Mn nodules, mottles, stains; 12 – gley mottles; 13 – alluvial silt

(an upper PK8.1 and a lower PK8.3) with a strongly calcareous PK8.2 in between. The PK8 is morphologically similar to PK4, but much thicker, has darker color as well as stronger vertic properties that are manifested by intersecting slickensides, forming parallelepipeds and angular blocks with shiny faces, and by manganese dendrites. The PK8 magnetic susceptibility values approach $100 \times 10^{-8}\,m^3\,kg^{-1}$, i.e. less than in PK4 but much higher than in PK5-PK7 (ca. $50 \times 10^{-8}\,m^3\,kg^{-1}$).

From ca. 20 m depth, soils acquire reddish hues and have abundant indurated, calcrete-like concretions that are mainly found in the subsoil. Notably, only PK11 and PK12 belong to truly red-colored paleosols with 5 YR Munsell readings. At ca. 30 m depth, a polyphase semihydromorphic soil PK13 shows red, ochre and green mottles on top of the alluvial loess. PK9 through PK13 have low magnetic susceptibility ($<50 \times 10^{-8}\,m^3\,kg^{-1}$).

In a second section on the opposite, north-facing bank of the gully we were able to recognize morphological variations of the lowermost PKs, attributable to catenary changes (Birkeland 1999). For example, below PK9, the non-soil sediments designated L10 through L12 are practically absent so that paleosols in fact merge into one another. Their diagnostic horizons however, are more strongly colored, e.g. PK11 enhances from 5YR in the major section to 2.5 YR red, while the lower PK12 has achieved 10R4/8 Munsell color.

2.2 Micromorphology

The micromorphological soil characteristics of reddish-brown and related paleosols from Novaya Etuliya are summarized in Table 1 and illustrated in Figs. 2 and 3. Micromorphological evidence of possible superimposition of slightly different soils within PK4 is provided by distinct microfabrics in the upper PK4.1 and an underlying PK4.2. The PK4.1 has partially compacted fine granular structure intergrading to spongy structure composed of simple subrounded pellets ranging from 50–200 µm in size, with abundant pore space (Fig. 2a). While continuous pores are lacking, there are occasional root channels of 150–400 µm and frequent tiny (less than 100 µm) sinuous pores (Fig. 2a) probably resulting from a partial collapse of the pristine fecal pellets and structural disintegration. Tiny charred pellets from completely decomposed and comminuted vegetal matter occur in the matrix, while occasional thin micritic coatings are found in pores. In the lower PK4-2 the microfabric is calcite free, stained by homogeneous dark humus, and with frequent continuous sinuous interaggregate pores, while lacking spongy microstructure (Fig. 2b and Table 1). Processes of biotic mixing and humus accumulation are widely documented in the East European Chernozems (Gerasimova et al. 1992, Gerasimova 2003). Although the welded PK4 soil clearly demonstrates a Chernozem-type isotropic biologically related microfabric, sporadic weak stress coatings were also observed (Table 1). In applying the concept of a higher-level micromorphological interpretation (Stoops 1994) to the paleosol identification, we may then propose to classify both PK4 soils as associated with

Table 1 Micromorphological features of A (AB) horizons of paleosols at Novaya Etuliya

Pedocomplex, depth, m	Micro-structure	B-Fabric and/or clay concentration features	Secondary carbonates			Fe/Mn impregnations	Formation/syndrome, after Stoops (1994)
			(Hypo) coatings	Soft masses	Nodules		
PK4-1, 7.15–7.20	S+Ch+V; Homogeneous	Isotropic, Weak oriented clay separations	•	–	–	–	Bio-related weakly calcareous
PK4-2, 7.65–7.70	Ch+V; Homogeneous	Isotropic, Weak oriented clay separations	–	–	–	–	Bio-related leached
PK5, 10.40–10.45	M+B Heterogeneous	Crystallitic, rounded non-calcareous aggregates	•	•••	••	•	Strongly calcareous aquic
PK8-1, 16.65–16.70	Ch+V+B Homogeneous	Isotropic, Strong oriented clay separations	•	–	–	–	Bio-related weakly calcareous vertic
PK8-3, 17.60–17.65	Ch+V+B Homogeneous	Isotropic, strong oriented clay separations	•	–	–	–	Bio-related weakly calcareous vertic
PK9-2, 19.0–19.05	M Homogeneous	Crystallitic	••	•••	–	•••	Aquic
PK10-1, 20.50–20.55	M Heterogeneous	Crystallitic, Spheroidal aggregates	–	•••	••	••	Vertic chromic calcareous aquic
PK10-2, 21.15–21.20	M+B Heterogeneous	Crystallitic, Non-calcareous zones	–	•	•	•••	Vertic
PK11-1, 22.65–22.70	M+B Heterogeneous	Crystallitic, aggregates with strial b-fabric	–	•••	••	••	Vertic chromic calcareous aquic
PK11-2, 23.10–23.20	M+B Heterogeneous	Crystallitic, Strong Birefringent clay	–	•	•	•	Vertic chromic
PK12-1, 24.40–24.45	M+B Heterogeneous	Crystallitic aggregates with strial b-fabric	–	•••	••	••	Vertic chromic calcareous aquic
PK12-2, 25.10–25.15	M+B Heterogeneous	Crystallitic, Strong Birefringent clay	–	•	•	•	Vertic chromic

Codes used: S, spongy; Ch, channel; V, vughy; M, massive; B, blocky; – = absent; • = rare; •• = common; ••• = abundant

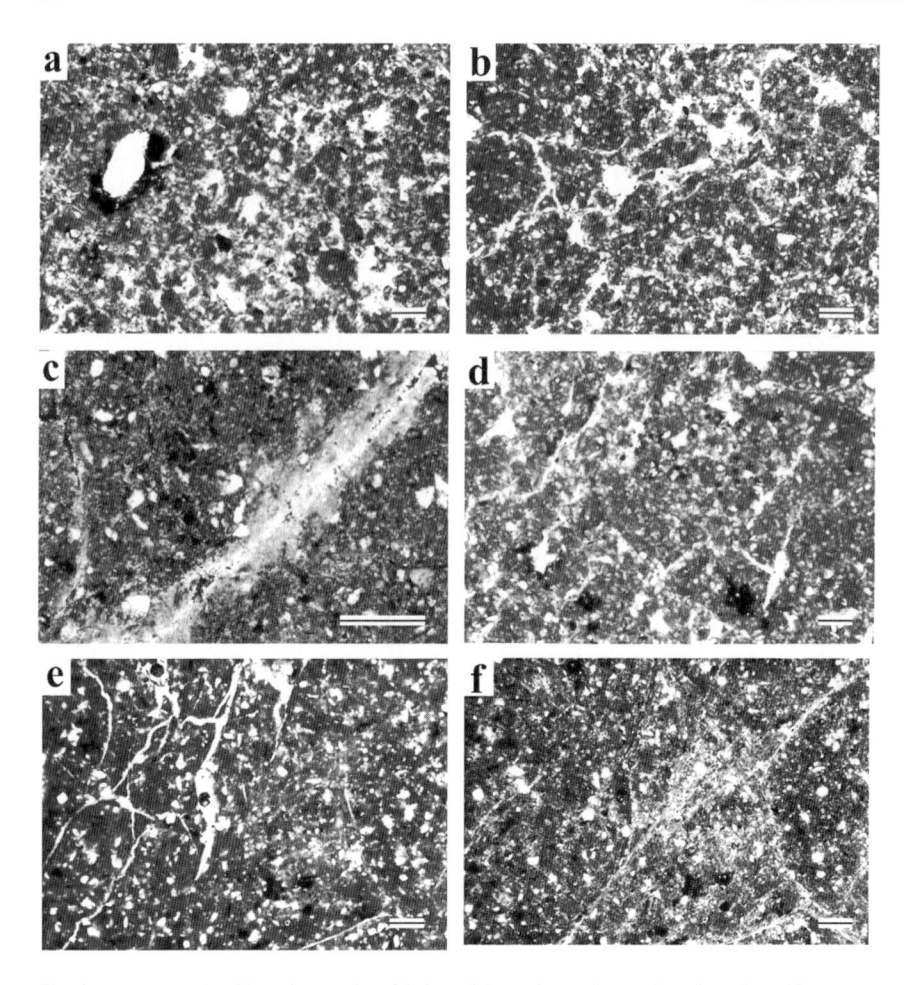

Fig. 2 Representative bio-related microfabrics with vertic syndrome in paleosols at Novaya Etu-liya (for comparison vertic fabric in terra rossa, Lower Galilee, Israel). Plane polarised light (PPL); crossed polarised light (XPL); scale bar = 200 μm; (**a**) Fine granular (spongy) structure in PK4.1, slightly compacted; some root channels and frequent sinuous pores between pellets and rounded aggregates (PPL); (**b**) Continuous interaggregate pores within isotropic humus-rich fine mass in PK4.2 (PPL); (**c**) massive microstructure with orthogonal planar voids coated with polyphase recurrently re-crystallized microsparitic/micritic calcite infillings (note microsparite growing inside the void) in PK8.2 (XPL); (**d**) Welded subangular blocky peds separated by incipient cracks (planar voids) in PK8.3; tiny opaque 20–100 μm pellets presumably originated from completely decomposed and comminuted vegetal matter are more frequent here than in PK4 (PPL); (**e**) Development of blocky pedality and numerous well expressed planar voids in a Holocene terra rossa in Israel (PPL), (**f**) same field of view as (**e**), note groundmass having strong strial b-fabric unaccommodated with cracks visible in (**e**) (XPL)

bio-related micromorphic formation. However, the upper one also shows a calcareous syndrome while the lower one relates to a leached type (the last column in Table 1). If PK4.2 is analogous to semi-humid steppe soils, then the younger PK4.1 soil may have developed under conditions of increased aridity (probably

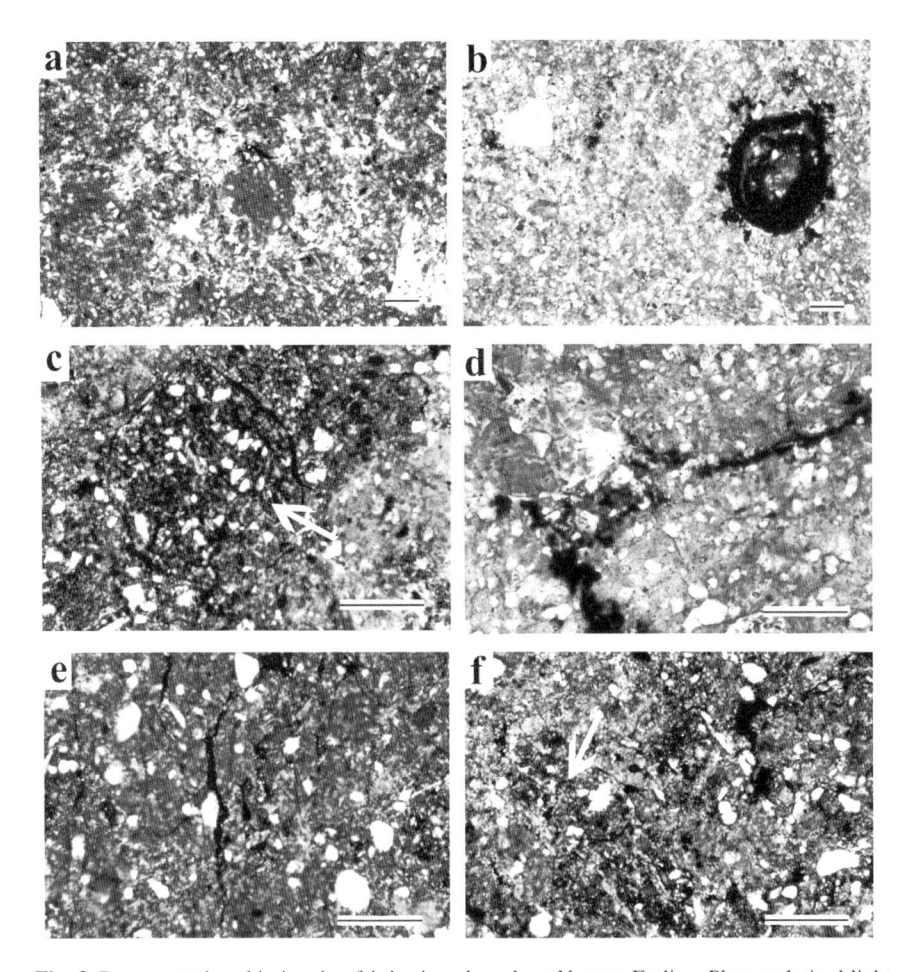

Fig. 3 Representative abiotic microfabrics in paleosols at Novaya Etuliya. Plane polarised light (PPL); crossed polarised light (XPL); scale bar 200 μm; **(a)** weak to strong impregnation of the matrix by micrite with 0.4 mm typical micritic nodule of irregular shape (center) in PK5 (XPL); **(b)** Very dense massive fabric with a 0.5 mm concentric Fe/Mn nodule with sharp boundary and a halo of tiny mottles in PK9.2 (XPL); **(c)** Rounded aggregate with stipple-speckled b-fabric (arrow) embedded within calcareous groundmass having crystallitic b-fabric in PK10.1 (XPL); **(d)** Compacted, blocky, reddish-brown fine mass with Fe/Mn infillings along bifurcated orthogonal planar voids in PK11.1 (PPL); **(e)** Massive structure with few cracks and poorly preserved striated b-fabric in PK11.2 presumed to originate from assimilation and aging of clay coatings in the fine mass (XPL); **(f)** Rounded aggregate (arrow) with stipple-speckled to striated b-fabric embedded within calcareous groundmass having crystallitic b-fabric in PK12.1; note the basic similarity of those features with **(c)** (XPL)

Kastanozem (WRB 1998)). As is shown later, both are magnetically enhanced in accord with the abundant data on the accumulation of pedogenic magnetite in Chernozem soils (Babanin et al. 1995).

Bio-related microfabrics are also characteristic of the much older PK8 (Table 1), where they are juxtaposed, although, with frequent stress coatings, compaction features and almost linear cracks (Fig. 2d). The latter set of features is related to processes of deep mechanical transformations as known in the clay-expandable Mediterranean soils, particularly in Vertisols (Yaalon and Kalmar 1978, Mermut et al. 1988, Wilding and Tessier 1988, Blokhuis et al. 1990, Kapur et al. 1997). In contrast, the PK8.2 demonstrates the development of a massive microstructure with orthogonal planar voids coated with polyphase and repeatedly re-crystallized microsparitic/micritic calcite infillings (Fig. 2c). A similar albeit distinct microfabric was found in a near-surface Terra Rossa (Lower Galilee, Israel), rich in smectite and finely dispersed hematite (Fig. 2e,f). Subangular blocky structure of a strongly colored fine mass (Fig. 2e) is mainly composed of unaccommodated or partially accommodated blocky peds with oblique and bifurcating planar voids (Fig. 2f), which are occasionally reminiscent of very fine incomplete wedges (Kapur et al. 1997). Terra Rossa demonstrates a much stronger development of strial b-fabric than that found in PK8, indicating stronger swell/shrink effects leading to the formation of slickensides. Both, Terra Rossa and PK8 show enhanced production of pedogenic Fe oxyhydroxides, as demonstrated below. In addition, similar soil microfabrics are found in surface chromic Vertisols (WRB, 1998 typological unit) on alluvial/colluvial clay-rich parent materials in semiarid Mediterranean areas in Turkey, in which self-mulching, mixing and very strong bioturbation by soil mesofauna dominate (Mermut et al. 2004).

In contrast, the paleosols from PK5 through PK7 and from PK9 through PK12 at Novaya Etuliya (Fig. 1) do not contain any biologically related fabrics, but instead show strong development of juxtaposed fabric, calcite and amorphous pedofeatures, which all exist in varied proportions in different paleosols (Fig. 3). The mostly ubiquitous are secondary carbonates that manifest themselves in PK5 as soft micritic masses (impregnations), responsible for the non-uniformly distributed crystallitic b-fabric (Bullock et al. 1985); or as typical micritic nodules ranging 0.2–0.4 mm of both regular and irregular shapes (Fig. 3a). Abundances of micrite soft masses and nodules increase in the A horizons of older paleosols while their coatings wane (Table 1). At the present time, it is not quite clear if this massive calcification was the leading process during soil formation or if it occurred at a later stage. The later stage seems more plausible because older calcareous paleosols with heterogeneous matrix show frequent ca. 0.5 mm rounded calcite-free aggregates with a stipple-speckled b-fabric peak (Table 1, Fig. 3c). A similar case was reported by Kemp et al. (2004) who explained the occurrence of rounded aggregates due to their higher resistance to posterior calcite precipitation from groundwater solutes than the bulk mass. If so, rounded aggregates may be interpreted as analogs of spherical aggregates in Mediterranean Vertisols (Yaalon and Kalmar 1978, Kapur et al. 1997) of an earlier developmental stage. Hence, older paleosols with cracked patterns, densely packed fine mass and rounded aggregates are related to vertic micromorphic formation, and in case of calcite impregnation, the term calcareous is attached (Table 1). Older paleosols exhibit abundant ferruginous pedofeatures, peaking in PK9 and PK10 (Table 1). Typical examples of Fe-Mn concretions nodules and infillings are provided in Fig. 3b and 3d. Abundant amorphic nodules with no hard core and a halo of tiny impregnation mottles are embedded in the micritic groundmass in PK9.2

(Fig. 3b). The crystallization of Fe-Mn nodules, mottles, stains and hypocoatings in the calcareous groundmass indicates that the ancient paleosol underwent recurrent changes in waterlogging conditions with fluctuating pH and Eh.

Apart from clay concentration features described within well delineated aggregates (e.g. Fig. 3f) we were also able to identify a very specific fine mass b-fabric, as exemplified in PK11.2 (Fig. 3e). It shows a dull quasi-isotropic bright-brown strial heterogeneous mass that is reminiscent of the aged ferriargillans in Mediterranean soils as reported and interpreted by Fedoroff (1997). A strong brown coloration in carbonate-free patches of the fine mass is also seen in the upper part of the PK11.1 (Fig. 3d). Although the Mössbauer data only allowed us to achieve a better understanding of the nature of this rubefication (see below), the 'argillic' fabric, characterized in Table 1 as chromic, is indeed remarkable. This fabric of optically oriented clay seems to fit Kubiena's (1970) Braunlehm/Rotlehm found in tropical soils. It is noteworthy that coalescence of crystallitic calcareous formations and the specific clayey b-fabric formation are confined to deeper parts of PK11.2 and PK12.2, whereas the upper parts PK11.1, PK12.1 are genuine calcisols affected by massive precipitation of soft carbonates. Since this later phase was probably strong enough to affect the lower solum with secondary carbonates, we propose at this point to distinguish a new type of pedogeomorphic development termed "transformational welding" (vs. the accretionary type where subtle alterations by a new soil-forming episode are assumed). Coexistence of clay patches with strongly calcareous hard patches is characteristic of Mediterranean polygenetic columnar calcretes, petric calcisols (Kapur et al. 1987, 1997), although the extent of carbonate induration at Novaya Etuliya is much less than required for petrocalcic horizons.

3 Magneto-Mineralogical Properties of Vertic Brown-Reddish Paleosols

Key magnetic properties of vertic reddish-brown- paleosols at Novaya Etuliya (Table 2) include maximum values of magnetic susceptibility, induced magnetization J_i, and remnant saturation magnetization J_{rs} from the topsoils of PK4, PK5-PK6, PK8, and PK11-PK12. The PK4 and PK8 belong to biorelated micromorphic formation and are clearly distinct from other older paleosols with vertic, aquic and calcareous syndromes. The normalized χ_{lf}, J_i and J_{rs} values (see bottom rows of Table 2) have been calculated with respect to the best preserved "old" loess layer L7 with average values $\chi_{lf} \approx 12 \times 10^{-8} m^3 kg^{-1}$, $J_i \approx 16 \times 10^{-3} Am^2 kg^{-1}$, $Jrs \approx 1.2 \times 10^{-3} Am^2 kg^{-1}$. It may be kept in mind that low-field susceptibility χ_{lf} and induced magnetization J_i are related to the total concentration of paramagnetic, superparamagnetic (SP) and ferrimagnetic minerals, while the remanent magnetization J_{rs} is only related to ferromagnetic minerals larger than the size of SP particles (>20 nm).

As shown in Table 2, both PK4 and PK8 with high susceptibility $\geq 100 \times 10^{-8} m^3$ kg^{-1} in topsoil also have elevated J_i and J_{rs}, in contrast to PK5-PK6 and PK11-PK12.

Table 2 Key magnetic properties of older paleosols at Novaya Etuliya (normalized values are calculated in respect to L7 loess)

	PK4	PK5-PK6	PK8	PK11-PK12
χ_{lf}, 10^{-8} m^3 kg^{-1}	135	60	95	40–60
J_i, 10^{-3} Am^2 kg^{-1}	53.2	30–40	55	25–44
J_{rs}, 10^{-3} Am^2 kg^{-1}	7.3	2.5–5.0	6.7	2.5–3.9
χ_{lf} (soil)/χ_{lf} (loess)	11.2	5.0	7.9	3.3–5.0
J_i(soil)/J_i(loess)	3.3	1.9–2.5	3.4	1.6–2.7
J_{rs}(soil)/J_{rs}(loess)	6.1	2.1–4.2	5.6	2.1–3.2

In the latter cases, susceptibility is as low as ca. 50×10^{-8} m^3 kg^{-1}, but still remains distinctly higher than the most pristine loess unit L7. The ratio χ_{LF} (soil)/χ_{LF} (loess) in PK5-PK6 equals 5.0 and in PK11-PK12 even less, but rises to distinctly higher values in PK4 and PK8 (Table 2). The same trend is found regarding the ratios J_i(soil)/ J_i(loess) and J_{rs}(soil)/J_{rs}(loess). Hence, the key magnetic properties of the PK4 and PK8 paleosols at Novaya Etuliya differ substantially from PK5-PK7 and PK11-PK12. It is important to remember that PK4 and PK8 were previously defined micromorphologically as humified, biologically active paleosols. They are also magnetically strongly enhanced, albeit to a different degree, than the rest of the reddish-brown palesols in the age range from 0.5 Ma through ca. 1.5 Ma. J_i values for separated clay fractions increase 2–2.5 times for PK4 and 1.2–1.7 times for PK8 compared with bulk samples.

The magnetic minerals in the paleosols at Novaya Etuliya were identified on the basis of their Curie points. Thermomagnetic curves for the clay fractions from PK4.1 and PK4.2 (Fig. 4a,b) show the presence of partly oxidized (maghemitized) magnetite with Curie point around 600°C. However, the shape of the curves, dependant

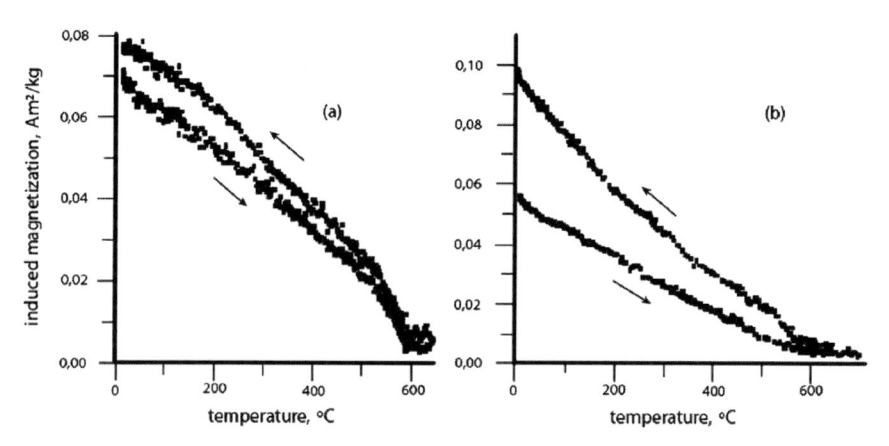

Fig. 4 Thermomagnetic curves of induced magnetization Ji(T) for clay fractions from PK4.1 **(a)** and PK4.2 **(b)**. Differences in ferromagnetic behavior are inferred from comparison between heating (arrow down) and cooling (arrow up) curves. Applied field 0.45T

on the grain size of ferromagnetic grains (e.g. Virina et al. 2000b), is different. For example, the heating curve of PK4.1 has a convex shape, in contrast to a concave one in the PK4.2 below, and is almost reversible during cooling. Curve shape and thermal stability imply a larger grain size of a pedogenically formed ferrimagnetic phase in PK4.1 than in PK4.2. If this is correct, then magnetic methods seem to be no less (and in case of PK4 even more) sensitive in their identification of soil stages within paleosols than micromorphology.

The ^{57}Fe Mössbauer analysis is particularly important for better understanding the nature of pigmentation in red-colored paleosols which is generally associated with Fe oxides and hydroxides and their crystallinity (Schwertmann 1993). An example of typical Mössbauer spectra at room temperature and at a liquid nitrogen temperature (80 K) from a separated clay fraction of PK4.1 is illustrated in Fig. 5. The spectra of bulk samples and clay fractions are principally the same except for reduced peaks intensity

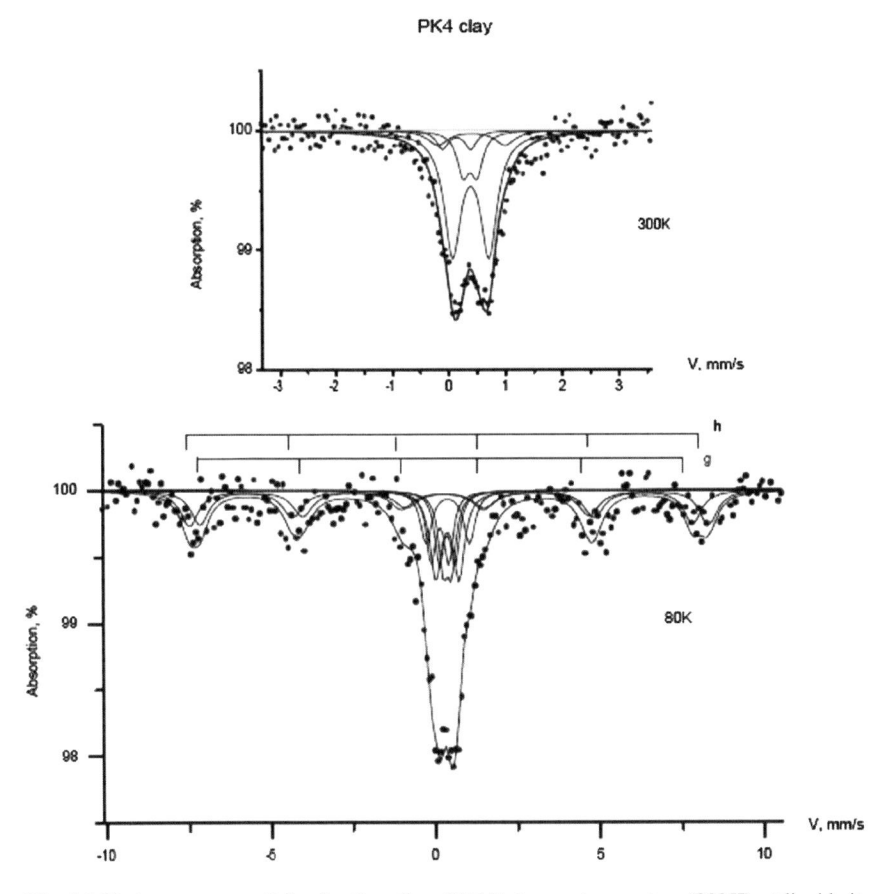

Fig. 5 Mössbauer spectra of clay fractions from PK4.2 at room temperature (300 K), at liquid nitrogen temperature (80 K). Points-experimental data, thin lines-fitted subdoublets, thick line-computer fitted sum of subdoublets, h – hematite; g – goethite. Note two sextets and decrease in the intensity of a doublet (center of a spectrum) at 80 K indicating hematite and goethite respectively

in bulk samples. Paleosols in general show substantial increase of spectra intensity compared to loess, albeit the total Fe content is nearly the same (Gendler et al. 2006). This is explained as the result of pedogenic transformation of detrital Fe-bearing compounds in loess. At room temperature (RT) the clay fraction demonstrates a characteristic broadened ferric doublet (Fig. 5), which was fitted into 4–5 sub-doublets related to different Fe^{3+} sites in the smectite crystal lattice and one sub-doublet from SP Fe oxyhydroxides (further explanation in Gendler et al. 2006). At 80 K the intensity of the SP doublet substantially decreases, and broadened sextets of magnetically ordered minerals are revealed (Fig. 5; lower spectrum). Because these magnetic phases are revealed in low-temperature conditions only, we may conclude that their size is less than 15 nm (i.e. superparamagnetic, SP) and they have poor crystallinity. The low temperature spectra can be fitted into two magnetically ordered sextets: one of them has a hyperfine field (H_{hf}) of 49.5 T which is due to finely dispersed hematite, while another sextet with 46.8 T H_{hf} is accounted for by either goethite or ferrihydrite or both (Fig. 5b). The ratio of the sextet areas in PK4.1, which helps assessing the relative content of hematite/hydroxide SP particles, is about 0.9. Hence, the reddish brown hue of PK4 is associated with subtle domination of SP goethite over hematite. The older paleosols from PK5 through PK11, all having similar Mössbauer spectra at room temperature, show an increasing proportion of SP hematite compared with Fe hydroxides. For example, the 5–15 nm SP hematite/iron hydroxides ratio in PK12 (with maximal redness and Fe-impregnated decayed clay coatings) is about 2, as exhibited by Mössbauer spectra at liquid nitrogen temperature (Gendler et al. 2006). In the lowermost paleosol PK13, Fe oxyhydroxides with grain size >5 nm are not observed at all. Although under the experimental conditions that were available here, the distinction between hematite and goethite is impossible, we may assume that the major Fe mineral phase in PK13 is ultra-fine ≤5 nm poorly crystallized SP hematite, responsible for red hue of the soils.

On the basis of the magneto-mineralogical data and prior to general discussion it seems relevant to address the question of genetic similarity between the reddish-brown paleosols at Novaya Etuliya and the hematite-rich Terra Rossa soils in Israel. According to our preliminary data, Terra Rossa soils show very high magnetic susceptibility of ca. $300 \times 10^{-8} m^3\ kg^{-1}$ (without clear depth functions), induced magnetization J_i up to $183 \times 10^{-3} Am^2\ kg^{-1}$ and remanence J_{rs} up to $22.5 \times 10^{-3} Am^2\ kg^{-1}$ (Table 2). Thermomagnetic curves indicate that nearly stoichiometric magnetite is the only ferrimagnetic phase. The Mössbauer room and low temperature spectra of Terra Rossa soils look similar to Novaya Etuliya paleosols and can be interpreted using the same model. The Terra Rossa shows an unambiguous dominance of SP hematite particles, as is the case for the oldest paleosols at Novaya Etuliya, while SP goethite/ferrihydrite slightly outweighs SP hematite in PK4 (see above).

4 Discussion

The integration of micromorphology, environmental magnetism and the Mössbauer effect for the older paleosols at Novaya Etuliya allowed us to reconstruct the key paleopedological processes and to enhance our understanding of the occurrence of welded paleosols. Welded pedocomplexes appear to be more common in loess sequences than was previously thought. Accepting Kemp's (2001) definition of welded pedocomplexes as paleosols whose covering sediments are insufficiently thick to isolate an earlier soil from the effects of a later one, we wish to stress that field observations do not always provide enough information in order to identify paleosol welding. In these cases, micromorphology and magnetic properties allow for a much more refined solution both in taxonomical paleosol classification and in the reconstruction of paleosol evolutionary trends (Table 3). Specifically, pedocomplexes effectively contain features of secondary alteration, primarily confined to the upper part of the profile that presumably originated during climatic deterioration at the time of interglacial-to-glacial transition. The geomorphic, sedimentary and water-controlled pedogenic processes operating at this stage may exert a strong impact on a soil. For reasons still unknown, the impact of this final pedogenic/geomorphic and early diagenetic stage was relatively slight on PK4 and PK8, while featured prominently on PK5-PK6 and PK11-PK12. Thus in terms of evolution, PK4 and PK8 are interpreted as accretionary with PK5-PK6 and PK11-PK12 classified as transformational and therefore less compatible with the earmarks of a modern analogue.

The PK4 and PK8 paleosols with bioturbated profiles, homogeneous and leached humus horizons, and strong magnetic enhancement are ascribed the highest paleoclimatic value (Table 3). PK4 with a double mollic horizon is defined as a type of Chernozem with signs of rubefication (i.e. more reddish than surface Chernozems and with slickensides), implying that the environmental conditions transitioned from temperate steppe to Mediterranean (Table 3). The PK8 is related to rubefied vertic self-mulching and biologically active soils of a wetter Mediterranean climate. Its strong magnetic enhancement is due to production and preservation of pedogenic, fine-grained, partly oxidized magnetite. As has been convincingly shown (Babanin et al. 1995, Maher 1998, Evans and Heller 2003), pedogenic ferrimagnets accumulate in modern well-drained, intermittently wet/dry soils with good buffering capacity and a sufficient supply of substrate Fe, whereas magnetic depletion characterizes soils in arid, cold or very moist climates and in waterlogged conditions. Secondary ferrimagnetic minerals are rapidly destructed by waterlogging and are associated with significant changes in bacterial populations (Dearing et al. 1996). Soils of the Mediterranean areas overwhelmingly contain fine-grained hematite, presumably formed from precursor iron hydroxide phases (Schwertmann 1993, Singer et al. 1998). However, our preliminary studies of Israeli Terra Rossa soils show the accumulation of substantial amounts of magnetite of varying sizes as well, and thus confirm the micromorphologically based analogies between PK8 soils at Novaya Etuliya and reddish-brown Mediterranean soils. This comparison has to be understood as tentative, considering an increasingly accepted view that surface soils indeed reflect an integration of modern dynamic processes superposed on the features from earlier soil-forming and geological stages (Targulian and Sokolova 1996), which is particularly widespread in hot climates (Paton et al. 1995, Bronger et al. 2000).

In contrast, PK5-PK6 and PK11-PK12 are related to an opposite class of magnetically depleted paleosols with immature bio-related features. In PK5-PK6, the later soil-forming occurrence of hydromorphic re-distribution of secondary carbonates was likely to have overprinted the vertic microfabric, which led to its crucial role in paleosol formation and preservation (transformational welded type). In PK11-PK12, the initial microfabric is interpreted as oriented clay heterogeneous, resulting from a complete reworking of red and yellow clay coatings and their incorporation into the matrix, with strong redness due to the formation of ultra-fine goethite/hematite pigment with less than 5 nm particle size. PK11-PK12 are also related to a transformational pedogeomorphic class, because at the final stage the paleosols were deeply reworked by secondary calcite precipitation as soft masses and calcrete-like polygenetic concretions.

It is, however, unclear at which stage pedogenic magnetite was destroyed, or whether it had accumulated at all. It is plausible that PK5-PK6 and PK11-PK12 are not identical, although both have undergone the same strong transformational evolution under the impact of massive secondary carbonate precipitation. We cannot discount the possibility that in PK11-PK12, with their birefringent fine mass microfabric, the earlier phase was even more humid than in PK8, which is detrimental to magnetite production. At the time of massive carbonate accumulation, tentatively termed "calcretization", soils also intermittently undergo waterlogging so that magnetite may have been destroyed at this stage.

Micromorphological evidence from "modern" calcrete, widespread on different geomorphic surfaces in arid and semi-arid climates soils (Kapur et al. 1987), is not identical, albeit not incompatible, with ancient paleosols at Novaya Etuliya. Micromorphic evidence from PK5-PK6 and PK11-PK12 (Table 2, 3) may suggest that these paleosols could have undergone recurrent micrite migration, partial leaching, re-precipitation as soft masses and hard concretions, in addition to the formation of iron/manganese impregnations and strong mechanical turbations in the shrink–swell zone, as is characteristic for many soils with calcrete (Mermut et al. 2004). Since "calcretization" implies wet/dry seasonal cycles, we may speculate that semihydromorphic and hence reduction conditions in a soil are the possible explanation for the posterior destruction of magnetite.

At Novaya Etuliya, as well as at Roxolany on the Lower Dniestre (Tsatskin et al. 1998), the extent of posterior secondary changes in older reddish-brown paleosols is not equal. It is suggested that the magnetically enhanced, bio-related PK4 and PK8 have preserved reliable morphology and Fe compounds mineralogy of their major soil forming episodes. Hence they are proposed as being the most reliable paleoclimate indicators. Under the constraints of paleomagnetism and other methods of absolute geochronology, PK4 and PK8 may certainly be extremely useful as key stratigraphic markers in South Eastern Europe altogether in the time interval ~0.5–~1.5 Ma. For example, PK4 (an equivalent of the Vorona pedocomplex of Velichko 1990) and PK8 (formed during the Jaramillo subchron) can easily be recognized also in the Khadzimus loess/soil sequence overlying the alluvium with the Tamanian (Late Villafranchian) faunal complex (Dodonov et al. 2006). In a broader sense, since a reasonably secure framework for European loess already exists, thoroughly defined paleosol types from well-ordered type loess sections have a good potential for distant, inter-regional stratigraphic correlation.

5 Conclusions

(1) Integration of micromorphology, environmental magnetism and Mössbauer spectroscopy along with the identification of pedogeomorphic classes of reddish-brown paleosols in the Black Sea area allowed us to improve the quality of paleopedological reconstructions. A higher level of micromorphological classification for ancient soils, which still needs further development by soil micromorphologists, is particularly useful for their taxonomical identification. However, not all paleosols share the same taxonomic faculty, either because of local geomorphic imprints, posterior transformation or partial convergence with other interglacial soils. At Novaya Etuliya, the best paleoclimatically resolvable soils in the time interval ~0.5– ~1.5 Ma are magnetically enhanced reddish Chernozems of the pedocomplex PK4, and red-brown vertic Mediterranean soils of pedocomplex PK8. The PK4 and PK8 are therefore proposed as the key stratigraphic markers in South Eastern Europe.

(2) The question of diagenesis, including the impact of the final stages of soil formation under changing sedimentary environment and degrading climate, is central to the understanding of buried Pleistocene paleosols. The hypothesis that the final stage of paleosol evolution may have been responsible for the rapid destruction of ferrimagnets during the reduction state in the paleosols with a low magnetic susceptibility signal cannot be ruled out.

(3) Since the magnetic signals in Mediterranean soils like elsewhere seem to be controlled by the formation of secondary ferrimagnetic minerals, which can be detected by rock magnetic methods, magnetic susceptibility measurements should be wider employed in the research of Mediterranean soils.

Acknowledgements We thank the reviewers for their useful comments. The work has been done as part of further development of the INTAS project 'Rock Magnetism of Loess Sediments from the Russian Plain: Palaeoclimatic and Environmental Aspects' (1995–1996), and is dedicated to the memory of our late friend Elena I.Virina.

References

Azzaroli A, Colalongo ML, Nakagawa H, Pasine G, Rio D, Ruggieri G, Sartoni S, Sprovieri R (1997) The Pliocene-Pleistocene boundary in Italy. In: Van Couvering JA (ed) The Pleistocene Boundary and the Beginning of the Quaternary. Cambridge, Cambridge University Press, pp. 41–155

Babanin VF, Trukhin VI, Karpachevskyi LO, Ivanov AV, Morozov VV (1995) Magnetism pochv [Soil magnetism]. YaGTU, Yaroclavl' (in Russian)

Birkeland P (1999) Soils and Geomorphology, 3rd edn. Oxford University Press, New York

Blokhuis W, Kooistra M, Wilding L (1990) Micromorphology of cracking clayey soils (Vertisols). In: Douglas L (ed) Soil Micromorphology: A Basic and Applied Science. Developments in Soil Science, vol. 19. Elsevier, Amsterdam, pp. 123–148

Bronger A (2003) Correlation of loess-paleosol sequences in East and Central Asia with SE Central Europe: Towards a continental Quaternary pedostratigraphy and paleoclimatic history. Quat Int 106–107: 11–31

Bronger A, Wichmann P, Ensling J (2000) Over-estimation of efficiency of weathering in tropical "Red Soils": Its importance for geoecological problems. Catena 41: 181–197

Bullock P, Fedoroff N, Jongerius A, Stoops G, Tursina T (1985) Handbook for Soil Thin Section Description. Waine Research, Wolverhampton

Catt JA (1988) Soils of the Plio-Pleistocene: Do they distinguish type of interglacial? Philos Trans R Soc, London B 318: 539–557

Catt JA (1990) Paleopedology manual. Quat Int 6: 1–95

Classification and Diagnostics of Soils of the USSR (1986). Washington DC

Courty M-A, Goldberg P, Macphail R (1989) Soils and Micromorphology in Archaeology. Cambridge University Press, Cambridge

Dearing JA, Hay KL, Baban SMJ, Huddleston AS, Wellington EMH, Loveland PJ (1996) Magnetic susceptibility of soil: An evaluation of conflicting theories using a national data set. Geophys J Int 127(3): 728–734

Dodonov AE, Zhou LP, Markova AK, Tchepalyga AL, Trubikhin VM, Alexandrovski AL, Simakova AN (2006) Middle–Upper Pleistocene bio-climatic and magnetic records of the Northern Black Sea Coastal Area. Quat Int 149: 44–54

Evans ME, Heller F (2003) Environmental Magnetism: Principles and Applications of Enviromagnetics. Academic Press, San Diego, CA

Faustov SS, Bol'shakov VA, Virina EI, Demidenko EL (1986) Application of rock magnetic and palaeomagnetic methods to Pleistocene studies. In: Kaplin PA (ed) Itogi Nauki i Tekhniki [Science and Technology Results] Palaeogeography 3. VINITI Press, Moscow (in Russian)

Fedoroff N (1997) Clay illuviation in red Mediterranean soils. In: Mermut AR, Yaalon DH, Kapur S (eds), Red Mediterranean Soils. Catena 28: 171–189

Gendler TS, Heller F, Tsatskin A, Spassov S, Du Pasquier J, Faustov SS (2006) Roxolany and Novaya Etuliya – key sections in the western Black Sea area: Magnetostratigraphy, rock magnetism and paleopedology. Quat Int 152–153C: 89–104

Gerasimova MI, Gubin SV, Shoba SA (1992) Mikromorfologiya pochv prirodnukh zon SSSR [Micromorphology of soils of the natural zones of the USSR]. Russian Academy of Science, Pushchino (in Russian)

Gerasimova MI (2003) Higher levels of description – approach to the micromorphological characterization of Russian soils. Catena 54(3): 319–337

Heller F, Liu TS (1982) Magnetostratigraphical dating of loess deposits in China. Nature 300: 431–433

Heller F, Liu TS (1986) Palaeoclimatic and sedimentary history from magnetic susceptibility of loess in China. Geophys Res Lett 13: 1169–1172

Heller F, Evans ME (1995) Loess magnetism. Rev Geophys 33: 211–240

Holliday VT (2004) Soils in Archaeological Research. Oxford University Press, Oxford

Kapur S, Cavusgil VS, FitzPatrick EA (1987) Soil-calcrete (caliche) relationship on a Quaternary surface of the Çukurova region, Adana (Turkey). In: Fedoroff N, Bresson LM, Courty A-M (eds) Micromorphologie des sols – Soil Micromorphology. AFES, Paris, pp. 597–603

Kapur S, Karaman C, Akça E, Aydin M, Dinç U, FitzPatrick EA, Pagliai M, Kalmar D, Mermut AR (1997) Similarities and differences of the spheroidal microstructure in Vertisols from Turkey and Israel. Catena 28(3–4): 297–311

Kemp RA (1998) Role of micromorphology in paleopedological research. Quat Int 51/52: 133–141

Kemp RA (2001) Pedogenic modification of loess: Significance for palaeoclimatic reconstructions. Earth-Sci Rev 54: 145–156

Kemp RA, Toms PS, King M, Kröhling DM (2004). The pedosedimentary evolution and chronology of Tortugas, a Late Quaternary type-site of the northern Pampa, Argentina. Quat Int 114: 101–112

Klassifikatsiya i diagnostika pochv Rossii [Classification and Diagnostics of Soils of Russia, 2nd ed] (2004). Oikumena, Smolensk (in Russian)

Kubiena WL (1970) Micromorphological features in soil geography. Rutgers University Press, Brunswick

Maher BA (1998) Magnetic properties of modern soils and Quaternary loessic paleosols: Paleoclimatic implications. Palaeogeogr, Palaeoclim, Palaeoecol 137: 25–54

Mermut AR, Sehgal J, Stoops G (1988) Micromorphology of shrink–swell soils. In: Kirekerur L, Pal D, Sehgal J, Deshpande S (eds) Classification, Management and Use Potential of Swell – Shrink Soils. Oxford Univ. Press and IBH, New Delhi, pp. 127–144

Mermut AR, Montanarella L, FitzPatrick EA, Eswaran H, Wilson M, Akça E, Serdem M, Kapur B, Ozturk A, Tamagnini T, Cullu MA, Kapur S (eds) (2004) 12th International Meeting on Soil Micromorphology, Excursion Book, 20–26 Sep. 2004, European Communities

Morozova TD (1995) Identification of paleosol types and their applicability for paleoclimatic reconstructions. Geoj 36 (2–3): 199–205

Nikiforova KV (1982) The boundary between Neogen and Quaternary (Anthropogene) Systems. In: Shantzer EV (ed) Stratigraphy of the USSR. Quaternary System, 1, Moscow, Nedra, pp. 95–110 (in Russian)

Nikiforova KV (1997) The Pliocene and Pleistocene of the European part of the Commonwealth of Independent States. In: Van Couvering JA (ed) The Pleistocene Boundary and the Beginning of the Quaternary. Cambridge, Cambridge University Press, pp. 221–226

Paton TR, Humphreys GS, Mitchell PB (1995) Soils: A New Global View. Yale University Press, New Haven

Retallack GJ (1990) Soils of the Past: An Introduction to Paleopedology. Allen & Unwin, London

Schwertmann U (1988) Occurrence and formation of iron oxides in various pedoenvironments. In: Stucki JW, Goodman BA, Schwertmann U (eds) Iron in soils and clay minerals. D.Reidel, Dordrecht, pp. 267–308

Singer A, Schwertmann U, Friedl J (1998) Iron oxide mineralogy of Terre Rosse and Rendzinas in relation to their moisture and temperature regimes. Eur J Soil Sci 49: 385–395

Sirenko NA, Turlo SI (1986) Razvitie pochv I rastitel'nosti Ukrainy v pliotsene I pleistotsene [Development of soils and vegetation in Ukraine during Pliocene and Pleistocene]. Naukova Dumka, Kiev (in Russian)

Spassov S (1998) Gesteinmagnetismus und Paläoklima: Das Lössprofil von Novaya Etuliya, Moldavien. Unpublished Diploma thesis, ETH Zürich

Stoops G (1994) Soil thin section description: Higher levels of classification of microfabrics as a tool for interpretation, In: Ringrose-Voase AJ and Humphreys GS (eds) Soil Micromorphology: Studies in Management and Genesis, Developments in Soil Science 22, Elsevier, Amsterdam-London-New York-Tokyo, pp. 317–325

Stoops G (2003) Guidelines for analysis and description of soil and regolith thin sections. Soil Science Society of America, Madison, WI

Targulian VO, Sokolova TA (1996) Soil as a biotic/abiotic natural system: A reactor, memory and regulator of biospheric interactions. Eurasian Soil Sci 29(1): 30–41

Tsatskin A, Heller F, Hailwood EA, Gendler TS, Hus J, Montgomery P, Sartori M, Virina EI (1998) Pedosedimentary division, rock magnetism and chronology of the loess/paleosol sequence at Roxolany (Ukraine). Palaeogeogr Palaeoclim Palaeoecol 143: 111–133

Tsatskin A, Heller F, Gendler TS, Virina EI, Spassov S, Du Pasquier J, Hus J, Hailwood EA, Bagin VI, Faustov SS (2001) A new scheme of terrestrial paleoclimate evolution during the last 1.5 Ma in the western Black Sea region: Integration of soil studies and loess magnetism. Phys Chem Earth 26: 911–916

Veklitch MF (1982) Paleoetapnost' i stratotipy pochvennukh formatsii Ukrainu [Paleogeographical stages and stratotypes of soil formations of Ukraine]. Naukova Dumka, Kiev (in Russian)

Velichko AA (1973) Prirodnui protses v pleistotsene [Evolution of Nature in the Pleistocene]. Nauka, Moscow (in Russian)

Velichko AA (1990) Loess-paleosol formation on the Russian plain. Quat Int 7/8: 103–114

Velichko AA, Akhlestina EF, Borisova OK, Gribchenko YN, Zhidovino NY, Zelikson EM, Iosifova YuI, Klimanov VA, Morozova TD, Nechaev VP, Pisareva VI, Svetlitskaya TV, Spasskaya II, Udartsev VP, Faustova MA, Shik SM (1999) Vostochno-Evropeiskaya ravnina [East-European plain], In: Velichko AA, Nechaev VP (eds) Climate and Environmental Change during the

Last 65 Million Years (Cenozoic: from Paleocene to Holocene). GEOS, Moscow, pp. 43–83 (in Russian)

Virina EI, Faustov SS, Heller F (2000a) Magnetism of loess-palaeosol formations in relation to soil-forming and sedimentary processes. Phys Chem Earth (A), vol. 25, No. 5: 475–478

Virina EI, Heller F, Faustov SS, Bolikhovskaya NS, Krasnenkov RV, Gendler TS, Hailwood EA, Hus J (2000b) Paleoclimatic record in the loess-paleosol sequence of the Strelitsa type section (Don glaciation area) deduced from rock magnetic and palynological data. J Quat Sci 15: 487–499

Wilding L, Tessier L (1988) Genesis of Vertisols: Shrink – swell phenomena. In: Wilding L, Puentes R (eds) Vertisols: Their Distribution, Properties, Classification, and Management. Texas A&M University Printing Center, College Station, pp. 55–79

World Reference Base for Soil Resources (WRB) (1998). FAO Report No. 84, Rome

Yaalon DH (1971) Soil-forming processes in time and space. In: Yaalon DH (ed) Paleopedology: Origin, Nature and Dating of Paleosols. International Society of Soil Science and Israel University Press, Jerusalem, pp. 29–39

Yaalon DH (1997) Soils in the Mediterranean region: What makes them different? In: Mermut AR, Yaalon DH, Kapur S (eds) Red Mediterranean Soils, Catena 28: 157–169

Yaalon DH, Kalmar D (1978) Dynamics of cracking and swelling clay soils: Displacement of skeletal grains, optimum depth of slickensides, and rate of intra-pedonic turbation. Earth Surf Process 3: 31–42

Ferricretes in Tamil Nadu, Chennai, South-Eastern India: From Landscape to Micromorphology, Genesis, and Paleoenvironmental Significance

Hema Achyuthan and N. Fedoroff

Abstract Ferricretes located in the coastal plain around Chennai, southeastern India, were analysed in thin sections using the concept of pedo-sedimentary sequence of events based on the hierarchy of sedimentary and pedological features and organisations. Field characteristics of investigated ferricretes are similar to those described in the literature. From top to bottom, they consist of a pisolithic crust, lying on hard ferruginous crust merging into a plinthite developed on mottled clays. Frequently the hard crust is hosted on ferruginous gravels. A sandy clay layer covers the ferricrete in one of the surveyed site. These ferricretes appear as inliers above the Late Pleistocene and Holocene riverbeds. Micromorphological analysis lead to conclude that the genesis of ferricretes studied have a complex succession of phases and attributes which were overlapping. Some of these phases are purely pedogenic, characterised by an intense iron oxide accretion, which occurred during episodes of landscape stability characterised by high ground water levels. Other phases reveal two different modes of erosion that have been recognised as, (i) deep erosion of plinthitic soils followed by a local deposition in the form of ferruginous gravel, (ii) surface erosion of hard ferruginous crust, long distance transportation and sedimentation in the form of a nodular layer. The sandy clay layers, which were found in one site, also confirmed the aeolian input in the area. A reconstruction of the local past landscapes is also proposed as well as a discussion on the environmental significance of the ferricretes studied.

Keywords Abrupt events · aeolian sedimentation · climate fluctuations · ferricrete · laterite · pedo-sedimentary sequences

Hema Achyuthan
Department of Geology, Anna University, Chennai 600 025, India

N. Fedoroff
Institut National Agronomique, 78850 Thiverval-Grignon, France,
e-mail: nicolas.fedoroff@wanadoo.fr

S. Kapur et al. (eds.), *New Trends in Soil Micromorphology*,
© Springer-Verlag Berlin Heidelberg 2008

111

1 Introduction

Laterites and ferricretes have been investigated mainly by geologists (e.g. Harrassowitz 1926, Maignien 1966, McFarlane 1976, Boulangé 1984, Bourman 1993b). Their studies have been somehow neglected by pedologists, probably because of their insignificance for agriculture. Consequently soil classifications (FAO-WRB 1998, US Soil Taxonomy 1999) are not well suited for these materials. The terms ferricrete and laterite are not mentioned in these classifications. FAO-WRB (1998) proposes only the soil group of Plinthosols in which a petroplinthite or a plinthite starts within 50 cm from the soil surface. The plinthite, a diagnostic soil characteristics in US Soil Taxonomy (1999), corresponds only to not hardened laterites, where vesicular and pisolithic ones are excluded from the definition of plinthite.

Consequently we have followed geologists (Tardy 1993, Delvigne 1998) for describing and analysing the profiles. According to Tardy (1993), the typic lateritic iron crusted profiles consist of: (i) an unweathered bed-rock, (ii) an isoalterite (the bed-rock is weathered but its texture is preserved), (iii) an alloterite (the bed-rock texture is no more recognisable), (iv) mottled clays, (v) an intermediate horizon enriched in iron oxides, (vi) the ferricrete, a hard, cemented, ferruginous crust, and (vii) a nodular surface horizon. In this chapter, we will consider ferricretes as a distinct horizon of lateritic iron crusted profiles.

Ferricretic laterites are common on the deeply weathered, stable cratons of the intertropical belt affected by monsoonal types of climate. Their extension in the Deccan peninsula is somehow limited where they occur in patches and also in specific height zones. They were first recognised and termed by Buchanan (1807). The most developed lateritic ferricretes occur on the Western Ghat plateau with an elevation of 1400 m, while others are found on the coastal plateaus at an elevation below 250 m (Dikshit and Wirthmann 1992). On the slopes of Western Ghats, ferricretes are absent, however incipient ferruginisations and inherited nodules are present (Peterschmitt 1993). More information about their distribution can be found in Indian soil survey reports (Raychaudhuri et al. 1963, Perur and Myrhyantha 1972). Most of the literature on laterites in India deals with their morphology and their chemistry; in Andhra Pradesh (Raman and Vaidyanadhan 1980), in Kerala (Karunakaran and Roy Sinha 1980, Ollier and Rajaguru 1989), and Tamil Nadu (Subramanian and Mani 1980).

There is no general agreement on the genesis of ferricretes and related horizons. Briefly the abundant literature on their genesis can be summarized as follows: (i) the linear model, the most commonly referred model, in which the ferricrete formation occurred through pedogenesis under the tropical monsoonal type of climate; their development is independent from climatic fluctuations (e.g. Tardy 1993), (ii) the non-linear model considers the ferricretes developing from the interactions of geomorphic and pedogenic processes in relation to climatic fluctuations (e.g. Bertrand 1998), (iii) some authors (Brimhall and Lewis 1992, Brimhall et al. 1994, Iriondo and Kröhling, 1997) suggest that aeolian sedimentation should also be taken in account in the genesis of lateritic soils.

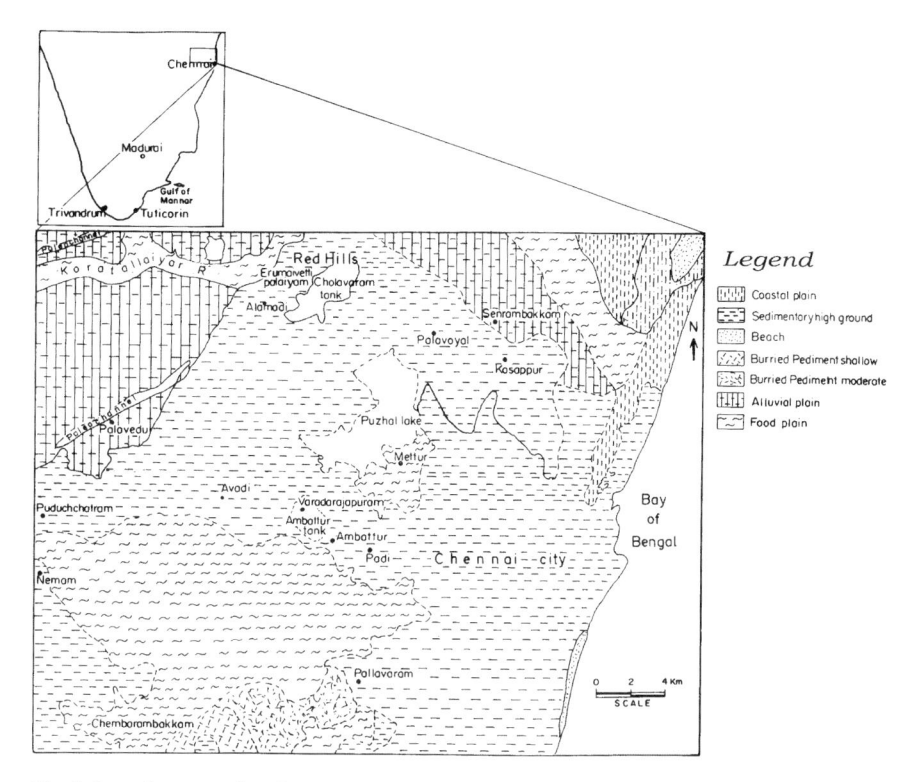

Fig. 1 Location map of study area

We present here the results of field works and micromorphological analyses of ferricretic profiles from the coastal zones of Chennai, south-eastern India (Fig. 1), previously investigated by Achyuthan (1996), Achyuthan et al. (2000), and Achyuthan (2004). Our main objectives were to: (1) apply the method of hierarchy to test the existing genetical models and (2) reconstruct the paleo-landscapes during ferricrete development and infer their paleo-climatic significance.

2 Geologic, Geomorphologic, and Climatic Setting

The bedrock around Chennai mostly consists of sandstone and shale of the Upper Gondwana age. Mostly early Cenozoic sequences (Paleogene), form a thick cover surrounding the Chennai region, leaving it exposed as inliers. Tectonically, a north-south trending fault traverses through the coastal part of the area With a relative uplift still active since the Proterozoic through to the late Cenozoic (Neogene) mainly along reactivated NNE–SSW, NNW–SSE, and N–S trending structures and alignments, consisting of tilted fault blocks. Uplifted parts of these blocks are dominated by charnockite, sandstone, and shale as bedrock exposures with a partial cover of different sediments (Subramanian and Selvan 2001).

The ferricretic laterite landscape is incised (10–20 m) by meandering rivers of which the riverbed is infilled by sandy sediments, free of ferruginous segregations. Presently only inliers of this landscape are preserved. A peat sampled further down in the delta of one of these rivers was dated to 6,590±120 yrs BP (Achyuthan 1996).

The present day climate in south-eastern India is monsoonal, characterised by an annual rainfall averaging 1,200 mm and very high evapotranspiration. Temperatures range from an average maximum of 40–42°C in the summer and average minimum of 16°C.

3 Methods

Samples collected along the vertical profiles from the different sites were analysed for major oxides with a spectrophotometer following the procedure of Shapiro and Brannock (1962). Clay minerals and fine fractions were identified with an X-ray diffractometer with CoK-alpha radiation.

Mammoth size thin sections (13 × 7 cm) were made according to Guilloré (1985), which served to determine the overall features present in ferricretes, some of which are centimetres in size or larger. Thin section descriptions were performed following Bullock et al. (1985), and Fedoroff (1994), as well as Delvigne (1998) and Eschenbrenner (1988) for weathering of the ferruginous features. Hierarchy of features and organisations was established according to Fedoroff and Courty (2002) to establish the sequence of pedo-sedimentary events.

4 Field Morphology

Two inliers covered by ferricretic laterites, dissected by erosion and also by intense quarrying, exist around Chennai; the Red Hills which are located north-west, and Chambarrambakkam lying west of the city (Fig. 1). Truncated lateritic profiles and redeposited lateritic materials are present on some of the charnockite uplifted blocks (not studied here).

4.1 Red Hills

A typic profile of ferricretic laterite developed on sandy white clays in Red Hills (Fig. 2a) consists of four layers:

1. A very hard pisolith crust, 20–30 cm thick, in which very abundant, rounded, hard, red and black pisoliths as well as fragments of the underlying crust are cemented by a black ferruginous matrix. Common tortuous, a few centimetres large channels, some coated by a black shiny material and also frequently infilled by pale brown, 10 YR 6/3 or white material. The upper layer is limited by horizon 2 with an abrupt boundary.

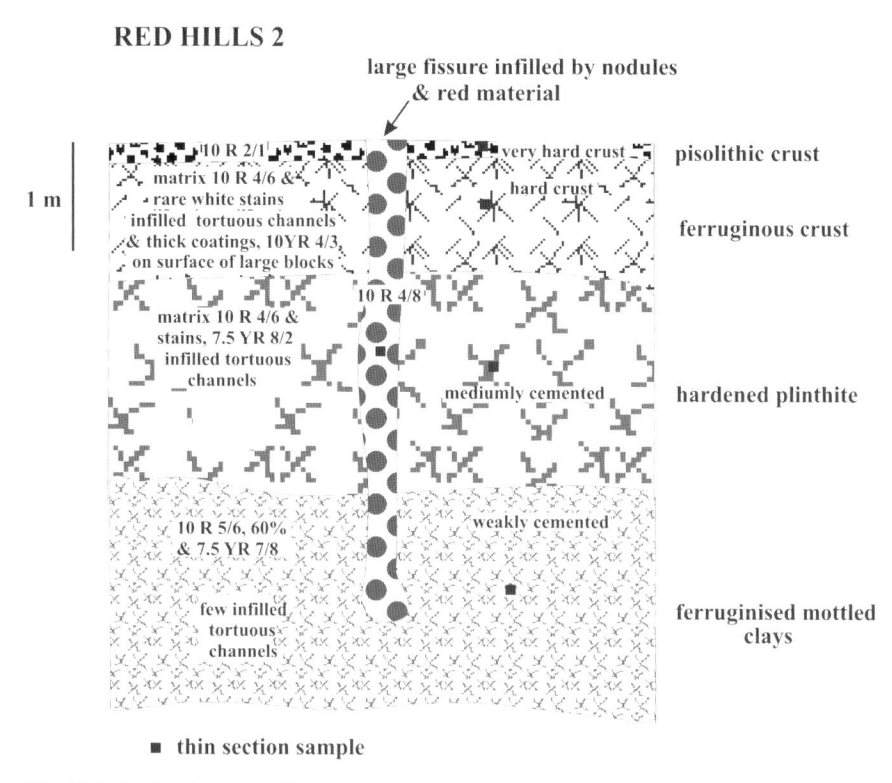

Fig. 2(a) Section 2 of Red Hills (Zone without gravel)

2. A thick (120 cm thick), hard ferruginous crust, red, 10R 4/6, with rare white stains, and with abundant tortuous channels and cavities (up to 25%), disjointed in a few meters large blocks coated by a dark brown, 10 YR 4/3, material.
3. A thick, (200 cm thick), medium hard, red, 10R 4/6, ferruginous material which responds to the definition of plinthite, with rather abundant white stains. Common channels and cavities coated or infilled by red material (10R 4/6).
4. Mottled clays (visible on 200 cm thick), weakly hardened, red, 10R/6, (60%), reddish yellow 7.5 YR 7/8 and pinkish white, 7.5 YR 8/2, mottles with common large (from 1 cm to few mm) tortuous channels coated by dark red (10R 4/4) material frequently infilled by red (10 YR 4/8) coarse sand grains.

Deep (reaching the mottled clays), large (10 cm wide) vertical cracks penetrate the ferricretic profile. They are infilled by coarse sand and even gravel grains embedded in a red (10R 4/8) matrix with cavities and channels which are infilled by dark red (10R 3/3) material with thin yellowish red stains.

Laterally, the profile studied merge into partially or totally redeposited material characterised by gravel lying on the mottled clays with an abrupt boundary. The gravel is embedded by a red (10R 5/8) matrix consisting of fragments of zones

1 and 2, present in the upper part of the gravel layer, while fragments of zones 3 and 4 exist throughout. This gravel outcrops frequently underlined by an abrupt boundary under which a hard ferruginous crust exists on a pisolithic crust.

4.2 Chambarrambakkam

The quarry of Chambarrambakkam (Fig. 2b) occurs at an elevation of 29 m, some metres lower than the Red Hills. Ferricretic profiles in this site are comparable to those of Red Hills, but they differ by: (i) a sandier parental material, (ii) thinner encrusted horizons, and (iii) a surface cover. As in Red Hills, below the upper encrusted horizons, a horizon consisting of fragments of ferricrete (7.5 YR 5/4, brown) coated by red (10 R 5/8) material was commonly observed. In some sections, it splits in two parts with an abrupt boundary, the lower consisting of finer gravel of weathered brown (10 YR 4/3) sandstone.

The surface is irregular in thickness, while in the elevated zone the thickness does not exceed 30 cm, however in depressions it reaches a few metres. A short profile description from a depression is given below:

1. A thin (20–30 cm) eluvial horizon, sandy loam, yellow (10YR 8/6), friable, rather diffuse boundary.
2. A thick (100 cm) illuvial horizon, sandy clay, yellowish red (5 YR 5/8), friable, porous, diffuse boundary.
3. A thick (100 cm), weakly gleyed illuvial horizon, with mottles, red (2.5 YR 5/8, 70%), and reddish yellow (7.5 YR 6/8, 30%), with diffuse boundary. This horizon lies always with an abrupt boundary on a ferricrete.

Fig. 2(b) Lateral distribution of ferricretes and associated layers in Chembarambakkam quarry

5 Results

5.1 Micromorphology

Micromorphological description of a typic profile of Red Hills is given below.

The pisolithic crust consists of pisoliths (20%) embedded in a ferruginous matrix. Isolated chambers and channels, a few mm to 1 cm in size, (10%) are rather common, most of them being infilled. The pisoliths are rounded to sub rounded, between $\frac{1}{2}$ and 1 cm in diameter, can be split according to their internal structure in two great groups: (i) black and opaque, (ii) partially translucent.

Black and opaque pisoliths can be subdivided according the host material and the degree of weathering of quartz grains. The following subtypes have been identified: (i) the host is an alloterite of a deeply weathered charnockite (Fig. 3a) (partially kaolinised micas) (frequency, 10%), (ii) homogeneous black and opaque with individual small vughs, some, which are occupied by ruiniform quartz grains (Fig. 3b), (iii) rare pisoliths are hosted on charcoal fragments (Fig. 3c).

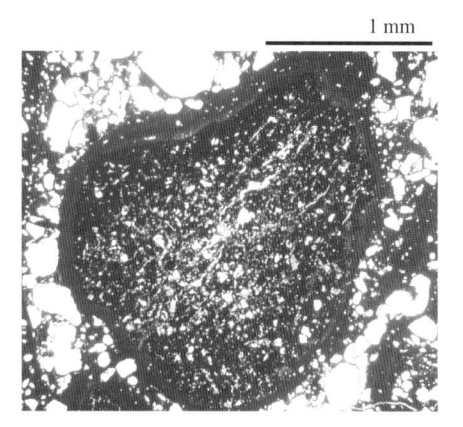

1 mm

Fig. 3(a) A pisolith core hosted on weathered charnockite

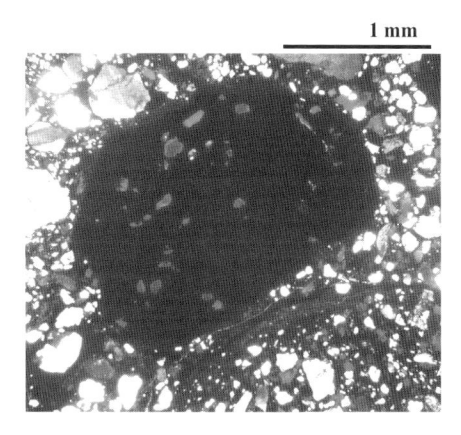

1 mm

Fig. 3(b) Microphotograph of an opaque pisolith core in which quartz can be recognised

Fig. 3(c) Microphotograph
of a pisolith core hosted on a
charcoal fragment

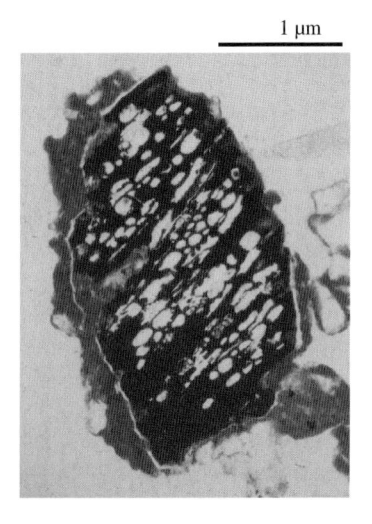

1 µm

The partially translucent pisoliths, frequently fragmented, are hosted on clayey matrices usually rich in clay coatings. The cortex of pisoliths are dark red, microlaminated, medium translucent, 50 µm thick in average, most are simple (Fig. 3d). Some are large (up to 150 µm) and exhibit laminae containing silt and sand grains.

The ferruginous matrix appears at low magnification fully opaque while at high magnification it becomes slightly translucent; some features, e.g. microlaminated clay coatings, partially masked by iron oxides then become visible. Ferrugination is somehow irregular. Rounded to sub rounded nodular forms are present as a result of circular iron oxide concentration, free of coarse fraction. Distribution of the coarse fraction in this matrix is not uniform. Some areas are enriched in coarse fraction while others are depleted. The vugh and channel infillings consist of whitish grey clay, in which embedded quartz grains exist.

100 µm

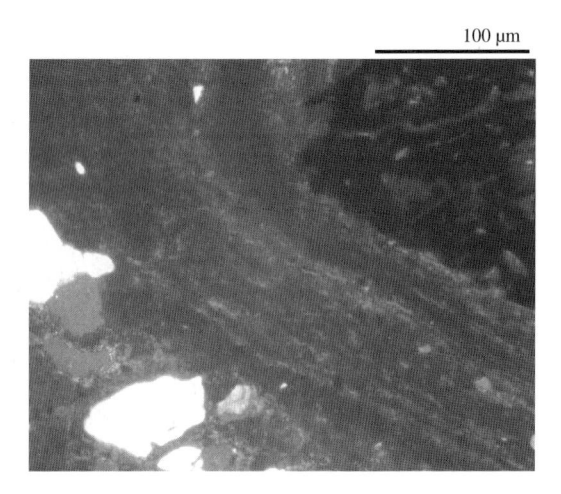

Fig. 3(d) Cortex of a pisolith

The hard ferruginous crust differs from the pisolithic crust by an absence of cortified nodules (pisolith), however some rounded to sub rounded, black and opaque nodules are present (5%) as well as rare ferruginised charcoal fragments. The ferruginous matrix is similar to the equivalent matrix of the pisolithic crust.

The chamber and channel infillings are varied. The following types have been identified: (i) a few are whitish grey as in the pisolithic crust, (ii) many transitions in colours exist between whitish grey type and a red one, (iii) the abundant coarse fraction, fluctuates with a few infillings consist only of a sand grain packing, (iiii) microlaminated clay coatings (from pale yellow to red) are common in secondary voids of type (ii) and (iii) infillings, (v) a few consist of an irregular packing of fragments of microlaminated clay features (Fig. 3e).

The plinthite appears as an intermediate phase between the hard ferruginous crust and the mottled clays. The irregular zones of weakly to moderately ferruginised whitish grey matrix alternates irregularly with the highly ferruginised zones enriched by coarse locally sorted particles.

The mottled clays appear as a dense, homogeneous, whitish grey, clay mass, almost without interference colours, in which quartz grains are embedded (Fig. 3f). Ferruginous segregations, opaque and black infill create a dense network of small fissures and commonly expand from these fissures in the ground mass (brownish black). Channels are infilled by whitish grey microlaminated to laminated clays and very pale grey interference colours. Some of these infillings are fragmented.

Quartz grain distribution and morphology. Three distinct classes of quartz grains are present throughout the profile: (i) coarse sand ($\sim 1.000–1500\,\mu m$), (ii) medium sand ($\sim 250–300\,\mu m$), and (iii) fine sand, and coarse silt ($\sim 50–100\,\mu m$). However their abundance and morphology vary significantly. In the mottled clays, the coarse fractions do not exceed 40%, while the coarse sand are dominant (25%) and the fine sand and silt are rare (5%). In the hard ferruginous crust, the coarse fraction reaches 70% and is dominant in the medium sand (30%) while the fine sand and silt are only 15%. The distribution of the coarse fraction in the ground mass of the

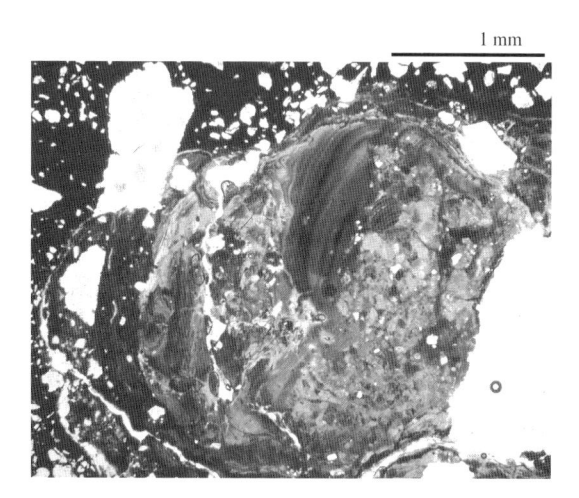

1 mm

Fig. 3(e) Irregular packing by fragments of microlaminated clay features of a large cavity of the hard ferruginous crust

Fig. 3(f) The mottled clay host material of ferruginised gravel. Note clayey textural features in the form of fragments

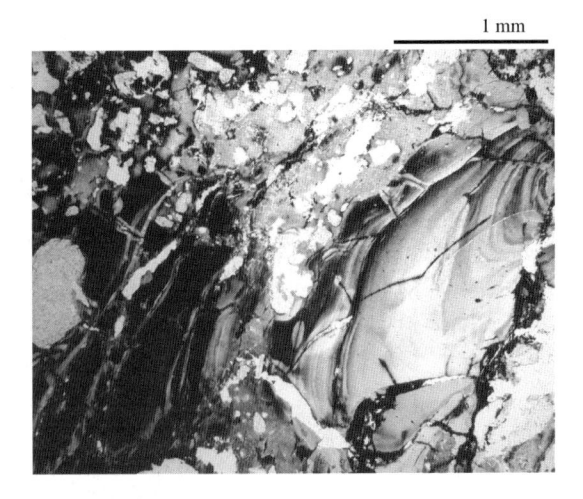

1 mm

pisolithic crust (excluding the pisoliths) is similar to the one of the ferruginous crust just below. The transition between the quartz grain distribution of the mottled clays and the one of the ferruginous crust occur progressively in the upper half of the plinthite, where the zone of medium and fine sand and silt concentrations progressively become more frequent.

The quartz grain morphology also varies in the coarse sand size quartz grains in mottled clays exhibiting irregular shapes with rough and fresh surfaces. In the nodules and pisoliths of the crusts, the coarse sand grains are corroded exhibiting a ruiniform aspect (corrosion voids are infilled by iron oxides) while some have been totally dissolved. Part of the medium sand grains are also ruiniform and ferruginised.

Lateral variations: The characters of the pisolithic crust remain constant throughout sites investigated, except the vugh and channel infillings which vary from a whitish grey matrix, as in the typic profile, to a red sandy clay matrix (Fig. 4a), which is characterised by a microstructure evolving from crumby, to spongy, and to fissure. Its coarse fraction is similar to the sandy clay of the Chambarrambakkam, containing common black and opaque ferruginous, as well as red, slightly translucent fragments. These infillings consist of microlaminated clay coatings and infillings frequently disturbed and even fragmented (and masked by the ferruginisation).

The gravely layers consist of ferruginous fragments of various size (from $\frac{1}{2}$ to some cm) and forms (most are irregular with a smooth boundary). The fragment host material is the whitish grey sandy clay, usually present in the mottled clays. The embedding, slightly birefringent, sandy clay, matrices vary from whitish grey to reddish yellow; they all contain abundant small fragments of clay coatings; merging locally to complex argillic organisations (Fig. 4b). The coarse fraction of these fabrics consists of the three grain size classes mentioned above, rather irregularly distributed. However, the fine sand and silt size particles are dominant, the medium sand is present while the coarse sand is almost absent. The large fissures locally

Fig. 4(a) Red sandy ground mass infilling large fissures of the pisolithic crust in Red Hills

1 mm

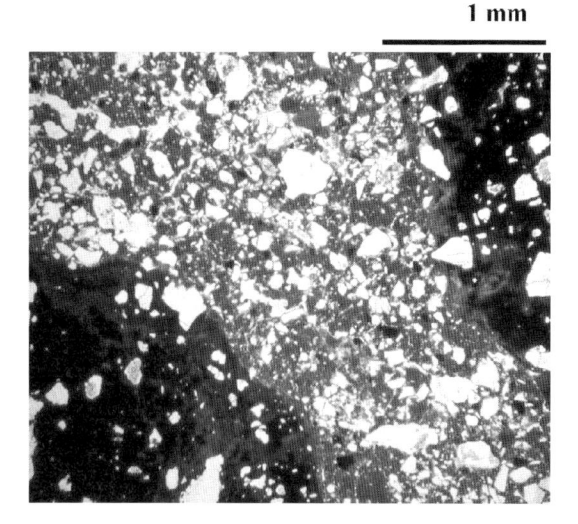

Fig. 4(b) Ground mass of mottled clays with embedded quartz

1 mm

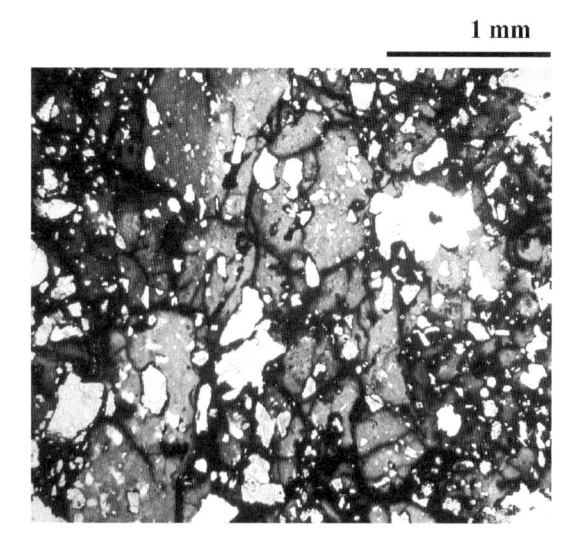

widening in vughs of these matrices are coated or infilled. The most complete infillings consist of the void wall of: (i) red microlaminated clays, (ii) covered by whitish grey clays, (iii) discordant red microlaminated clays, (iiii) poorly sorted red fine silt and clay.

The sandy clay illuvial horizon in the Chambarrambakkam is characterised by a channel and fissure microstructure. It consists of a red, clayey mass in which the embedded, regularly distributed, juxtaposed argillic organisation, is characterised by common, thin, microlaminated red clays (5%), (Fig. 4c). It is moderately birefringent, which coats and mostly infills a dense, regular, network of thin channels and chambers.

Fig. 4(c) The red, clayey
mass of the sandy, clay
illuvial layer, voids of which
are coated or infilled by
microlaminated clays

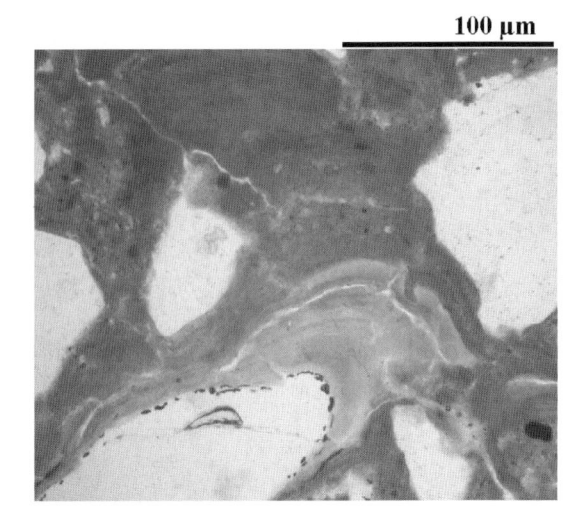

100 μm

The red clayey mass is characterised by a speckled b-fabric and it contains, rather
common, small (<5 μm), black and opaque and dark red ferruginous fragments,
some muscovite flakes and a few charcoal micro-fragments (Fig. 4d).

The coarse fraction as in the ferricretic horizons can be sub-divided in
three classes: (i) coarse sand (~ 1.000 μm) of about 15–20%, (ii) medium sand
(~ 250–500 μm) 20–25%, and (iii) fine sand and coarse silt (~ 100–50 μm) of
about 10%. Some coarse grains are rounded and smooth while most of them are
rough and angular. Quite a few splinters are present in the fine class (Fig. 4e).
This coarse fraction consists of: (i) quartz grains (80%), most of them fresh,
some are corroded while a few exhibit a ferruginous coating, (ii) feldspar (10%),

100 μm

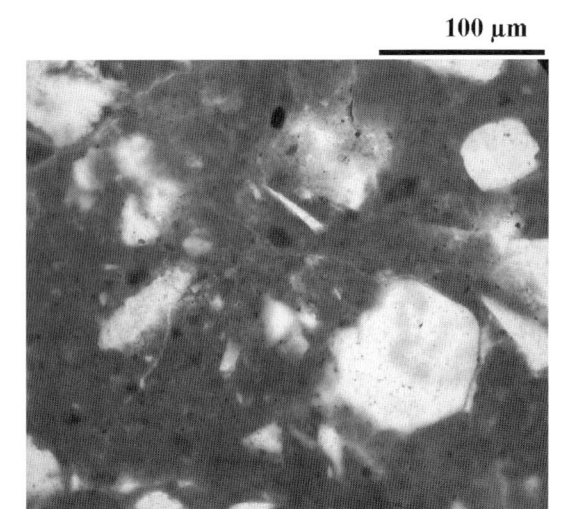

Fig. 4(d) Quartz splinter in
the red, clayey mass

Fig. 4(e) Small charcoal
fragments in the red, clayey
mass

200 µm

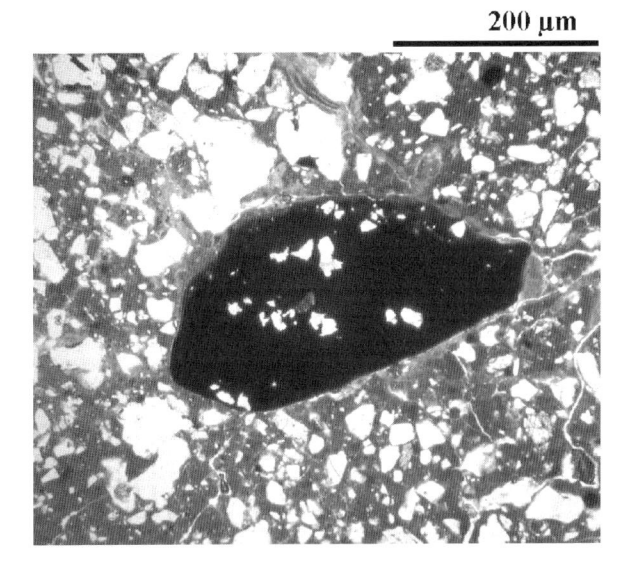

weakly to commonly altered, (iii) rounded to elongated, smooth ferruginous nodules present only in the coarse class, most are black and opaque while a few are red.

The argillic fabric (5–7%) consists of microlaminated, red, clay features, polarising in the yellow of the first order, predominantly in the form of infillings, regularly distributed. Some of these features are partially bleached.

XRD analysis have shown only few weatherable minerals: around 1–5% feldspars and 1–5% phyllosilicates which are dominated by muscovites. Clay minerals are dominated by kaolinite, but illite was also identified. The oxides are chiefly, hematite, goethite, limonite and magnetite.

Micro-sampling and XRD analysis of the concentric zones within individual slabs of ferricretes and pisoliths demonstrate variations in mineralogy. Kaolinite and gibbsite tend to be concentrated in the rim of pisoliths, while the cores tend to be richer in iron oxides and quartz.

6 Discussion

Ferricretes studied belong undoubtedly to ferricretic laterites as defined by Tardy (1993) and other authors. They possess the same succession of horizons, however as they are developed upon a quartzic clayey sediment, the weathering zones are missing. The hierarchy of features and organisations that we were able to establish, lead us to built a sequence of pedo-sedimentary events (Figs. 5 and 6). We will first discuss the processes involved in this sequence, then the succession of events.

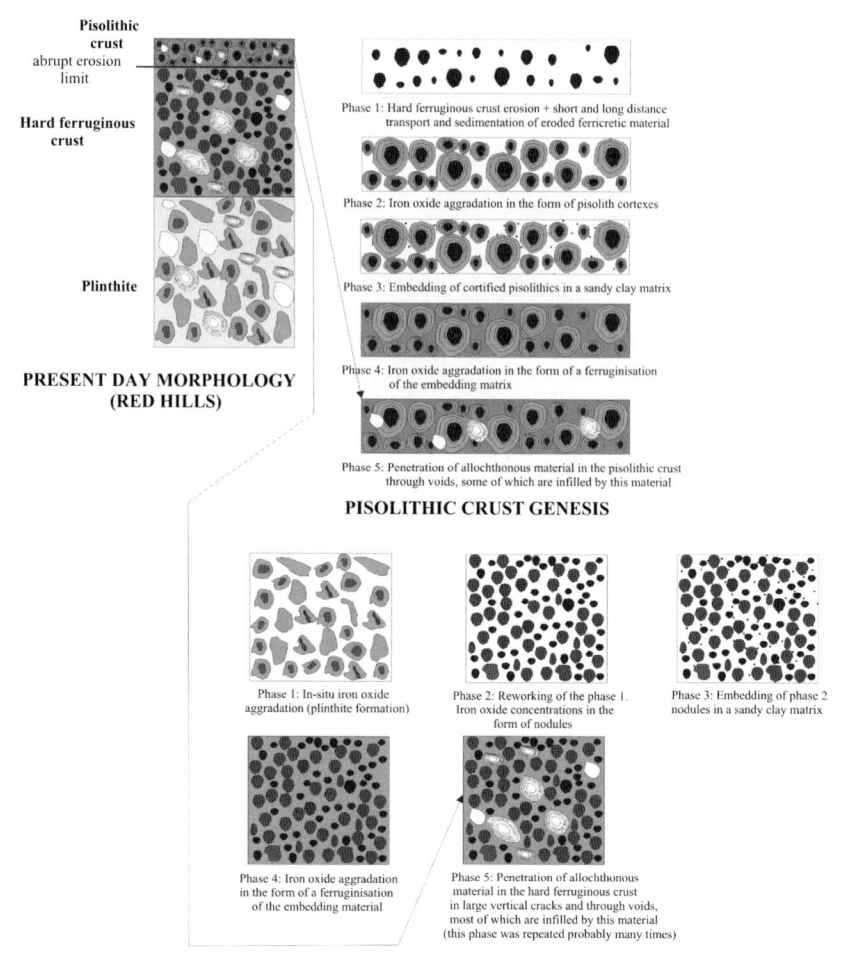

Fig. 5 Re constructed sequence of pedo-sedimentary events forming a cycle in ferricretes of northern Tamil Nadu

6.1 Evidence of Ferricrete Reworked by Water Erosion

Ferricrete reworked by water erosion is evidenced by: (i) the gravel layers, (ii) the varied host materials of cores of pisoliths present in the pisolithic crust, (iii) the cracks, penetrating in to the ferricrete, infilled by coarse sand and gravel.

The size of gravels indicate a deposition by turbulent waters. Their abrupt limit with underlying horizons confirms this inference. The composition of gravels, only ferruginised fragments hosted on quartzic clays, points for a unique origin. Gleysols and Plinthosols were developed on these clays. Frequent fragments of clay coatings

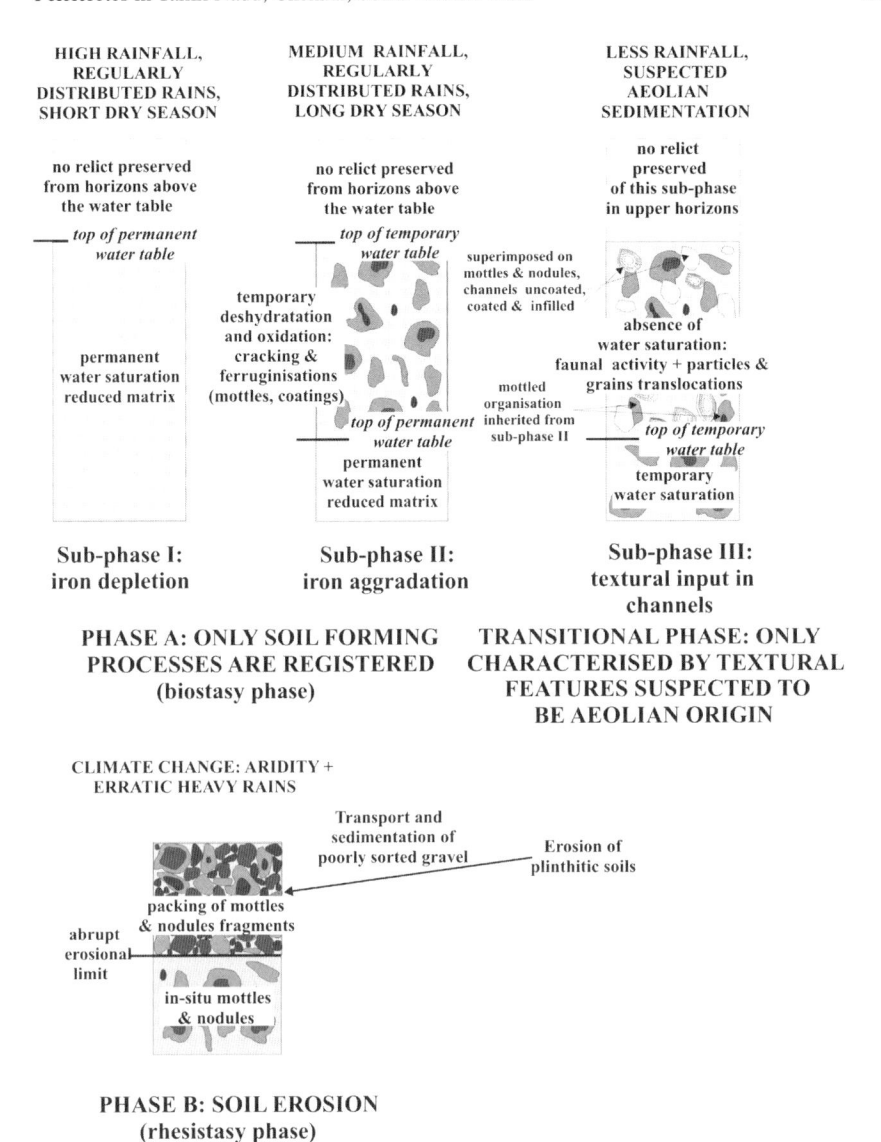

Fig. 6 Tentative reconstruction of a cycle in investigated ferricrete

in the gravel indicate also that these soils possessed illuvial characteristics. As these quartzic clays are only present in the coastal plain, we think that erosion had stripped only the local soil cover. The sub angular morphology of gravels confirms the short distance of transportation. The preservation of these ferruginised gravels during erosion and transportation in turbulent waters causes the formation of a hard and dry plinthite. This erosion took place only in plinthitic horizons and the episode was not unique, e.g. two clearly differentiated layers were observed in the Chambarrambakkam site.

We have considered that, prior to the erosional episodes, a poorly drained coastal alluvial plain existed on which the Gleysols and Plinthosols were developed. Erosion was likely preceeded by the lowering of the ground waters and drying and hardening of the plinthites. This could be associated with either drought, or the regression of the sea. Then the coastal plain was affected by very severe floods which eroded the soils down to a few meters and transported to short distances, attested by the varied shapes of fragments and their poor sorting. The wide distribution of the gravel layers indicate a series of events. The coarse sand and gravels which infill the cracks are likely associated with floods which have affected the surface of the hard ferruginous crust.

Pisolith cores in the pisolithic crust suggest a different process of erosion and sedimentation than of the underlying ferruginous gravels. Cores hosted on weathered charnockite originate from the charnockite horsts, which exist in the coastal plain, some of which still bear signs of a deep weathered mantle. These charnockitic cores in both sites studied suggest a rather long distance transportation, at least some tens of kilometres. Other cores, because of the high degree of weathering of their quartz grains (floating quartz, totally dissolved quartz), have no direct pedogenic relationships with the underlying hard ferruginous crust in which quartz grains are only weakly to moderately weathered.

Consequently we think that these cores have to be traced to older very mature ferricretic crusts that existed earlier and were completely eroded. However, the cores partially translucent appear directly related to underlying horizons. Consequently we have to conclude that the cores in the pisolithic crust are some kind of a residuum of earlier ferricretes which were eroded and of which the most resistant part was distributed randomly, probably by floods.

On another scale, the fragments of clay coatings in embedded matrices should be interpreted as a consequence of a disruption of an argillic organisation followed by a mass movement. This means abrupt, very heavy rains saturating the top soil, which thus became thixotropic and consequently was able to move laterally. The infilling of some large channels by fragments of clayey features resulted in an intense percolation, forming clay coatings in the lower horizons.

6.2 Evidence of Aeolian Sedimentation

The sandy clay ferricretes in the Chambarrambakkam area possess characteristics of an aeolian sediment. This materials occur between the sandy clay cover and the underlying ferricrete with a smooth boundary. This supports the hypothesis that there have been two distinct geogenic formations of different origin. In the field, the thickness of this cover appears topography dependant, very thin on elevated points and thick in depressions, which is another argument in favour of an aeolian origin. Its homogeneity, except the differentiation of elu-illuvial horizons, which is post-depositional is also in favour of such an origin. The following micromorphological characteristics are also in favour of the aeolian origin: (i) some coarse grains are rounded and smooth, (ii) a few splinters are present in the fine sand and coarse silt class. The clay mineralogy points to an enrichment in illites.

This sandy clay cover has never been described earlier in northern Tamil Nadu. However, further south in this state and in north western Sri-Lanka on both sides of the gulf of Mannar exist a thick (from 1 to 30 m), red (2.5 YR 4/8 to 10 R 3/6), clay rich (3–29%) dunes which cover wide surfaces (Gardner 1995) and could be a lateral equivalent of the Chambarrambakkam sandy clay cover. Origin of the embedding red clays of Mannar gulf is disputable.

Foote (1883) and Ahmad (1972) consider that the clayey sands were derived from eroded lateritic soils from the Eastern Ghats and were later wind reworked. Gardner (1981) on the other hand states that the clays of these dunes result of an in-situ weathering of primary minerals brought by wind together with quartz grains. We are in favour of the detrital hypothesis for the clays in the Chambarrambakkam area. Such a hypothesis is supported by: (i) The absence of a weathering profile as well as of any preserved partially or totally weathered primary mineral and (ii) the complete homogeneity of sandy clay cover (except the post depositional eluvial-illuvial differentiation).

Various hypotheses for explaining the genesis of unconsolidated, friable (e.g. oxic or argillic) horizons, such as those in the Chambarrambakkam sandy clay cover which overlie ferricretes, have been proposed for other continents, e.g. in western Africa and Brazil. Some authors consider that ferricretes and overlying horizons are formed simultaneously (e.g. Lucas 1989 in Brazil). Others believe that these overlying horizons result of a chemical weathering of the underlying ferricrete (Costa 1991). A detrital hypothesis was also suggested (Sombroek 1966). Iriondo and Kröhling (1997) suggested that these tropical surface horizons in Northeast Argentina are aeolian in origin, which is an equivalent of mid-latitude loess.

The genesis of the Chambarrambakkam sandy clay cover could be the following: (i) red lateritic soils from the well drained part of the landscape as well as to a lesser extent ferricretes from poorly drained ones were eroded during an erosional phase, (ii) pedo-sediments were deposited on the continental shelf and also in river beds, (iii) these pedo-sediments were later wind eroded during a low marine stand, i.e. a glacial maximum, (iv) they were transported as sand grains and clayey pseudo-sands near the soil surface and deposited on higher surfaces of the coastal zone. The few, rounded to elongated, smooth ferruginous nodules which are similar by internal morphology and composition to pisoliths cores and nodules were eroded from ferricretes.

The above described sandy clay cover is absent in the Red Hills, however some voids of the pisolithic and ferruginous crusts are infilled by a material similar to the one present in Chambarrambakkam sandy clay cover which leads to consider that such a cover was also deposited in the Red Hills, but later eroded. This sandy clay cover corresponds undoubtedly to an unique episode, as testified by the homogeneity of the sediment and monophased character of the clay coatings and infillings through the whole aeolian cover. However, the origin of the abundant infillings and clay coatings present in voids of ferricretes and especially in the gravely layers is puzzling. Two origins can be envisaged for these features, their material could be: (i) brought by floods loaded with sediments which after the flooding have penetrated inside the crust or the gravel letting the materiel to settle in the form of infillings and of coatings as in argillic horizons, (ii) aeolian deposited at the surface, as the Chambarrambakkam sandy clay cover and later translocated inside the ferricrete or the gravel layer. The second hypothesis has been proposed for laterites in south-western Australia by Brimhall et al. (1994).

6.3 Significance of Ferruginised Charcoal Fragments

Ferruginised charcoal fragments are quite common in ferricretes, however, they are detectable only in thin sections. In the field they appear as nodules. They are mentioned in the literature on ferricretes only by authors practising thin section analysis (e.g. Eschenbrenner 1988, Delvigne 1998). However, their significance has never been fully discussed. In the studied ferricretes we were not able to establish a hierarchy between these fragments and other features. The presence of these fragments are likely associated with forest fires. Soil reworking due to erosion and redeposition after these fires was powerful enough to bury the charcoal fragments down to subsurface horizons.

6.4 Soil Forming Processes Identified in Ferricretes

Weathering of primary minerals: As the parental material was already a pedo-sediment consisting dominantly of kaolinite and quartz, the weathering sequence is restricted to quartz grain dissolution. The quartz dissolution remains incipient, characterised by few internal and peripheral pores through the whole profile, except in nodules and pisolith cores of the pisolithic crust in which three degree of quartz weathering exist: (i) quartz grains in most nodules and a few pisoliths exhibit wide cracks and cavernous cavities coated by iron oxides, (ii) in some pisolith cores, residual quartz fragments are floating in the former quartz grain frame, (iii) in other few pisolith cores only the frame of the quartz grain is recognised, as discussed above.

Iron oxide accretion: Ferruginous segregations in mottled clays have infillings described by Bullock et al. (1985), while the bleached clays appeared totally depleted in iron oxides. The ferruginisation of the matrix of the hard ferruginous and of the pisolithic crusts affect the whole host material, which consequently has acted as a cementing agent. It differs from the iron oxides accretion in mottled clays by its continuity and high content in iron oxides.

Such ferruginous organisations differ from a common gleying observed in mid latitude hydromorphic soils only by a higher percentage of iron oxides and the higher abundance of ferruginous features. This high content of iron oxides in investigated ferricretes and more generally in ferricretes under the tropics has to be explained by a monsoonal climate. The rainy season in the tropics is associated with high temperatures, which favour in higher microbiological activity while in mid latitudes, e.g. the Western Europe, soils get water saturated during the winter. Monsoon is characterised by very high precipitations during short periods, which implies important seepage from upland soils to down valley floors, which become rapidly water saturated. At the end of the monsoon, the ground water level decreases, favoring soil oxygenation and consequently accretion of ferric iron. Such an accretion of iron oxides creates an acidic vegetation cover supplying pseudo-soluble chelates.

Iron oxide depletion: The parental clays were undoubtedly deposited free of ferric iron, as at depth their greyish colour is continuous without any mottling. In mottled clays and plinthite, iron oxide depleted features are rather rare. In the embedding material of gravel, the alternating iron oxide accretions and depletions shows that episodes, some characterised by oxidation due to temporary water saturation and others by reduction due to permanent water saturation have occurred alternatively. The depleted whitish grey matrices in plinthites are different from the parent material by their moderately to well sorted coarse fraction. We considered these as the remnants of episodes of permanent water saturation.

Cortification of pisoliths: Pisoliths are very common attributes in ferricretes, frequently complex (consisting of various layers, commonly discordant) (Delvigne 1998). Genesis of pisoliths, especially of the cortex is controversial. Briefly some scientists such as Tardy and Nahon (1985) and Muller and Bocquier (1986) consider that pisoliths as well as their cortex are formed in situ while others (Milnes et al. 1985, Bourman et al. 1987) regard them as resulting from the physical breakdown of ferricretes and associated horizons, a water transport and a post-depositional alteration.

We suppose that eroded pisolith cores were deposited in the form of loose sediments, which latter became seasonally saturated, by water-laden chelates rich in ferrous iron and some clay minerals in suspension (kaolinite and even illite). XRD analysis reveals an admixture of clay minerals to the cortexes vs the highly ferruginised cores indicating an external input of these clay minerals. Such enrichment in clay minerals of the pisolith cortexes indicates different geochemical conditions than those prevailing during genesis of pisolith cores.

Aggradation in the form of microlaminae on pisolith cores indicates probably a seasonal phenomenon, which occurred at the beginning of the dry season when the pisolithic layer started to dry out. The unique cortex in investigated pisoliths suggest one phase of cortification, while the polyphased and even polycyclic complexity of cortexes were discussed in literature (e.g. Delvigne 1998).

Eluviation-illuviation: These processes are clearly expressed in the Chambarrambakkam sandy clay cover where a well expressed eluvial horizon lies upon a typic, well developed (approximately 5% of clay coatings and infillings). The significance of textural features present in crusts, plinthite and in gravely layers is questionable. The infillings consisting of unsorted matrices, whitish grey to red, originate from friable material were present on the surface of the ferricrete. This material was probably transported as a mud through the network of chambers and channels and deposited in it. The importance of the friable material lying on top of the ferricrete cannot be estimated.

The colours of the infillings varied, from whitish grey to red, resulting from the fluctuation of redox conditions in the soil matrix. The sand grain packings were located in main, large voids through which most of soil water flow circulated. Microlaminated clay features which usually coat and infill secondary voids of whitish grey to red matrices were deposited in a latter stage when the matrices were stabilised. Altogether the textural features in crusts, plinthite and in gravely layers can be compared to the set of textural features which infill old, silted-in tile drainage pipes (Sole-Benet 1979).

6.5 Reconstruction of a Sequence of Episodes

The ferricretes we studied evolved discontinuously. The varied phases were succeeded, some of which were pedogenic of iron oxide accretion, while others were dominated by water erosion or aeolian sedimentation. We suggest the following phases, from the most recent to the oldest.

I.   Present day functioning: This is characterised by progressive erosion on hedges of ferricretic laterite inliers, presently reinforced by intensive human quarrying and the process of elu-illuviation of the sandy clay cover where it exists.

II. Sedimentation in valleys. During this phase, valleys were infilled with unweathered materials to which ferricretic laterites have not contributed significantly.

III. Aeolian sedimentation: This is characterised by the Chambarrambakkam sandy clay cover. This phase occurred probably during a dry episode with strong winds during a low marine stand. The material deposited on the continental shelf during the preceding phase was redeposited by the wind on the coastal plain.

IV. Severe erosion episode: Valleys were deeply incised which led to erosion of most of the ferricretic cover and to fossilisation of remaining ferricretic laterites in the form of inliers. This phase have probably occurred during low marine stands.

V.   Last phase of iron oxide aggradation: Cortified pisoliths became cemented by iron oxides during a phase of a high water stand.

VI. Cortification of pisolith transported during phase VII.

VII. Erosional episode. Erosion affected only the surface of ferruginous crusts as only well developed ferruginous cores are present in materials reworked during this phase (nodules originated from plinthites are excluded). These ferruginous cores, at least some of them, were transported on long distances, of which testify the nodules hosted on charnockite.

VIII. Iron oxide aggradation: The hard ferruginous crust became encrusted as a result of a dense and compact ferruginous impregnation of its matrix. Such an aggradational phase should correspond to a long, stable, humid episode during which the coastal plain was flooded during the rainy season and covered by an acidic vegetation, degradation of which produced abundant chelates.

IX. Illuviation: The infillings present in pores of the gravel embedding matrix correspond to a phase of clay translocation during which the soil water regime has fluctuated. At the beginning, the profiles were well drained as shown by the red microlaminated clays, then the whitish grey clays indicate an impervious drainage. The red microlaminated clays which lay in discordance on whitish grey clays correspond to a restoration of well drained conditions while the poorly sorted red fine silt and clay indicate an erosional phase. The origin of translocated clays is questionable. We have supposed that they could be of an aeolian origin as suggested by Brimhall et al. (1994) for the Australian laterites.

X.   Iron oxide depletion: The dominant whitish grey colour of the gravel embedding matrix means a post-depositional iron depletion corresponding to high ground water stand with no or little inter-seasonal fluctuation.

XI. Erosion of the hydromorphic soil cover down to the plinthite: Eroded gravels were transported on short distances as they have a local origin and sedimentation appeared in the form of gravely layers. The sandy clay matrixes in which are embedded in the gravels was deposited probably simultaneously. The abundant small fragments of clay coatings in this matrix indicate an erosion of probably sub-surface illuvial horizons.

XII. Iron oxide aggradation: Ferruginous segregations in the mottled clays and in the plinthite of which are presently only preserved the mottled clays and a variable part of the plinthite. Open water of the sedimentation of the parental sandy clays was followed by a fluctuating water table which induced a clay shrink which was responsible of the fissure network in which water loaded with iron chelates penetrated. Iron oxide accretion occurred when the subsurface water evaporated and the soil was oxidising at the beginning of the dry season.

XIII. Sedimentation of the parental sandy clays: This process has occurred under a permanent reduced water level.

Phases of pedogenesis (phases I, V, VI, VIII, IX, X and XII) have alternated approximately with episodes of erosion and sedimentation (phases II, III, IV, VII, XI and XIII). Pedogenic phases were discontinuous, depending on the soil water regime. The following sub-phases have been recognised: (i) sub-phase of iron depletion due to permanent soil water saturation and reduced conditions, (ii) subphase of iron aggradation which occurred during periods of fluctuating soil high ground water, (iii) phase of iron oxide stability characterised by a congruent translocation of iron oxides and clays which means the absence of major water saturation of soils in some part of the year. We have to consider that sub-phases (i), (ii) and (iii) occurred in this order and that each succession (i), (ii) and (iii) forms a cycle together with an erosional episode (Fig. 6).

Such a phase succession fits with the theory of bio-rhexistasy proposed by Erhart (1956) in which the soil covers and sediments are the result of an alternation of two phases, the first named biostasy, during which a stable humid climate favour forest and soil development and the second named rhexistasy during which the vegetation cover is destroyed and soils are eroded.

6.6 The Age of Northern Tamil Nadu Ferricretic Laterites

Many problems exist in establishing the age of lateritic ferricretes. Bourman (1993a) considers that critical investigations should include: (1) the age of underlying materials as well of overlying soil or sediment, (2) whether the ferricrete has formed at the surface or in subsurface and (3) the ferricrete and bedrock weathering are genetically coupled or uncoupled. Our studies, answered only part of the above mentioned questions. Dating laterites with ^{10}Be has been attempted (Braucher et al. 2000). However, applying this method, because of its complexity, seems to be doubtful.

The perched position just above the present day valleys of ferricretic laterites, in the form of inliers indicates that these profiles are inherited from a past landscape which should be part of a low terrace, rather recent, Middle to Late Pleistocene. Radiogenic dates for valley upper infills range from Holocene to Very Late Pleistocene (Achyuthan et al. 2000) which lead us to conclude that the valley incisions followed by their infill occurred during the last glacial cycle.

Absence of neoformed gibbsite in ferricretic laterites points for a Pleistocene origin of these laterites which is consistent with Subramanian and Selvan (2001). The aeolian episode of the Chambarrambakkam sandy clay deposit could be contemporaneous of the Mannar gulf red dune sedimentation. Gardner (1995) proposed the age to be the Late Glacial maximum (26.000 BP). Singhvi et al. (1986) obtained similar dates. Middle–Upper Palaeolithic artefacts, which occur on the surface of the pisolithic crust at the Red Hills, should mean that this crust existed already some 150,000 years ago. We would conclude that the ferricretes studied date from the Pleistocene and were formed during a time span covering some glacial cycles with complex alternation of erosive and pedogenic phases.

6.7 Reconstruction of Past Landscapes and Climate

The vegetation and soil covers as well as sub-surface water flows, of Tamil Nadu northern coastal plain during iron oxides aggradation was undoubtedly different from the present day conditions (Subramanian and Selvan 2001). For a long period (Early and Middle Pleistocene, including Pliocene), this plain was marshy, without significant valley incisions. Upland waters moved towards the sea mainly as seepage water in the form of sub-surface aquifers which fluctuated from a high level during the rainy season to a lower level during the dry season. This induced the development of hydromorphic soils, a monsoonal type of climate and acidic vegetation favouring high iron oxide accretion. Ferricretic laterites are considered to form on laterally continuous, 'flat' or extremely low-relief land surfaces, with gentle slope 5–8° (Woolnough 1927, Wopfner and Twidale 1967).

Erosional episodes occurred during this period but without a significant modification of the landscape. At first, they were affected only the local soil cover, while during the last phase hard ferruginous crusts were destroyed of which fragments were transported on rather long distance.

A drastic environmental change occurred during the Late Pleistocene characterised by intense erosion. Landscape was deeply modified. Valleys were incised and later infilled with unweathered sediments. Only inliers of the former landscape were preserved. The causes of this drastic change has to be searched in an uplift of the Deccan peninsula probably associated with climatic fluctuations marked by dry periods with strong, erratic rainfalls leading to forest degradation.

An aeolian episode occurred when iron oxide accretion was already accomplished, sometime during the valley incision. This episode is connected most probably with a marine low stand and was characterised by strong winds and relative aridity. The continental shelf was exposed and red pedo-sediments originating dominantly from erosion of lateritic soils, deposited earlier were blown away and deposited on the coastal plain. Sediments belonging to earlier aeolian episodes have not been found, however some textural features present in the ferricrete could be relict of such episodes.

Discordances between clayey features in infillings and the variability of these features in the ferricrete points for severe fluctuations of the soil water regime probably in relation with climatic changes.

7 Conclusions

Classical micromorphological analysis (e.g. Bullock et al. 1985), combined with the micromorphology of ferruginous features under the tropics (Delvigne 1998), associated with identification of erosional and sedimentary features and organisations and coupled with the concept of sequency of events based on hierarchy of features and organisations (Fedoroff and Courty 2002) enabled to decipher and to interpret the complex organisations of investigated ferricretes. Analysis of large thin sections ($13 \times 7\,cm$) appears also quite essential for recognising the whole of most organisations present in ferricretes.

All pedological attributes present in these ferricretes could be divided in elementary pedological features and corresponding organisations. They all correspond to well known soil forming processes, e.g. elementary ferruginous features and organisations correspond to iron oxide accretion, due to alternating reducing and oxidising conditions. Well expressed evidence of water erosions and sedimentation of ferricretic fragments can be observed already in the field, e.g. plinthitic gravel layers and latter analysed in detail under the polarising microscopic, e.g. identifying the material on which ferruginous nodules are hosted.

The aeolian sedimentation already supposed in the field could be clearly demonstrated using thin sections. The concept of hierarchy enabled to establish sequence of events that consisted of pedogenic phases vs erosional-sedimentary ones which alternate more or less regularly. Consequently, ferricretes should be considered as a very compressed sequence of soil developments and of erosion and sedimentation and thus has to be compared with extended sequences such as those of loess and paleosols e.g. the well known sequence of the Loess Plateau of China described by Guo Zhentang (1989).

The compression in studied ferricretes is expressed by the overlapping of features and organisations generated during different phases. Consequently only some phases can be identified, most of the relicts being erased. The crusts, the hard ferruginous, and the pisolithic materials act as a barrier through which only particles and more rarely grains are translocated. Reconstructing the post-crust history can only be performed by analysing and comparing various textural features present in crusts and underlying horizons.

Counting cycles as it is done in loess-paleosol sequences is a task almost impossible in ferricretes. A cycle in studied ferricretes could consist in a very first approximation of the following phases: (i) an iron oxide depletion, (ii) an accretion of iron oxides, (iii) an erosional phase, and (iv) aeolian sedimentation. The iron oxide depletion occurred during a period characterised by a humid climate with a short dry season during which the lowlands were drained almost the whole year. Such a climate should have taken place during interglacial optima. The iron oxide accretion is due to a still humid climate, but with a long dry season which could have taken place during transitions from interglacial to glacial episodes. Erosional phases occurred probably during pleniglacials which correspond to sea low stands and likely lower and erratic rainfall. However, tectonic uplifts cannot be excluded. Aeolian sedimentation phases took also place during arid period of pleniglacials.

The high concentration of iron in ferricretes studied is explained by: (i) an abundant source of iron through the weathering of mafic bedrocks, e.g. charnockite, (ii) a monsoonal climate favoring high rates of chemical reactions, (iii) a vegetation cover producing abundant chelates on one hand and favouring seepage on the other. Their hardening occurs due to: (i) a high content in iron of the ferricretic horizons, (ii) a continuous ferruginous impregnation only perforated by channels and cavities, and (iii) a post functional drying.

Controversies on the genesis of ferricrete can be explained by:

1. Lack of soil micromorphology and different concepts used by many investigators. Field observations supported by analyses of bulk samples do not provide sufficient information to solve the complexity of ferricretes and consequently infer their genesis. Micromorphological investigations should be carried out systematically, i.e., (i) the sampling for thin sections should be performed in order to follow at microscopic level the evolution of all features and organisations from top to bottom of the lateritic profile as well as their lateral variations, (ii) large size thin sections are compulsory for the analyses of ferricretes.
2. The concept of hierarchy is vital for reconstructing the sequences of events registered in ferricretes. Using this concept leads to conclude that development of ferricretes was non-linear, in the form of juxtaposed phases.
3. Erosional and sedimentary features and organisations should be identified. They have been ignored by many investigators. Studying of ferricrete should be multidisplinary, including the viewpoints of pedologists, specialists of weathering and sedimentologists.

References

Achyuthan H (1996) Geomorphic evolution and genesis of laterites around the east coast of Madras, Tamil Nadu, India. Geomorphology 16:71–76

Achyuthan H (2004) Paleopedology of ferricrete horizons around Chennai, Tamil Nadu, India. Rev Mex Cienc Geol 21:133–143

Achyuthan H, Ramasubramaniyam S, Nagalakshmi T (2000) Formation of red soils around Chennai, Tamil Nadu, India. Indian Geogr J 75:17–36

Ahmad E (1972) Coastal Geomorphology of India. Orient Longman. New-Delhi

Bertrand R (1998) Du Sahel à la forêt tropicale. Clés de lecture des sols dans les paysages ouest-africains. Repères. CIRAD, Montpellier

Boulangé B (1984) Les formations bauxitiques latéritiques de Côte d'Ivoire. Les faciès, leur transformation, leur distribution et l'évolution du modelé. Travaux et Documents. ORSTOM. Paris

Bourman RP (1993a) Perennial problems in the study of laterite: a review. Aust J Earth Sci 40:387–401

Bourman RP (1993b) Modes of ferricrete genesis: Evidence from southeastern Australia: Z Geomorphol 37:77–101

Bourman RP, Milnes AR, Oades JM (1987) Investigations of ferricretes and related surficial ferruginous materials in parts of southern and eastern Australia. Z Geomorphol supplementband 64:1–24

Braucher R, Siame L, Bourlès D, Colin F (2000) Utilisation du [10]Be cosmogénique produit in-situ pour l'étude de la dynamique et de l'évolution des sols en milieux latéritiques. B Soc Geol Fr 171:511–520

Brimhall GH, Lewis CJ (1992) Bauxite and laterite ores. In: Encyclopedia of Earth system science, Academic Press Inc

Brimhall GH, Compston W, Williams LS, Reinfrank RF, Lewis CJ (1994). Darwinian zircons as provenance tracers of dust-size exotic components in laterites: mass balance and SHRIMP ion microprobe results. In: Ringrose-Voase AJ, Humphreys GS (eds) Soil Micromorphology: Studies in Management and Genesis. Dev Soil Sci 22, Elsevier, Amsterdam

Buchanan F (1807) A journey from Madras through the countries of Mysore, Kanara and Malabar. East India Co, London 2:436–461

Bullock, P, Fedoroff N, Jongerius A, Stoops G, Tursina T (1985). Handbook for Soil Thin Section Description: Waine Research Publication, Wolverhampton, UK

Costa ML (1991) Aspectos geologicos des lateritos da Amazonia. Rev Bras Geociencias 5:146–160

Delvigne JE (1998) Atlas of micromorphology of mineral alteration and weathering. Can Mineral (Special publication 3) and ORSTOM, Ottawa and Paris

Dikshit KR, Wirthmann A (1992) Strip planation in laterite – a case study from western India. Petermann Geogr Mitt 136:27–40

Erhart H (1956) La genèse des sols en tant que phénomène géologique. Esquisse d'une théorie géologique et géochimique. Biostasie et rhésistasie. Masson, Paris

Eschenbrenner V (1988) Les glébules des sols de Côte d'Ivoire. Nature et origine en milieu ferrallitique. Modalités de leur concentration. Rôle des termites. ORSTOM, Paris

Fedoroff N, Courty MA (1994) Organisation du sol aux échelles microscopiques. In: Bonneau M, Souchier B (eds.), Pédologie. Masson, Paris

Fedoroff N, Courty MA (2002) Paléosols et sols reliques. In: Géologie de la Préhistoire. J. C. Miskovsky Ed. Géopré et Presses universitaires de Perpignan

Food and Agriculture Organization (FAO) (1998) World reference base for soil resources. Rome, Italy

Foote RB (1883) On the geology of Madurai and Tinnevelly districts. Memoirs of the Geological Survey of India. Soils of India 20

Gardner RAM (1981) Reddening of dune sands – Evidence from Southeast India. Earth Surf Process 6:459–468

Gardner RAM (1995) Red dunes and Quaternary Palaeoenvironment in India and Sri Lanka. Mem Geol Soc India 32:391–404

Guilloré P (1985) Méthode de fabrication mécanique et en série des lames minces INA P-G, Grignon. France

Guo ZT (1990) Successsion des paléosols et des loess du centre-ouest de la Chine. Approche micromorphologique. Thèse Univ. Pierre et Marie Curie. Paris

Harrassowitz H (1926) Laterit. Fortsschr. Geol Pal 4:253–565

Iriondo M, Kröhling D (1997) The tropical loess. In: An Z, Zhou ZW (eds) Quaternary geology Proceedings, 30th Inte. Geol Congress Rotterdam

Karunakaran C, Roy Sinha S (1980) Laterite profile development linked with polycyclic geomorphic surfaces in South Kerala. In Krishnaswamy VS (ed) Lateritisation processes. Oxford and New Delhi

Lucas Y (1989) Systèmes pédologiques en Amazonie brésilienne:équilibre, déséquilibres et transformation. Thèse Univ Poitiers. France

Maignien R (1966) Review of research on laterites. UNESCO. Nat resources res 4

McFarlane MJ (1976) Laterite and Landscape. San Diego, Academic Press

Milnes AR, Ludbrook NH, Lindsay JM, Cooper BJ (1985) Field relationships of ferricretes and weathered zones in southern South Australia. Aust J Soil Res 23:441–465

Muller J, Bocquier G (1986) Dissolution of kaolinites and accumulation of iron oxides in lateritic-ferruginous nodules. Geoderma 37:113–136

Ollier CD, Rajaguru RW (1989) Laterite of Kerala (India). Geografia Fisica e Dinamica Quaternaria 12:27–33

Perur NG, Mithyantha S (1972) Soils of Mysore. In: Alexander TM (ed) Soils of India.

Peterschmitt E (1993) Les couvertures ferrallitiques des Ghâts occidentaux en Inde du Sud. Caractères généraux et dégradation par hydromorphie sur le revers. Institut français de Pondichery

Raman PK, Vaidyanadhan R (1980) Laterites and laterisation in Eastern Ghats and Coastal Plain in North Coastal Andra Pradesh, India. In: Krishnaswamy VS (ed) Lateritisation processes. Oxford and New Delhi

Raychaudhuri SP, Agarwal RR, Datta Biswas NR, Gupta SP, Thomas PK (1963) Soils of India. ICAR New-Delhi

Shapiro L., Brannock WW (1962) Rapid analyses of silicates, carbonates and phosphate rocks. US Geol Survey Bull 1114A:1–56

Singhvi AK, Deraniyagala SU, Sengupta D (1986) Thermoluminescence dating of Quaternary red sand beds: A case study of coastal dunes in Sri Lanka. Earth Planet Sci Lett 80:139–144

Sole-Benet A (1979) Contribution à l'étude du colmatage minéral des drains. Une démarche expérimentale basée sur la micromorphologie pour étudier les transferts solides dans les sols. CEMAGREF Antony France

Sombroek WG (1966) Amazon soils: a reconnaissance of soils of the Brazilian amazon region. Centre for Agriculture, Wageningen, The Netherlands

Subramanian KS, Mani G (1980) Genetic and geomorphic aspects of laterites on high and low landforms in parts of Tamil Nadu. In: Krishnaswamy VS (ed) Lateritisation processes. Oxford and New Delhi

Subramanian KS, Selvan T A (2001) Geology of Tamil Nadu and Pondicherry. Geol Soc India

Tardy Y (1993) Pétrologie des latérites et des sols tropicaux. Masson, Paris

Tardy Y, Nahon D (1985) Geochemistry of laterites, stability of Al-goethite, Al-hematite and Fe(III)-kaolinite in bauxites and ferricretes: an approach to the mechanism of concretion formation. Am J Sci 285:865–903

US Department of Agriculture. Natural resources conservation service (1999). Keys to Soil Taxonomy. 8th edn Washington D.C.

Woolnough WG (1927) The duricrust of Australia. J Proc Royal Soc New South Wales 61:24–53

Wopfner H, Twidale CR, (1967) Geomorphological history of the Lake Eyre Basin, In: Jennings JN, Mabbutt JA (eds) Landform Studies from Australia and New Guinea. Aus National University Press, Canberra

Palygorskite Dominated Vertisols of Southern Iran

A. Heidari, S. Mahmoodi, and G. Stoops

Abstract Vertisols are generally known as smectite-dominated soils with prominent shrink-swell properties, although Vertisols with mixed, kaolinitic, illitic or other non-smectitic mineralogical compositions have also been reported. Palygorskite-dominated Vertisols have not yet been described in detail. From about 70,000 ha of cultivated Vertisols in Iran, 16,000 ha is located in the Fars Province (southern Iran). These soils formed on calcareous sediments in the lowlands, are classified as Sodic Haplusterts, Aridic Gypsiusterts, Aridic Haplusterts, Halic Haplusterts and Aridic Haploxererts. X-ray diffraction and electron microscopy indicate that the smectite content is very low to almost absent in most soils, while palygorskite and chlorite are dominant silicate clay minerals, especially in the fine clay fraction. A good relationship was observed between COLE and fine clay/total clay ratio. Strongly developed angular blocky to prismatic structure with wedge-shaped peds is dominant in the subsurface horizons. Due to the high micritic carbonate content of the groundmass, the b-fabric is mostly crystallitic, but decalcified zones (after HCl treatment) show a stipple-speckled b-fabric. Fe/Mn hydroxide impregnations are omnipresent, which is an indication of seasonal hydromorphic conditions. In some part of the profiles prismatic and lenticular gypsum crystals occur as loose continuous infillings of channels, reflecting mineral formation in brine-filled macropores.

A. Heidari
Department of Soil Science, Agriculture College,
University of Tehran, Daneshkadeh St. Karaj,
Iran, e-mail: aheidari82@yahoo.com

S. Mahmoodi
Department of Soil Science, Agriculture College,
University of Tehran, Daneshkadeh St. Karaj, Iran,
e-mail: smahmodi@chamran.ut.ac.ir

G. Stoops
Department of Geology and Soil Science, Ghent University,
Krijgslaan 281, S8, B-9000 Ghent, Belgium

S. Kapur et al. (eds.), *New Trends in Soil Micromorphology*,
© Springer-Verlag Berlin Heidelberg 2008

Keywords Chlorite · clay mineralogy · COLE · micromorphology · plasticity

1 Introduction

Historically, Vertisols are known as smectite dominated soils. Their typical morphological and micromorphological characteristics are supposed to be all related to clay type and content, in combination with the action of climate (Dixon 1982, Dudal and Eswaran 1988, Blokhuis et al. 1990, Coulombe et al. 1996). Their macro-and micromorphological characteristics, such as granular and wedge shaped structures, slickensides, striated b-fabrics and void pattern are mainly stress related features.

High variability in the soil forming factors, particularly the strong diversity in their parent materials (Mermut et al. 1996) and the different climatic condition, influence the properties of Vertisols, including the complexity of their mineralogy (Coulombe et al. 1996). Apart from typic Vertisols with smectitic composition, also soils with mixed, kaolinitic, illitic and non-smectitic mineralogical composition have been reported to have a vertic behaviour (Probert et al. 1987). Kaolinite is an abundant constituent of some Vertisols in El Salvador (Yerima et al. 1987), Sudan, Hawaii and Australia (Coulombe et al. 1996). The relative abundance of illite (mica) is also quite variable from one Vertisol to another. Dixon (1982) has reported that mica generally occurs in small amount in Vertisols from Texas. Norrish and Pickering (1983) and Hubble (1984) have shown that illite is abundant in south Australian Vertisols, or is co-abundant with kaolinite in many Vertisols in Northern Australia. Palygorskite and chlorite are reported as minor components of many Vertisols, generally originating from the parent material (Feigin and Yaalon 1974, Shadfan 1983), but very little is known about palygorskite dominated Vertisols in the world (Jeffers and Reynolds 1987). Palygorskite is common in many soils of Iran (Abtahi 1974, Khademi and Mermut 1998; 1999) and in the soils overlying Pleistocene calcretes in Central Anatolia and the Mediterranean coast of neighboring Turkey (Kapur et al. 1990).

According to Wilding and Tessier (1988), a clay fraction with a high surface area will influence positively the shrink-swell phenomena regardless of the type of clay minerals. Also Coulombe et al. (1996) mentioned that the characteristics of each mineral need to be considered in order to identify their influence on physical and chemical properties of Vertisols. Thus, mineralogy of Vertisols is not as simple as usually considered in the past (Coulombe et al. 1996) and the shrink-swell property, which is responsible for the genesis (Anderson et al. 1973) and behaviour of Vertisols, is a complex, dynamic but incompletely understood set of processes (Soil Survey Staff 1999). Recently, in Soil Taxonomy (Soil Survey Staff 2003), mixed and other non-smectitic mineralogical families have been recognised.

Due to the high clay contents in Vertisols and vertic subgroups, the fabric of the fine material in the groundmass requires more attention in micromorphological studies. Swelling processes in soils result in micro-shear deformations. Such stresses reorient the individual clay plates into lineated zones with face to face clay alignment (Wilding and Tessier 1988). Stress related fabrics such as porostriated and granostriated b-fabrics are therefore commonly observed in Vertisols. Planar voids and porphyric c/f related distribution patterns are the other common micromorphological features in these soils (Heidari et al. 2005).

The objectives of this work are: (i) to study soils with vertic features formed on very fine calcareous and marine deposits with a silicate clay fraction dominated by palygorskite and only minor amounts or no smectite, and (ii) to discuss, based on mineralogical and micromorphological characteristics, whether they can be classified as Vertisols.

2 Materials and Methods

2.1 Study Area

In Iran, Vertisols occupy about 70,000 ha of agricultural lands. The largest Vertisol areas being in the Fars province, southern Iran, were selected for this study (Fig. 1). Fars province is part of Post- Tethyan Sea environment which is rich in evaporites (salts and gypsum) in most southern parts. The southern part of this province is formed by sedimentary rocks of Tertiary age and younger (Asmari, Jahrom, Mishan, and Bakhtyari Formations), mainly composed of limestone, dolomite and marl. These sedimentary rocks contain considerable amounts of palygorskite which

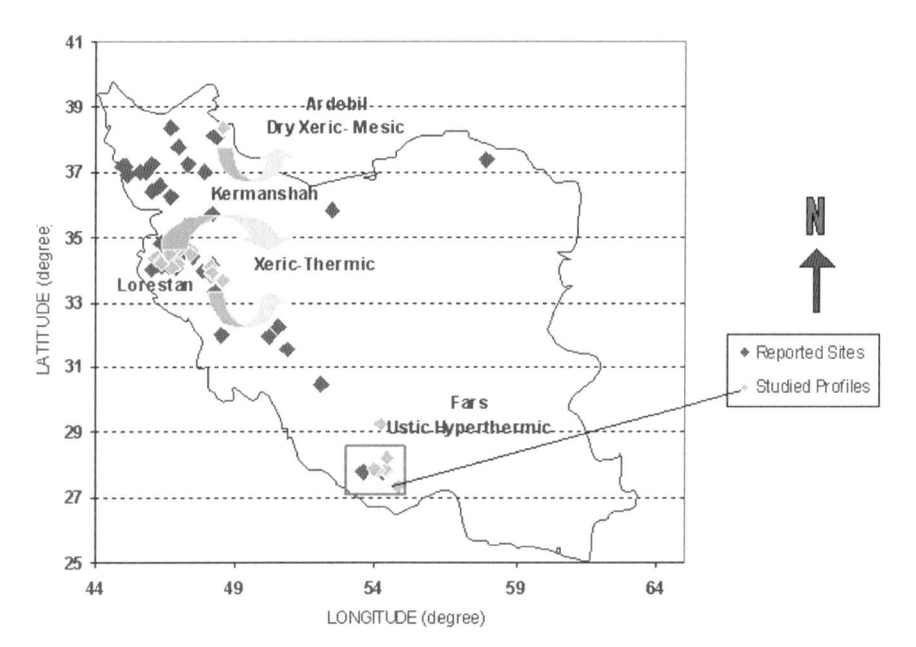

Fig. 1 Distribution of Vertisols and vertic subgroups in Iran (according to the Soil and Water Institute of Iran) and location of the profiles studied

can be inherited into the soils. Neoformed palygorskite has also been reported in gypsiferous soils of southern arid regions of this province (Khormali 2003).

The soils studied are formed on very fine calcareous and marine deposits. The soil climate is ustic-hyperthermic, with a mean annual air temperature above 23°C and mean annual rainfall of about 248 mm. Main land uses of these soils are occasional grazing land and extensive dry or irrigated farming of wheat, barley and forb. Nine pedons were described (Soil Survey Staff 1993) and sampled for further detailed laboratory analysis. Two pedons were selected for this paper, an Aridic Gypsiustert and a Halic Haplustert (Soil Survey Staff 1999).

2.2 Laboratory Studies

2.2.1 Physicochemical Analysis

After air drying at room temperature samples were ground to pass through a 2 mm sieve. The saturated extracts prepared by vacuum suction, were used for measurement of pH and electrical conductivity. Particle size distribution was determined by hydrometer method (Gee and Bauder 1986). The carbonate content was determined by calcimetery method (Salinity Laboratory Staff 1954). Cation exchange capacities of soil and clay samples were determined by sodium acetate method (Chapman 1965). The amount of fine and total clay was determined after removal of carbonates, organic matter, and Fe oxyhydroxides and subsequent fractionation for mineralogical analysis (Kunze and Dixon 1986). The gypsum content was determined by the acetone method (Salinity Laboratory Staff 1954). COLE values were determined on soil pastes (Schafer and Singer 1976). Plastic limit and liquid limit were determined according to Atterberg tests.

2.2.2 Mineralogical Analysis

The mineralogy of the clay fraction was studied by XRD analysis after Mg-saturation, Mg-saturation and glycerol solvation, K-saturation, and K-saturation followed by heating to 550°C, using a Siemens XRD, D5000, with CuK_α radiation. Pretreatments were carried out according to Kunze and Dixon (1986). Some samples were HCl treated (1 N, 80°C) to differentiate chlorite from kaolinite (Moore and Reynolds 1989). Fine and coarse clays were analysed separately for some samples.

Transmission electron microscopy (TEM), LEO 906E, was used for further characterisation and semi-qualitative determination of fibrous clays, by visual estimation of their relative abundance on TEM images.

2.2.3 Thin Section Preparation

Air-dried oriented clods were impregnated under the vacuum with an acetone diluted polyester resin (50/50 ratio). Thin sections with an area of about 30 cm² were prepared using standard techniques (Murphy 1986), and described following the guidelines of Stoops (2003). In order to observe b-fabrics masked by calcium carbonate, thin sections were HCl treated (2 N, 3 h), before covering.

3 Results

3.1 Morphological Properties

The studied soils are all deep to very deep, with an ochric epipedon, and usually a thin crust at the surface. Soil colour varies between 10YR3/3 and 10YR5/4 throughout the profiles. Structure of the surface horizons is fine, angular to subangular blocky (Table 1 and Fig. 2). Subsurface horizons show a moderately to strongly developed, medium to coarse prismatic structure with slickensides (wedge-shaped peds) grading to angular blocky in the lower part of some profiles. Many cracks, 1.5–3.5 cm wide at the surface, a gilgai relief and fox holes are the most important morphological features in these Vertisols (Fig. 3). Carbonate nodules, mottling and some of gypsum crystals have been observed in a number of profiles.

3.2 Physical and Chemical Properties

The pH varies between 7.1 and 8.0. Electrical conductivity generally increases with depth and reaches a maximum of 24.4 dSm^{-1}. The CaCO$_3$ content is usually high (25.2–33.7%) and slightly higher in the surface horizons than in lower parts. Cation exchange capacity of fine earth (8.3–11 cmol$^+$kg^{-1}) and of clay fractions (25.5–44.6 cmol$^+$kg^{-1}) is markedly low in comparison with normal smectitic Vertisols. Some profiles contain significant amounts of gypsum and were classified as Aridic Gypsiustert. Despite the high ratio of fine to total clay (0.48–0.68) the coefficient of linear extensibility (COLE) is rather low (0.06–0.12, Table 1). Yet, there is a significant relationship (R^2=0.8911) between the fine clay/total clay ratio and COLE values (Fig. 4). Plasticity indices for the studied profiles range between 10 and 20%, being lower in the surface horizons. A fairly good relation exists between the plasticity and total clay content.

3.3 Mineralogical Characteristics

The fine and total clay fractions are dominated by palygorskite and chlorite. Illite and kaolinite are accessory phases in the coarse clay fraction (Fig. 5 and Table 2). Relatively sharp and distinct 1.04 nm peaks together with lower intensity 0.64 nm peaks (Fig. 5) show the predominance of palygorskite over the other silicate clay minerals as confirmed by TEM micrographs (Fig. 6). The sharp and distinct 1.42 nm peak which does not change by Mg-Gly treatment and disappears on HCl treatment is considered as an

Table 1 Physicochemical and structural properties of the studied pedons

Horizon	Depth (cm)	Structure[a]	Color	Particle size (%)			pH_e	EC_e (dSm^{-1})	$CaCO_3$ (%)	CEC[b] (cmol$_c$ kg^{-1})		FC/TC[c]	COLE	LL[d]	PL[e]
				cl	si	sa				CEC_s	CEC_c				
6. Aridic Gypsiustert															
A_{p1}	0–12	m3abk	10YR5/2	69.0	27.8	3.2	7.9	2.83	33.70	9.36	–	–	0.12	39	29
A_{p2}	12–22	m2pr	10YR5/2	69.0	27.6	3.4	8.0	1.56	31.70	9.36	–	–	0.09	44	29
B_{ss}	22–48	m3pr	10YR4/3	68.8	27.6	3.6	7.6	3.67	28.60	8.28	39.20	0.61	0.08	45	28
B_{kyss1}	48–70	m2pr	10YR4/3	78.8	15.6	5.6	7.7	3.90	26.70	8.28	32.59	0.64	0.07	46	31
B_{kyss2}	70–90	m3pr	10YR4/4	80.8	15.8	3.4	7.2	4.26	28.90	8.82	–	–	0.08	46	29
B_{kyss3}	90–110	m3pr	10YR4/4	80.8	15.8	3.4	7.7	4.22	27.90	8.82	–	–	–	–	–
B_{kyss4}	110–150	m3pr	10YR4/4	76.8	17.6	5.6	7.0	9.10	27.90	8.28	–	–	0.09	47	30
7. Halic Haplustert															
A_1	0–10	f2sabk	10YR5/4	75.0	20.0	5.0	7.2	3.87	27.30	10.51	26.59	0.51	0.08	44	30
A_2	10–23	m2abk	10YR5/4	75.0	19.6	5.4	7.4	5.57	28.00	10.51	–	0.48	0.07	46	29
AB	23–45	m3pr	10YR3/4	85.6	11.2	3.2	7.4	8.20	26.00	11.10	–	0.53	0.08	47	29
B_{ss1}	45–60	c3pr	10YR3/4	85.6	10.8	3.6	7.3	9.56	26.60	9.93	–	–	0.09	47	29
B_{ss2}	60–80	c3pr	10YR3/3	81.6	11.2	7.2	7.2	13.0	25.20	10.51	25.46	0.64	0.09	46	28
B_{ss3}	80–110	c3pr	10YR3/3	77.6	10.8	11.6	7.1	24.4	25.30	9.36	–	0.55	0.06	46	29
B_{ss4}	110–150	c3pr	10YR3/3	81.6	11.2	7.2	7.2	18.1	26.60	10.51	44.56	0.68	0.10	49	29

[a] soil structure description according to Soil Survey Staff (1993),
[b] CEC of soil (CEC_s) and clay (CEC_c),
[c] ratio of fine clay to total clay,
[d] LL, Liquid limit,
[e] PL Plastic limit.

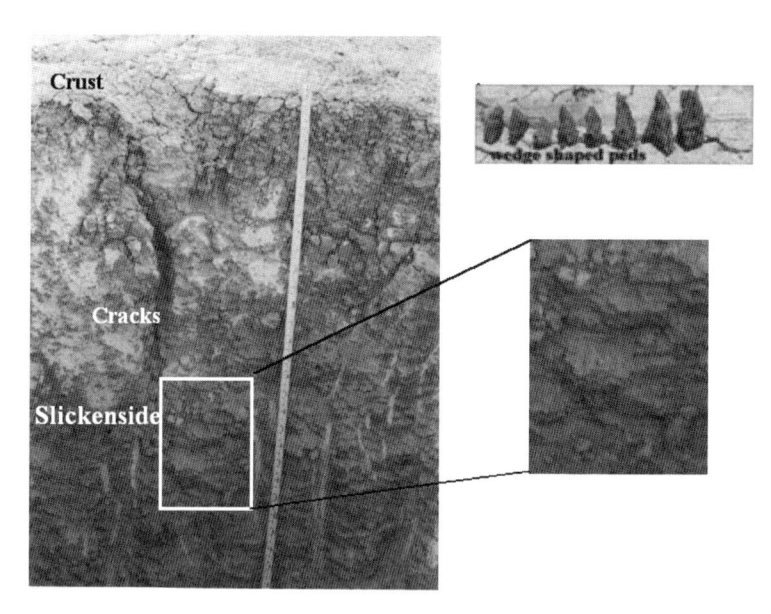

Fig. 2 Morphological features of Pedon 7: coarse wedge shaped peds (*right up*) and distinct slickensides (*right down*)

Fig. 3 Polygonal cracks, round gilgai (<15 cm height) (*up*) and fox holes (*down*) at the surface of Pedon 7

Fig. 4 Relationship between COLE and fine clay/total clay ratio in Pedon 7

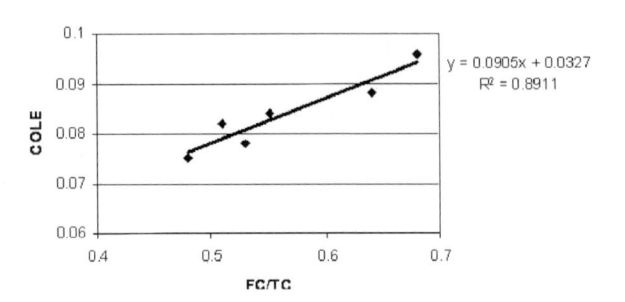

Fig. 5 X-ray diffraction patterns for the total **(a)** and fine **(b)** clay fractions of horizon B_{ss3} of Pedon 7

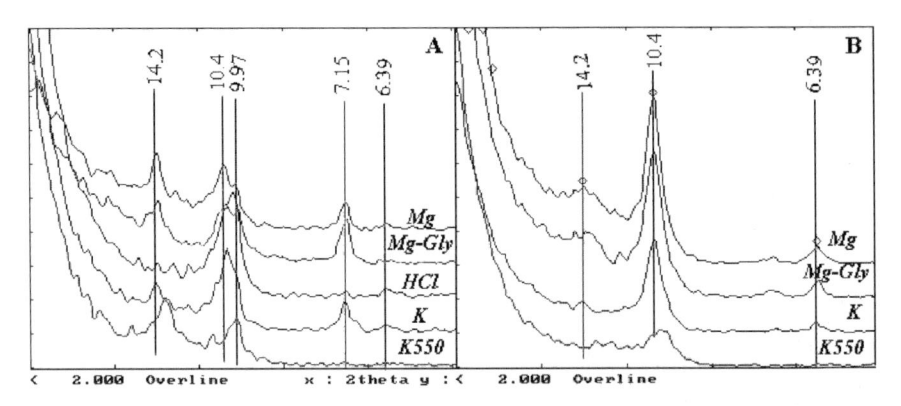

indicator for important amounts of chlorite in these samples. The relative intensities of 1.04, and 0.64 nm compared to 1.42 nm peaks are much higher in the fine clay fraction (Fig. 5), indicating the concentration of fibrous palygorskite in this fraction. Scanning and transmission electron microscopy confirmed these results (Fig. 6).

3.4 Micromorphology

The microstructure of the surface horizons is moderately to strongly developed angular and sub-angular blocky with intrapedal vughs (Fig. 7a). The subsurface microstructure is angular blocky and prismatic with dominantly planar interpedal

Table 2 Mineralogical composition of clay fraction in the pedons studied

Profile No.	Horizon	Depth (cm)	Fraction	Mineralogy[a]
6	B_{ss}	22–48	Total clay	Pal.>Chl.>Ill.>Kao.
	B_{kyss}	48–70	Total clay	Pal.>Chl.>Ill.>Kao.
7	A_1	0–10	Total clay	Pal.>Chl.>Ill.>Kao.
	A_2	10–23	Fine clay	Pal.>Chl
			Coarse clay	Chl.>Pal.>Ill.>Kao.
	AB	23–45	Fine clay	Pal.>Chl
			Coarse clay	Chl.>Pal.>Ill.>Kao.
	B_{ss1}	45–60	Fine clay	Pal.>Chl
			Coarse clay	Chl.>Pal.>Ill.>Kao.
	B_{ss2}	60–80	Total clay	Pal.>Chl.>Ill.>Kao.
	B_{ss3}	80–110	Fine clay	Pal.>Chl
			Coarse clay	Chl.>Ill.>Pal.>Kao.
	B_{ss4}	110–150	Total clay	Pal.>Chl.>Ill.>Kao.

[a] Pal. palygorskite, Chl. chlorite, Ill. illite, Kao. kaolinite, Ver. vermiculite, Smec. Smectite

Fig. 6 Transmittance electron microscopy (TEM) showing fibrous palygorskite minerals. Fine clay fraction (*left*) and total clay fraction (*right*) of B_{ss1} horizon of Pedon 7

voids (Fig. 7b). Related distribution patterns are fine monic to open porphyric. The micromass is yellowish brown and dotted, showing a calcitic crystallitic b-fabric which after HC treatment becomes porostriated, mosaic speckled and reticulate striated (Table 2 and Fig. 7). The coarse fraction is dominantly composed of sparitic limestone fragments. In most profiles Fe/Mn hydroxide impregnations are common, in the form of nodules, hypocoatings and impregnations of limestone fragments. Lenticular and microcrystalline gypsum crystals are the other important features observed in some pedons. Prismatic gypsum crystals occur as loose continuous infillings on channels of some pedons (Fig. 8).

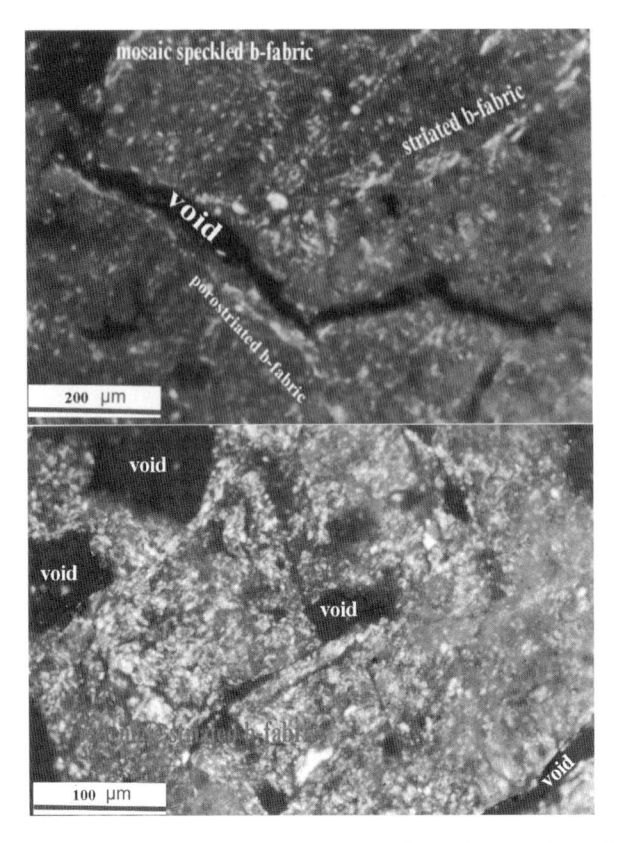

Fig. 7 HCl treated thin sections showing the masked mosaic speckled, porostriated and reticulate striated b-fabrics in Pedon 7, A^2 horizon (*up*) and Bss3 horizon (*down*). Cross polarizer

Fig. 8 Crystallitic b-fabric and infilling of channels with lenticular gypsum in Pedon 6. Crossed polarisers

4 Discussion

Key requirements of Vertisols (Soil Survey Staff 1999), including cracks, slickensides, wedge shaped aggregates and high clay content even as gilgai micro-topography are clear for the studied pedons (Fig. 3). Although these morphologies are fully related to shrink-swell phenomena and are mostly tied to smectitic clays (Blokhuis et al. 1970, Anderson et al. 1973), the clay fraction of the pedons studied is composed dominantly of low active and non-interlayer expansive clays like palygorskite, chlorite, and illite with little or no smectite (Table 2).

The behaviour of Vertisols can best be predicted by examining a combination of physical, chemical, and mineralogical soil properties in order to integrate them in a shrink-swell model. No single property however accurately predicts shrink-swell potential for all Vertisols. Shrink-swell phenomena in soils and sediments can not be explained at unit cell level, and one has to take into account changes in the microfabric and interparticle (interaggregate pores or pores between quasi-crystals) and intraparticle (pore organised in clay matrix or between sub-units of quasi-crystals and the interlayer space) porosity of clays (Coulombe et al. 1996).

The CECs of the studied pedons is quite low (Table 1), compared to the ranges mentioned in literature for Vertisols, which is in agreement with the presence of low activity clay minerals. The rather high fine clay/total clay ratios, particularly in B_{ss} horizons (Table 1) have probably a significant effect on shrink-swell forces in these soils. The effect of total surface area in relation to shrink-swell phenomena has been indicated in the past by several authors (Yerima et al. 1985, Yousif et al. 1988, Probert et al. 1987, Shirsath et al. 2000, Bhattacharyya et al. 1997). The relatively low COLE values in the palygorskite and chlorite dominated soils (Table 1) indicate a more limited degree of swell-shrink behaviour in these soils.

In addition to clay content and critical moisture content, the swelling potential of expansive soils is a function of the plasticity index. A wide range of results have been reported for plasticity index of smectitic minerals (e.g. plasticity index of montmorillonite ranges between 50 and 800%). The index varies between 25 and 60% for illite, 60 and 110% for attapulgite and about 8% for chlorite (Olson et al. 2000). The same authors reported a plasticity index of 33–46% for a fine, smectitic, mesic, vertic Hapludalf, whereas Kamara and Haque (1988) reported a range of 6–18% for surface Vertisols from the Ethiopian highlands.

Plasticity indices of the studied pedons vary between 10 to 20%, are located between the lower and higher limits of above mentioned results. This is agreement with the results obtained for COLE that vary between 0.06 and 0.12 which are lower than smectitic Vertisols.

The microstructure is mainly strongly developed angular blocky. Despite pedo-turbation related to vertic behaviour, channels and vughs commonly occur. The organisation of silicate clays in the micromass is masked by the presence of micritic calcite (crystallitic b-fabric); after HCl treatment porostriated, mosaic speckled and reticulate striated b-fabric appear (Table 3 and Fig. 7). Blokhuis et al. (1970),

Table 3 Micromorphological description of selected pedons

Horizon	Depth (cm)	c/f-rd[a]	c/f ratio	b-fabric[b]		Fe/Mn nodules %	Gypsum %
				Before HCl treat.	After HCl treat.		
6. Aridic Gypsiustert							
A$_{p2}$	12–22	fm	5/100	cr	ps,ms,rs	2	
B$_{kyss1}$	48–70	fm	5/100	cr	ps,ms,rs	5	5–10
B$_{kyss3}$	90–110	fm	5/100	cr	ps,ms,rs	5	5–10
7. Halic Haploustert							
A$_2$	10–23	fm to op	5/100	cr	ps,ms,rs	2	
B$_{ss3}$	80–110	fm to op	5/100	cr	ps,ms,rs	5	

[a] fm fine monic, op open porphyric,
[b] cr crystallitic, ps porostriated, ms mosaic speckled, rs reticulate striated

Kooistra (1982) and Blokhuis et al. (1990) have also indicated the presence of granostriated b-fabrics in Vertisols, with grains larger than 4 mm.

The most frequent pedofeatures in almost all horizons are Fe/Mn hydroxide nodules and hypocoatings (Table 3). According to Blokhuis et al. (1990), Mn or Fe/Mn hydroxide features are more common than carbonate nodules in Vertisols. However, in the soils under study carbonate nodules and hypocoatings, mostly fragmented and elongated parallel to the voids, are very common. Infillings of mainly subhedral prismatic gypsum crystals in the deeper part of some pedons (Table 3 and Fig. 7) reflect mineral formation from brine filling macropores (Mees 1999).

5 Conclusions

Clayey soils of the Fars Province, south Iran, show the macro- and micromorphological features characteristic for Vertisols, although their clay mineralogical composition is not dominated by smectitic clays, but by low activity palygorskite and chlorite. The vertic characteristics have to be explained by the large surface area of the clay fraction with high ratios of fine clay/total clay. Therefore, we concluded that, based on particle size, the interparticle pore spaces play a more significant role than the interlayer spacing of phyllosilicates in the development of vertic features in the soils studied.

The high amount of disseminated fine calcite crystals in the calcareous Vertisols makes it difficult to recognize in thin sections the striated b-fabrics that are typical for vertic materials. After HCl treatment, porostriated, mosaic speckled and reticulate striated b-fabrics appear, proving that moderate to strong shearing forces are involved in the formation of these soils. Pedofeatures include Fe/Mn hydroxide nodules and hypocoatings, compatible with periodic hydromorphic conditions.

References

Abtahi A (1974) Contribution to the knowledge of the soils of Shiraz area (Iran) Ph.D. Thesis State Univ. Ghent, Belgium

Anderson JU, Elfadil Fadul K, O'Connor GA (1973) Factors Affecting the Coefficient of Linear Extensibility in Vertisols. Soil Science Society of America Proceedings. 37: 296–299

Bhattacharyya T, Pal DK, Deshpande SB (1997) On kaolinitic and mixed mineralogy classes of shrink-swell soils. Australian Journal of Soil Research. 35: 1245–1252

Blokhuis WA, Slager S, Van Schagen RH (1970) Plasmic fabric of two Sudan Vertisols. Geoderma. 4: 127–137

Blokhuis WA, Kooistra MJ, Wilding LP (1990) Micromorphology of cracking clay soils (Vertisols). In: Douglas LA (ed.) Soil Micromorphology: A Basic and Applied Science. Developments in Soil Science 19. Elsevier, Amsterdam. pp. 123–148

Chapman HD, (1965) Cation exchange capacity In: Black CA (ed.) Methods of Soil Analysis, Part 2, Chemical and Microbiological Properties. Agronomy Monograph No. 9. American Society of Agronomy, Madison, WI. pp. 891–900

Coulombe CE, Dixon JB, Wilding LP (1996) Mineralogy and chemistry of Vertisols. In: Ahmad N and Mermut A (eds.) Vertisols and Technologies for their Managements. Developments in Soil Science 24. Elsevier, New York. pp.115–200

Dixon JB (1982) Mineralogy of Vertisols. In: Vertisols and Rice Soils of the Tropics, Symposia Papers II. 12th ICSS, New Dehli (India). pp. 48–59

Dudal R, Eswaran H (1988) Distribution, properties and classification of Vertisols. In: Wilding LP and Puentes R (eds.) Vertisols: Their Distribution, Properties, Classification and Management. Texas A&M University Printing Center, College Station, TX|

Feigin A, Yaalon DH (1974) Non-exchangeable ammonium in soils of Israel and its relation to clay and parent materials. Journal of Soil Science. 25: 384–397

Gee GW, Bauder JW (1986) Particle size analysis. In: Klute A (ed.) Methods of Soil Analysis, Part 1, Physical and Mineralogical Methods, 2nd edn. Agronomy Monograph No. 9. American Society of Agronomy, Madison, WI

Heidari A, Mahmoodi SH, Stoops G, Mees F (2005) Micromorphological characteristics of Vertisols of Iran, including nonsmectitic soils. Arid Land Research and Management. 19: 29–46

Hubble GD (1984) The cracking clay soils: Definition, distribution, nature, genesis and utilization of cracking clay soils. Review in Rural Sciences. 5: 3–13

Jeffers JD and Reynolds RC (1987) Expandable palygorskite from the Cretaceous-Tertiary boundary, Mangyshlak Peninsula, U.S.S.R. Clays and Clay Minerals. 35(6): 473–476

Kamara CS, Haque I (1988) Charactersitics of Vertisols at ILCA research and outreach sites in Ethiopia. Revised version. Plant Science Division Working Document B5. ILCA (International Livestock Centre for Africa), Addis Ababa, Ethiopia

Kapur S, Çavuşgil VS, Şenol M, Gürel N, Fitzpatrick EA (1990) Geomorphology and pedogenic evolution of quaternary calcretes in the northern Adana Basin of southern Turkey. Zeitschrift für Geomorphologie. 34: 49–59

Khademi H, Mermut AR (1998) Submicroscopy and stable isotope geochemistry of carbonates and associated palygorskite in selected Iranian Aridisols. European Journal of Soil Science. 50: 207–216

Khademi H, Mermut AR (1999) Source of palygorskite in gypsiferous Aridisols and associated sediments from central Iran. Clay Minerals. 33: 561–578

Khormali F (2003) Mineralogy, micromorphology and development of the soils in arid and semi-arid regions of fars province, southern Iran. Ph.D. Thesis, Shiraz University, Iran

Kooistra MJ (1982) Micromorphological Analysis and Characterisation of 70 Benchmark Soils of India, Part III. Netherlands Soil Survey Institute. Wageningen, the Netherlands

Kunze GW, Dixon JB (1986) Pretreatments for mineralogical analysis. In: Klute A (ed.) Methods of Soil Analysis, Part 1, Physical and Mineralogical Methods, 2nd edn. Agronomy Monograph No. 9. American Society of Agronomy, Madison, WI. pp.91–100

Mees F (1999) Distribution patterns of gypsum and kalistrontite in a dry lake basin of the southwestern Kalahari (Omongwa pan, Namibia). Earth Surface Processes and Landforms. 24: 731–744

Mermut AR, Padamanabham E, Eswaran H (1996) Pedogenesis. In: Ahmad N, Mermut A. (eds.) Vertisols and Technologies for their Managements. Elsevier Publ. Netherlands. pp.43–57

Moore DM, Reynolds RC (1989) X-Ray Diffraction and the Identification and Analysis of Clay Minerals. Oxford University Press Inc. Oxford

Murphy CP (1986) Thin Section Preparation of Soils and Sediments. AB Academic Publishers, Berkhamsted

Norrish K, Pickering JG (1983) Clay minerals. In: Soils, an Australian Viewpoint. Commonwealth Scientific and Industrial research Organization, Melbourne. Academic Press. London. pp. 281–308

Olson CG, Thompson ML, Wilson MA (2000) Phyllosilicates, In: Sumner ME (ed.) Handbook of Soil Science, pp. F77–F123, CRC Press, Boca Raton, FL

Probert ME, Fergus IF, Bridge BJ, McGarry D, Thompson CH, Russell JS (1987) The Properties and Management of Vertisols. CAB International, Wallingford, Oxon

Salinity Laboratory Staff (1954) Diagnosis and Improvement of Saline and Alkali Soils. Agriculture Handbook 60. U.S. Department of Agriculture, Washington, DC

Schafer WM, Singer MJ (1976) A new method of measuring shrink-swell potential using soil pastes. Soil Science Society of America Journal. 40: 805–806

Shadfan H (1983) Clay minerals and potassium status in some soils of Jordan. Geoderma. 31: 41–6

Shirsath SK, Bhattacharyya T, Pal DK (2000) Minimum threshold value of smectite for Vertic properties. Australian Journal of Soil Research. 38: 189–201

Soil Survey Staff (1993) Soil Survey Manual. U.S. Department of Agriculture Handbook No. 18. U.S. Government Printing Office, Washington, DC

Soil Survey Staff (1999) Soil Taxonomy a Basic System of Soil Classification for Making and Interpreting Soil Surveys. 2nd edition. USDA & NRCS, Washington, DC

Soil Survey Staff (2003) Key to Soil Taxonomy. 9th edition. USDA & NRCS, Washington, DC

Stoops G (2003) Guidelines for Analysis and Description of Soil and Regolith Thin Sections. Soil Science Society of America, Madison, WI

Wilding LP, Tessier D (1988) Genesis of Vertisols: Shrink-swell phenomena. In: Wilding LP, Puentes R (eds.) Vertisols: Their Distribution, Properties, Classification and Management. Texas A&M University Printing Center, College Station, TX pp. 55–81

Yerima BPK, Calhoun FG, Senkayi AL, Dixon JB (1985) Occurrence of interstratified kaolinite-smectite in El Salvador Vertisols. Soil Science Society of America Journal. 49: 462–466

Yerima BPK, Wilding LP, Calhoun FG, Hallmark CT (1987) Volcanic ash-influenced Vertisols and associated Mollisols of El Salvador Vertisols: Physical, chemical and morphological properties. Soil Science Society of America Journal. 51: 699–708

Yousif AA, Mohamed HHA, Ericson T (1988) Clay and Iron minerals in soils of the clay plains of Central Sudan. Journal of Soil Science. 39: 539–548

Contribution of Micromorphology to Classification of Aridic Soils

Maria Gerasimova and Marina Lebedeva

Abstract The function of micromorphology is perceived as the interpretation of phenomena and aid in recognizing diagnostic horizons in the major soil classification systems of the world. The micromorphological contribution to the diagnostics of the horizons in soil classification comprises the complementary characteristics and/or confirmation of the genetic implications on a horizon, the clarification and identification of the horizon with its differentiating criteria. Hence, the application of micromorphology to soils of (sub) arid regions is promising in discriminating the pedogenic properties from the inherited ones, assessing the development and degradation of structure, the re-arrangement of soil solids by pedofauna and other agents, explaining the nature of clay-enriched subsoil, as well as the behavior of carbonates and gypsum. These functions are helpful in solving problems related to aridic soils, whose diagnostics comprises the identification of yermic, takyric, cambic, calcic, (paleo) argic horizons and features derived from accumulation of salts.

Keywords Russian soil classification system · horizons · microstructure · aeolian features

1 Introduction

In many publications concerning the contribution of micromorphology (MM) to the development of major world soil classification systems (1981, 1985, 1991, 1998), the function of MM is perceived as the interpretation of phenomena and aid in recognizing diagnostic horizons. Although the horizons are declared to be identified

Maria Gerasimova
Moscow Lomonosov University, Faculty of Geography Leninskie Gory 119992, Russia

Marina Lebedeva
Dokuchaev Soil Science Institute, Pyzhevskiy pereulok, 7, Moscow 109017, Russia,
e-mail: mverba@mail.ru

S. Kapur et al. (eds.), *New Trends in Soil Micromorphology*,
© Springer-Verlag Berlin Heidelberg 2008

by properties, basically morphological ones, the choice of properties, their ranking and some other procedures are controlled by the concepts of soil genesis.

According to L. Wilding MM is a «marker for genesis» and soil classification (1985). Hence, the application of MM to soils of (sub) arid regions is promising in discriminating the pedogenic from the inherited properties, assessing the development and degradation of structure, the re-arrangement of soil solids by pedofauna and other agents, explaining the nature of clay-enriched subsoil, as well as the behavior of carbonates and gypsum. These functions are helpful in solving problems related to aridic soils, whose diagnostics comprises the identification of yermic, takyric, cambic, calcic, (paleo) argic horizons and features derived from accumulation of salts.

It is worth emphasizing that the imprints of structural and textural events are well preserved in the arid climate; this was a challenge for soil scientists in the beginning of the 20th century and now, facilitated by the MM studies (Allen 1985). Another noteworthy particularity of aridic soils is that their topsoils are "fair mirrors" of the present-day environment unlike their subsoils, which have a "long memory".

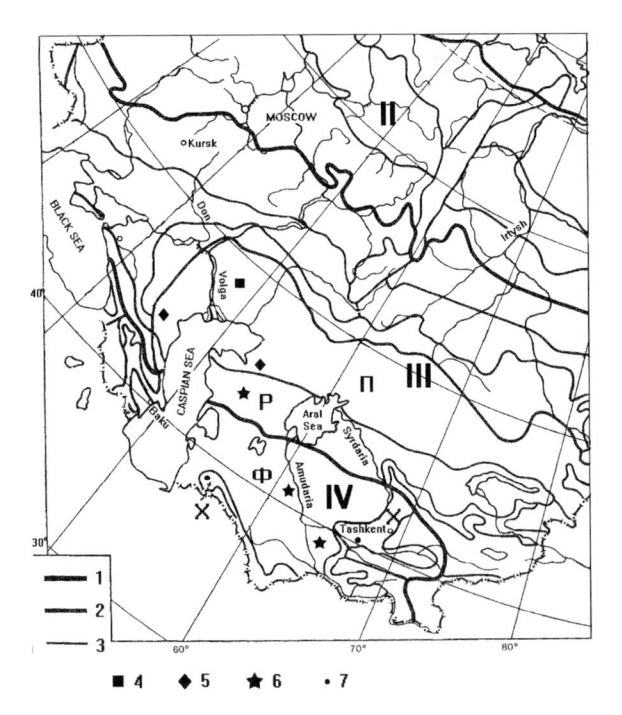

Fig. 1 Study objects – location of profiles. Fragment of the Map of Soil-geographic Zoning of the USSR, Authors: compiled by G.V. Dobrovol'skiy, N.N. Rozov and I.S. Urusevskaya, 1984. Conventions: 1 – boundaries between belts; 2 – boundaries between broad regions (facies); 3 – boundaries between zones. Soils: 4 – light chestnut, 5 – brown semidesert, 6 – gray-brown desert soils, 7– serozems. Designations on the map: II – boreal belt, III – subboreal belt, IV – subtropical belt; П – zone of light chestnut and brown semidesert soils; P – zone of grey-brown soils of subboreal desert; Ф – zone of grey-brown soils of subtropical desert X – zone of serozems of the foothill semidesert

Several new horizons were introduced in the last version of the Russian soil classification system (Shishov et al., 2004) for the soils of arid and subarid areas, which are rather small in area Russia (Fig. 1). These are:

- light-humus AJ horizon – separated from the light-coloured humus horizon in the previous approximation for its alkaline or neutral reaction and some other features produced by the arid climate;
- xerohumus AKL horizon – topsoil in the most arid environments, new horizon;
- xerometamorphic BMK horizon – subsoil in several soils, its definition is broader than that in the former version of the Russian system.

These horizons roughly correspond to the alkaline variant of ochric, to yermic[1] and to a part of cambic horizons in the WRB, respectively.

The objective of this research was to confirm that the micromorphology can considerably help to classify soils especially in aridic regions.

2 Materials and Methods

Soil types in the new Russian classification system (1997–2004) are identified by individual sets of diagnostic horizons. The recently introduced diagnostic horizons are ingredients of the profiles of the following soils of Russia: AJ – light chestnut soils (Calcic Kastanozems) and solonetzes; AKL and BMK – brown semidesert soils (Aridic Cambisols). However, these new horizons may be diagnostic for a broader set of soils in Central Asia – grey-brown desert soils (Aridi-Gypsiric Cambisols), serozems (Haplic Calcisols), takyrs (Takyric Regosols).

In the present research, only natural (without human-produced modifications) soils with texture not lighter than loamy sand were used in our studies. Thin sections were examined from the profiles of light chestnut and brown semidesert soils from the Caspian Lowland Kalmykia and Volgograd oblast, grey-brown desert soils and light serozems from Uzbekistan and Turkmenia, solonetzes from several localities in the Caspian Lowland (Fig. 1). Micromorphological properties were described on the basis of the Bullock et al. (1985), and some details were added for a better insight into the differences among our study objects.

3 Results and Discussion

The topsoils are presented by three diagnostic horizons. The AJ – light-humus horizon fits into the "aridic part" of the international ochric, and is well developed in light chestnut soils and serozems. The horizon is light in colour with moderate pedality, and weak to moderate crumb structure mostly related to the root systems;

---

[1] The horizons' names are given according to the FAO/ISRIC/IUSS, (1998).

the effervescence varies from none to moderate, evidences of faunal activity are common. The latter are more conspicuous in serozems, where insect chambers and channels contribute to high porosity and low density.

In light chestnut soils, a fine platy structure may be identified in the uppermost part of the horizon between the Festuca bunches, under which crumbs are clearly recognized. The MM properties fully confirm the field diagnostics by the weak to moderate pedality, granular microstructure with many packing voids, root and faunal channels, coprolites, few organic residues, absence of carbonates and clay plasmic features or the occasional to rare impregnation of the matrix by carbonates (Fig. 2a,b). The variation in MM properties among soils is due to texture, intensity of the biological activity and genesis of the parent material. Serozems on loess have stronger pedality and higher content of carbonates, the aggregates are rounded and rather small, which may have been, affected by the "loessic microaggregation" described by Minashina (1966, 1973).

The xerohumus AKL horizon, recently introduced as a diagnostic one, is light grey in colour (chroma<3), poor in humus and comprises two very thin (1–2 to 4–5 cm), and fragile subhorizons. They are different in structure – porous to vesicular K and very fine platy or scaly L. The third ingredient of the horizon index – A, attests to its surficial position in the profile, presence of organic matter and of some biogenic features, e.g. of a very thin subhorizon of algae

(a)

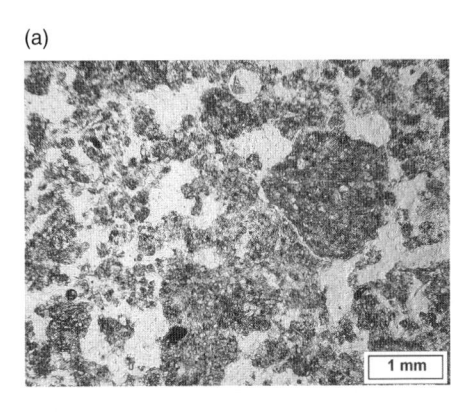

Fig. 2a AJ horizon of the light chestnut soil,
PPL. Moderate pedality,
rounded aggregates,
one coarse coprogenic
aggregate, many packing voids

(b)

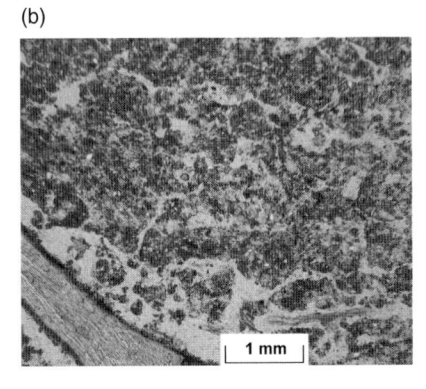

Fig. 2b AJ horizon of
the light serozem, PPL.
Moderate pedality, fine
aggregates, no apedal
material in the micromass, a
coarse plant residue in the
left corner

remnants. Since the thickness of both K and L subhorizons vary within few centimeters, and the sampling for preparing thin sections was mostly done without differentiation into the subhorizons, the MM data on their individual properties are scarce.

In thin sections, the AKL horizon of brown semidesert and grey-brown desert soils is rather heterogeneous in respect to the proportion of coarse and fine material, and its arrangement. Nevertheless, its upper part (K, 1–3 cm thick) has a vesicular habitus in contrast to the lower one (L) with its platy microstructure, where this differentiation is more advanced in grey-brown than in brown soils according to our observations and the literature data (Gubin 1984).

Vesicular microstructures in the upper part of the profiles of these soils were described by many authors (Lobova 1960, Romashkevich and Gerasimova 1977, Gubin 1984; Shoba et al., 1996) and their origin was adequately explained by the carbon dioxide emission during a very short period when the subhorizon is moist. Special MM observations enabled us to reveal some details in the microfabrics of K and L subhorizons.

The K subhorizon displays some differences in pedality, shape and arrangement of aggregates indicative of their origin. According to these differences, it seems to be possible to identify three fabric variants within the vertical section. The first variant – aeolian fine silty with wind-reworked embedded aggregates-is almost apedal, its micromass is composed of fine silt, and there are few fine sand grains. Some of the latter have patches of clay coatings on their surfaces and the boundary with the underlying material is clear (Fig. 2c). Rounded aggregates are common (Fig. 2d), they differ in size (0.3–0.8 mm) and composition: some are almost completely micritic with fragmentary clay coatings; others are composed of carbonate-impregnated clay with very few fine papules, which are presented by limpid transparent clay with crescent-shaped microlaminae. The kind of contact between these aggregates and the underlying one, along with the shape and composition of the rounded aggregates indicate an aeolian origin for the substrate. This material indeed corresponds to an early stage of aeolian dust accumulation on a more compacted "rough" surface retaining the detached coarser particles, namely, aggregates of different soil horizons. Vesicular pores are few and made by CO_2 emission.

(c)

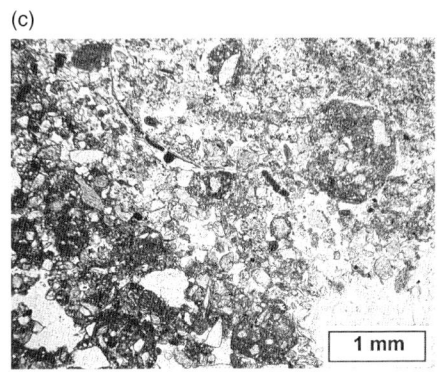

1 mm

Fig. 2c AKL horizon of the gray-brown soil, uppermost part, PPL. A clear boundary between the light-coloured apedal aeolian material with ventifacts and a darker material similar to that in Fig. 2e

The second fabric variant is presented by less compact microzones, comprising rounded or welded aggregates, some of 1 mm in diameter. Strongly birefringent clay coatings occur in aggregates and also on coarse skeleton grains (Fig. 2e). Some of these wind-reworked aggregates display elements of concentric striated b-fabric. The apedal silty material is of minor importance, but in microzones where it occurs, it also contains embedded aggregates. The packing and vesicular voids are closed, thus creating a spongy appearance of the micromass. Therefore, this variant was defined as aeolian-spongy with closed pores.

The third fabric variant of the K ingredient of the xerohumus horizon was termed as crumb with open voids (Fig. 2f). The microstructure is composed of rather rounded aggregates –crumbs (1–0.5 mm) with packing voids and few biogenic pores that, in distinction of the former variant, create an open-pore system. The

(d)

Fig. 2d The same horizon, XPL, higher magnification. Rounded aggregates – ventifacts

(e)

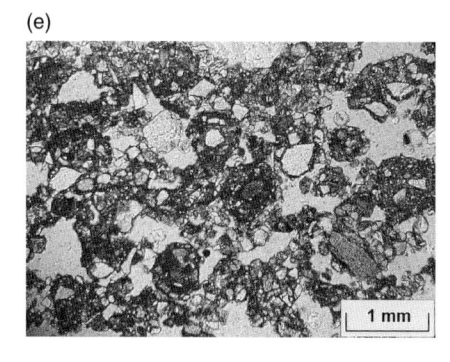

Fig. 2e K subhorizon of the gray-brown soil, PPL. *Aeolian-spongy* variant of fabric: aeolian and other kinds of aggregates form a closed-pore system

(f)

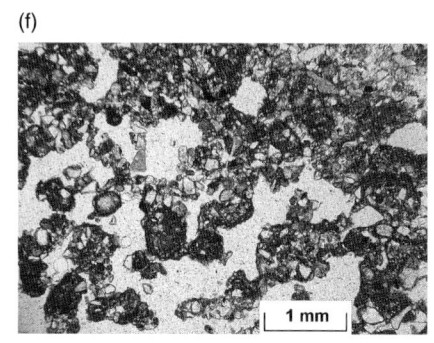

Fig. 2f The same horizon, PPL. Aggregates – *crumbs and the open-pore system*

aggregates are composed of organic-clay or carbonate-impregnated clay plasma binding the fine-silt particles. The aggregates are single, or bridged; features of their reworking by wind (ventifacts) are less distinct. This fabric variant may be regarded as an advanced stage of the aeolian deposit transformation, when the material is more re-arranged by physical processes, carbon dioxide emission, faunal activity and than accumulated.

The described variants do not display any strict preference to soils with the AKL horizon, although the first and second variants with more pronounced aeolian features are more common in grey-brown desert soils than in brown ones, which is in good agreement with stronger aridity and more active deflation in the desert environment. The L part of the AKL horizon is more homogeneous in fabric and microstructures. Groups of granular or crumb aggregates are rather clearly arranged in platy-like microstructures, so that voids acquire a subparallel orientation (Fig. 2g,h). The rounded aggregates are similar to the above-described, but they are less diverse, the carbonate-impregnated clay plasma is an important ingredient of aggregates. Few earthworm pellets were also encountered, presumably of Curculionidae and Scarabaeidae (Consultations with Prof. B. Striganova, Fig. 2h) together with scarce plant residues.

Presumably, the fine-platy L subhorizon has originated by the transformation of the K subhorizon upon its burial under the recent wind-blown material and subsequent rearrangement of the buried subhorizons by cryogenic mechanisms contributing to such stratifications (Gubin and Gulyaeva 1997, Konishchev et al. 2005).

(g)

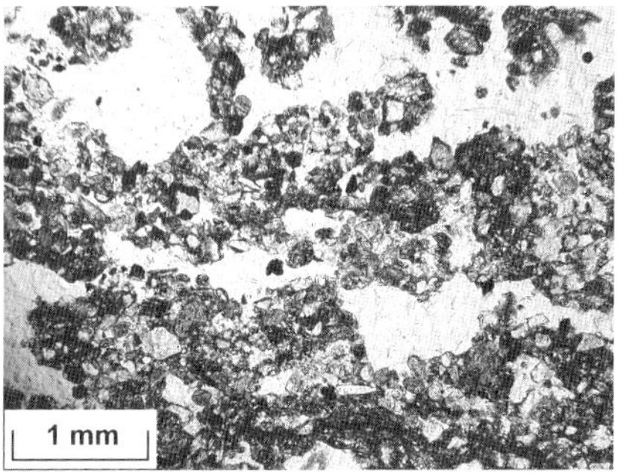

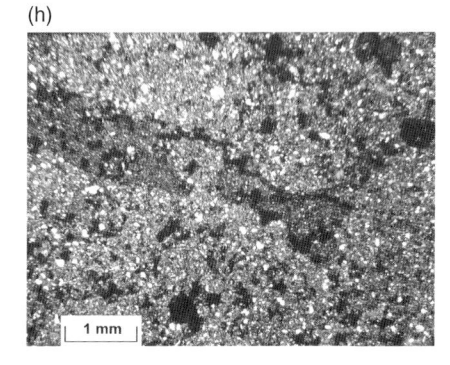

Fig. 2g L subhorizon of the gray-brown soil, PPL. Rounded aggregates tend to subparellel orientation forming platy microstructures

(h)

Fig. 2h The same horizon, PPL higher magnification, light-coloured insect casting in the horizontal void

Interpretation of the variants of the fabric types described in the K subhorizon and the MM properties of the L subhorizon permits to propose the following sequence of events, which comprise the successive stages of pedogenic reworking of the mixture of the aeolian fine silt, sand, and transported aggregates.

The diversity of the aggregates, in the initial vesicular materials (rather compact and apedal, with "fresh" embedded aggregates), is explained by the blowing out and the subsequent aeolian deposition of the fragments different horizons of the more recent soils, paleosols or sediments. Despite its fragility, that is due to its intrinsic properties, and susceptibility to external impacts of the wind, the inblowing of sand into voids, and the freezing-thawing cycles it is transformed into a more stable spongy material, which, in turn is either directly re-arranged into plates, or acquires weak crumb microstructure. The latter is also transient, and may easily evolve into platy microstructure, thus producing the L ingredient of the AKL horizon.

In all likelihood, the platy microstructures of the L subhorizon derive from any of the above-mentioned fabric variants. Although they are also fragile and unstable, they exist almost in all soils studied; hence, the mechanism of their formation seems to be universal and should be related to a seasonal freeze-thaw cycle, as suggested by many authors. Scarcity of rainfall and the continental climate may be regarded as limitations of this process. On the other hand, it is promoted by frequent alternation of cycles owing to the severity of the climate, slow infiltration of the atmospheric water due to the compact subsoil, and by the easy-to-rearrange substrate. The proposed mechanism is in agreement with the perception of the AKL horizon as a genetically unique system composed of interrelated subhorizons.

In terms of horizon diagnostics, the AKL horizon corresponds to the yermic horizon (WRB 1998) without a desert pavement, which is not an obligatory criterion for this horizon. The yermic horizon is identified by its "…vesicular crust above a platy A horizon" (WRB 1998). A similar combination of layers is diagnostic for the aridic epipedon in the Chinese Soil Taxonomy (2001), where a very thin vesicular crust is underlain by a platy horizon. It is worth noting that the depth requirements to the platy horizon are high – up to 10 cm, which matches with the explanation of its dependence on the severity of climate.

There are some other types of surface formations in aridic environments, which may be complementary to the above-discussed recent modifications of aeolian material. They may reflect other evolutionary sequences caused by prominent environmental changes, or occur in other topographic positions. They may be combined in time and space with the AKL horizon, and their general and MM properties may be helpful in explaining some processes and transformations in aridic environments. Thus, the true crusts of the K horizons of the typical takyrs – a horizon in WRB-98, or takyric property in the Russian system – have a clear set of MM characteristics related to the plasmic elementary fabric, non-pedality, low porosity, and vesicles and planes with sandy infillings. Properties of takyric crusts vary in accordance with the evolution of the takyr (Gerasimova 2003), where the evolutionary stages were not once described by pedologists in Uzbekistan and Turkmenia.

In salic or solonchakous horizons (WRB and Russian system), the fine-earth salty crusts significantly differ from both K and AKL horizons. They are rather homogeneous, have "lacy porosity" and "phantom" voids. They are also transient and may be destroyed or formed again in the course of (de)salinization (Lebedeva-Verba and Golovanov 2004). The new xerometamorphic BMK horizon in the Russian system basically corresponds to the cambic horizon of the WRB in its "arid wing", and is identified by pedogenic microstructures, although these structures are not specific and differ among soils (Aurousseau et al. 1985).

Thus, in light chestnut soils and serozems, the BMK horizons have common features like the moderate to high pedality, abundant coprogenic aggregates, predominance of packing and biogenic voids among the macropores, and dominant crystalline b-fabric with varying abundance of carbonates (Fig. 3a,b,c). In serozems, the horizon is dominated by carbonate accumulation (with nodule formations) and the activity of insects resulting to the more abundant development of voids – channels and chambers, than aggregates, although castings and pedotubules are infrequent. The rock-inherited "loessic microaggregation" (Minashina 1966) contributes to high microporosity and strong to moderate microstructure in serozems. However, the pedogenic microstructures of carbonate-redistribution and faunal origin are dominant in the horizon (Fig. 3d,e).

The BMK horizon in the light chestnut soils may be more compact the plasma is carbonate-free with few evidences of its mobilization recorded as weak poro- and granostriated b fabrics (Fig. 3.d,e).

(a)

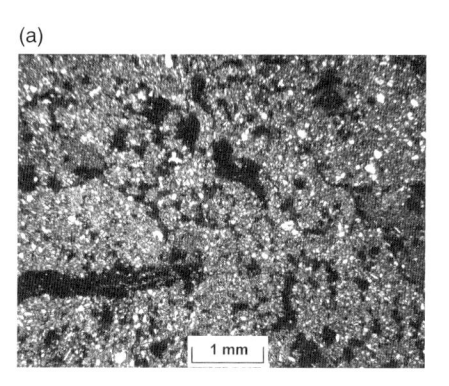

Fig. 3a BMK horizon of the light chestnut soil, XPL. Moderate pedogenic microstructure, many biogenic features, carbonate-impregnated clay plasma

(b)

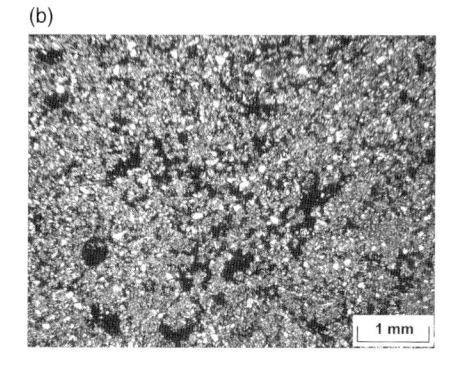

Fig. 3b BMK horizon of serozem, XPL. Same properties, less conspicuous and smaller size of aggregates

Fig. 3c BMK horizon of
grey-brown soil, XPL. In
the upper part – rounded
aggregates with birefringent
clay plasma, in the lower
part – coarse fragment
of micritic nodule with
fragments of clay coating on
its upper surface and in the
void

(c)

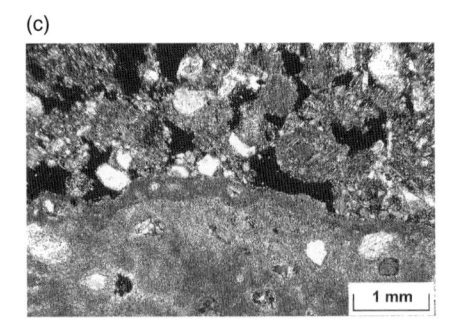

Fig. 3d BMK horizon of grey-
brown soil, PPL. An aggregate
with concentric clay laminae

(d)

Fig. 3e BMK horizon of grey-brown
soil, XPL. The same aggregate; on
the right – a skeleton grain with fine
clay coating around it

(e)

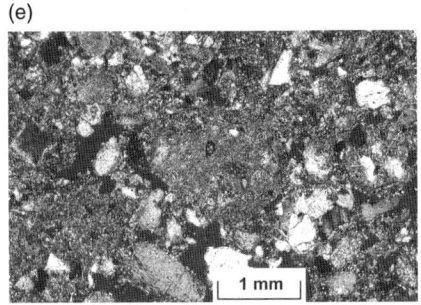

The latter phenomenon, along with few planar voids, indicates the development of a former or recent weak solonetzic features.[2] The xerometamorphic horizons in grey-brown and brown soils have low to moderate pedality, crystalline b-fabric, micritic segregations, and very few features related to pedofauna activity.

There are clay plasma separations and iron oxides pedofeatures, which are more prominent in grey-brown soils, so that they were interpreted as indicative of likely paleo clay illuviation and rubefaction (Lobova 1960). Gubin (1984) described very fine dispersed iron oxides on skeleton grains and in the micromass in the B horizon of both soils similar to our fine dispersed iron oxides on skeleton grains, and very specific clay features – pure clay pendants. Fragmentary grain

---

[2] Light chestnut soils occur in association (microcatenas) with solonetzic variants and solonetzes.

coatings sometimes with admixtures of micrite, or clay papules are related to coarser particles, wheresome are overlain by carbonate-clay coatings – the phenomenon mentioned by Stoops and Seghal according to Allen (1985). Individual rounded aggregates are composed of clay plasma with crystallitine b-fabric. In brown semidesert soils, seasonally fluctuating solonetzic mechanisms (Budina and Medvedev 1966, Glazovskaya and Gorbunova 2002) might have left their imprint on the behaviour of clay – signs of its weak mobility as porostriated and microzones with striated b-fabrics.

The occurrence of diverse aggregates and fragments of micritic nodules in gray brown soils under study, as well as clay pedofeatures, hardly compatible with the aridic environment, may be related to the much older age of these soils occupying the most elevated topographic positions – remnants of old plantation surfaces. They may contain fragments of calcrete, carbonate nodules, or paleoargillic horizons as was mentioned by Lobova (1960) and Gubin (1984).

4 Conclusion

1. The MM contribution to the diagnostics of the horizons comprises: (i) complementary characteristics and/or confirmation of the genetic implications on a horizon; (ii) clarification and/or ascertaining the identification of the horizon, or its differentiating criteria. Most efficient were the studies performed for substantiating the diagnostic AKL horizon recently introduced in the Russian soil classification system. Basically, it is in good correlation with the yermic horizon of the WRB system and aridic epipedon in the Chinese Soil Taxonomy.

2. The "humus-crusty-subcrusty" AKL horizon is composed of three ingredients: the uppermost vesicular aeolian crusty subhorizon – K, scaly or fine-platy "cryogenic" subhorizon – L, and the "background" A poor in humus and with few biogenic features; the microfabric characteristics were determined for each subhorizon. The two subhorizons are interrelated (conjugated), therefore, they form an entity – a diagnostic horizon; the mechanisms of their integral formation were revealed micromorphologically. The K subhorizon, although named crusty, is different in properties and mechanisms of true crusts: takyric – compact vesicular, salty – very porous "lacy", and diverse sedimentation and irrigation crusts that were beyond the scope of this study.

3. In contrast to the subsoils, the majority of the topsoils (AKL horizon, true crusts) are dynamic formations: aeolian and sedimentation processes with their seasonal cycles are responsible of their transient, unstable character, hence, their microfabric shows heterogeneity. In the subsoil – the xerometamorphic BMK horizon, some features testifying to paleosol- or rock-inherited features are preserved.

Acknowledgement The work was supported by the Russian Foundation of Basic Research, Project # 05-40-49098-a.

References

Allen B (1985) Micromorphology of Aridisols. Soil Micromorphology and Soil Classification, Soil Science Society of America, Madison, 15: 197–217.

Aurousseau P, Curmi P, Bresson LM (1985) Microscopy of the Cambic Horizon. Soil Micromorphology and Soil Classification, Soil Science Society of America, Madison, 15: 49–62.

Budina LP, Medvedev VP (1966) Brown semi-desert soils. In: Genesis and Classification of Semi-Desert Soils, Moscow, pp. 23–59 [in Russian].

Bullock P, Fedoroff N, Jongerius A, Stoops G, Tursina, T Babel U (1985). Handbook for Soil Thin Section Description. Waine Research Publications, Albrighton.

Cooperative Research Group on Chinese Soil Taxonomy (2001) Chinese Soil Taxonomy. Science Press, Beijing.

Dobrovol'skiy GV, Rozov NN, Urusevskaya IS (1984).Fragment of the Map of Soil-geographic zoning of the USSR.

FAO/ISRIC/ISSS (1998) World reference base for soil resources. Rome: World Soil Resources Rep. 84. FAO.

Gerasimova M (2003) Higher levels of description – approaches to the micromorphological characterization of Russian soils. Catena, 54: 3, pp. 319–337.

Glazovskaya MA, Gorbunova IA (2002) Genetic analysis and classification of brown arid soils. Pochvovedenie, 11: 1287–1297 [in Russian].

Gubin SV (1984) Micromorphological diagnostics of brown and grey-brown soils of northern Ust'-Urt. In: Nature, Soils and Reclamation of the Ust'-Urt desert, Pushchino, pp. 127–134. [in Russian].

Gubin SV, Gulyaeva LA (1997) Dynamics of soil micromorphlogical fabric. In: Soil Micromorphology. Studies on Soil Diversity, Diagnostics, Dynamics, Wageningen, Moscow, pp. 23–28.

Konishchev VN, Lebedeva-Verba MP, Rogov VV, Stalina EE (2005) Cryogenesis of Recent and Late Pleistocene Sediments of Altay and Periglacial Regions of Europe. GEOS, Moscow [in Russian].

Lebedeva-Verba MP, Golovanov DL (2004) Micromorphological diagnostics of elementary pedogenetic processes for the purposes of desert soils classification. In: Soil Classification-(2004) Petrozavodsk, pp. 53–54.

Lobova EV (1960) Soils of the Desert Zone of the USSR, Moscow [in Russian].

Minashina NG (1966) Micromorphology of loess, serozem and hei-loo-too and some problems of their paleogenesis. In: Micromorphological Method in Soil Genetic Research, Moscow, pp. 76–93 [in Russian].

Minashina NG (1973) Micromorphological fabric of desert takyric soils. In: Micromorphology of Soils and Sediments, Moscow, pp. 45–53 [in Russian].

Romashkevich AI, Gerasimova MI (1977) Major aspects of micromorphology of arid soils in the USSR. In: Aridic Soils, their Genesis, Geochemistry, Reclamation, Moscow, pp. 239–254 [in Russian].

Shishov LL, Tonkonogov VD, Lebedeva II, Gerasimova MI (2004) Classification and Diagnostics of Soils of Russia, Smolensk p. 342 [in Russian].

Shoba S, Gerasimova M, Miedema R (1996) Soil Micromorphology: Studies on Soil Diversity Diagnostics Dynamics. Printing Service Centre Van Gils, Wageningen.

Orientation and Spacing of Columnar Peds in a Sodic, Texture Contrast Soil in Australia

P.G. Walsh and G.S. Humphreys

Abstract The orientation and size of very coarse columnar peds in a sodic, texture contrast soil in Australia is largely influenced by the inherited fracture system from the underlying bedrock. The study is based on field measurement of fracture metrics including aperture (width), depth of penetration and orientation revealed along 36 m of trenches and from of over 100 columnar peds displayed in a 50 m^2 area where the topsoil had been removed. Four classes of fractures (mega, major, minor and incipient) were used to explore these associations more fully. The first three classes are continuous and define the columnar peds, whilst the incipient fractures represent a partial fracture across a dome. An orientation analysis indicates that the fracturing of the saprolite into large polygons is not random, but follows preferred orientations that coincide with jointing in the underlying bedrock. These joints define mega blocky peds (>1 m^3) of which the pentagonal to hexagonal columnar peds are a subset. These features indicate that brittle fracturing is the main mechanism involved in the initiation of columnar peds at the study site. A tentative model for the development of polygonal fractures in the saprolite is proposed in which relict joints in the bedrock provide the zones of weakness along which initial fracturing of the saprolite occurs. These fractures relieve tension produced by seasonal desiccation. Superimposed on these widely-spaced deep fractures is a smaller, shallower more random system of fractures that relieve surficial tension developed by rapid desiccation after rain. However, the presence of flow structures on the surface of domes and between domes indicates that ductile deformation occurs too, presumably when the soil expands on wetting and the apparent explosive nature

P.G. Walsh
Department of Physical Geography, School of Environmental and Life Sciences, Macquarie University, NSW, Australia; Forests NSW, Department of Primary Industries, PO Box 100, Eden, NSW, Australia, e-mail: peterwa@sf.nsw.gov.au

G.S. Humphreys
Department of Physical Geography, School of Environmental and Life Sciences, Macquarie University, NSW, Australia

of some of these may signify the release of confining pressure even though the capacity for expansion is not particularly high (COLE 7.7). It is possible that this behaviour may assist in the development of domes (the convex-up top of columnar peds) though a full explanation of this feature has yet to be developed. It is suggested that brittle fracturing is the prime factor involved in the development of several ped types including columnar, prismatic, blocky and parallelepiped and that this mechanism needs to be included with aggregation into a general theory of peds. In contrast gilgai, that also shares a similar polygonal structure to columnar peds, is normally viewed as responding to ductile deformation and associated shearing.

Keywords Jointing · planar voids · ped theory · saprolite · structural inheritance · texture contrast soils

1 Introduction

The inheritance of certain structural features of parent rock/material, including bedding fractures (such as joints) and cleavage, is as an important feature of many soils (e.g. Smith and Cernuda 1951, Brewer 1976, Hunt et al. 1977, Koo 1982, Beavis 1985, Hart et al. 1985, Hart 1988, Paton et al. 1995, Ollier and Pain 1996). Smith and Cernuda (1951) suggested that the exfoliation of tuffaceous and andesitic rocks led to angular, medium and loosely packed aggregates: horizontal cleavage in tuffaceous shales yielded a dominant horizontal planar void system in the subsoil whereas rounded, porous subsoil aggregates were associated with weathered quartz diorite. Similar associations were detected in other soils.

In a laterite profile, near Sydney, Hunt et al. (1977) found that well defined subhorizontal planar voids in the duripan layers had the similar sense and spacing of bedding planes described above in the underlying sandstone and that near vertical jointing was common to both units. A similar procedure was employed by Hart et al. (1985) to explore the link between planar voids in horizontally bedded and vertically fractured micaceous sandstone and the blocky peds in the overlying subsoil. Other studies have also attributed part of the planar void pattern to an inherited fracture system (Koo 1982) and ped shape to both fractures and bedding planes (Hart 1988, Paton et al. 1995).

Collectively these studies indicate that the inheritance of bedding, fractures and cleavage from the parent rock/material has an important influence on soil structure. If valid, these findings should be incorporated into a general theory of ped formation, a theory that is long overdue. The present study is a contribution to this theme. In particular the study seeks to test whether or not a pattern of structural inheritance also applies to very coarse columnar peds found in a sodic, texture contrast soil. This ped type was considered intriguing since on first inspection no obvious association to underlying fractures was evident – a situation that concurs with literature in Australia and elsewhere, including standard soil texts. The soil type is also intriguing since previous work in the study area focused on explaining the texture contrast (Hallsworth and Waring 1964) rather than the columnar ped structure.

2 The Site

2.1 Location and Geomorphic Setting

The study area is situated in north-western New South Wales within the Pilliga State Forest north of Coonabarabran (Fig. 1). Drainage is north and northwest to the Namoi River with the majority of creeks rising from the Warrumbungle Range to the south. The climate is warm temperate, subhumid, with a mean annual rainfall of 625 mm and a summer maximum. Annual rainfall is highly variable with both drought and extreme wet periods common. Daily temperatures vary between an average summer maximum of 33°C and an average winter minimum of 2°C. The study site is located on a gentle (<0.3°) broad spur at about 250 m a.s.l. (149° 10.377' east,

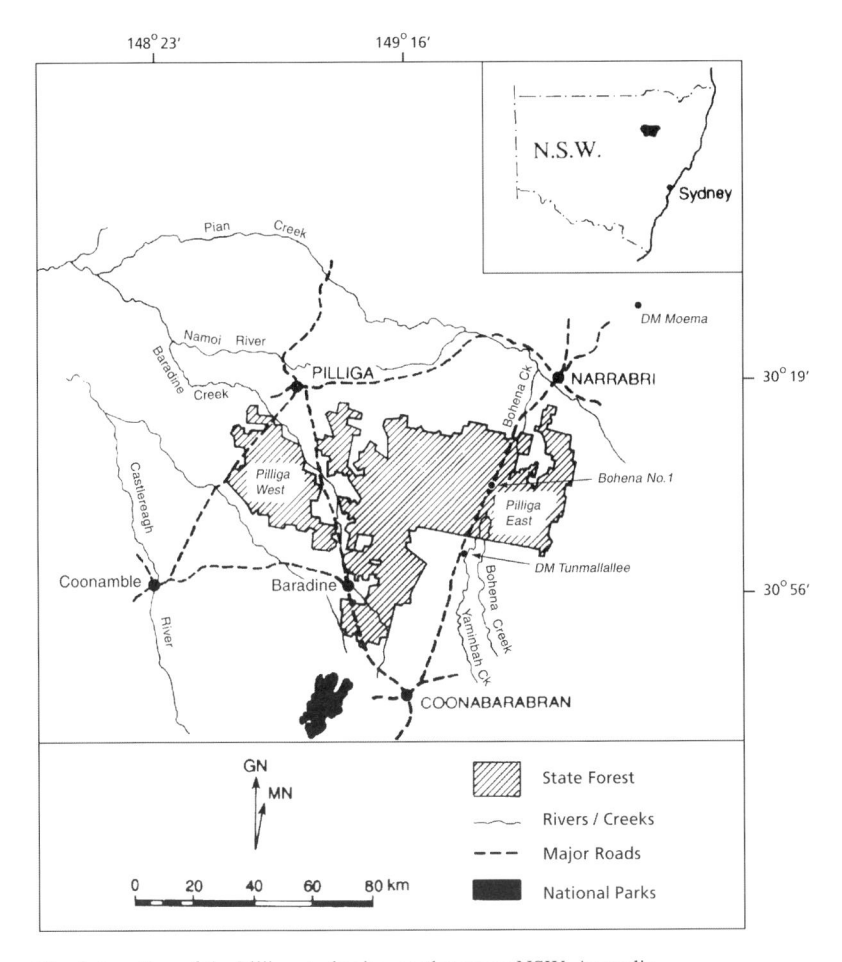

Fig. 1 Location of the Pilliga study site, northwestern NSW, Australia

30° 39.351' south) and supports mallee vegetation dominated by Eucalyptus viridis (Hesse and Humphreys 2001, Humphreys et al. 2001). Pilliga Sandstone (Jurassic) outcrops in the area and is dominated by massive and cross-bedded sandstone and pebbly conglomerate beds with occasional lenses of fine lithic sandstone, siltstone and shale (Arditto 1982).

2.2 The Study Soil

Apart from the texture contrast the most pronounced feature of the study soil is the presence of very coarse columnar peds (or cobblestone structure), mostly 25–35 cm wide (Fig. 2). This soil has a sandy loam to sandy clay loam topsoil 10–30 cm thick and a very thin (<1 cm) pulverscent, bleached sandy loam E, with scattered pebbles (stone layer), covering the tops of the domes (the convex-up tops of columnar peds) and extending down between them to a depth of 45 cm and sometimes through to the sandstone at greater depths (Appendix A). The pebbles, present in the A and E horizons but absent in the underlying lithic sandstone, are dominated by well rounded vein quartz, chert and jasper, and can be traced 50 m upslope to a pebbly conglomerate (part of the Pilliga Sandstone) that outcrops as a discontinuous low bench across the slope.

Fragments of lithic sandstone are common toward the base of the columns but are absent in the topsoil. These fragments often exhibit an oblique orientation to the surface, and to the flat-lying beds in the underlying lithic sandstone (Fig. 2). Adjacent to joint faces and at the interface between bedrock and saprolite the lithic sandstone is highly weathered. The lithic sandstone unit is white to grey in colour with some iron-oxide staining, but it contains thin (1–5 mm) dark brown clay bands that are mostly parallel to bedding (Fig. 2).

The soil is acid throughout (pH between 4.8 and 5.8) but has a sodic subsoil (ESP increases to 18–23% with depth) with magnesium the dominant exchangeable cation. The ESP depth trend is consistent with soloths elsewhere in Australia (e.g. Powell et al. 1994). The matrix of the columnar peds consists of medium clay with a fine sand component but with a dense arrangement (i.e. dense plasma support fabric cf. Paton et al. 1995).

Slight stickiness and slakiness yield a harsh quality to the material. Abandoned infilled channels (pedotubules), identified by their characteristic tubular form and contrasting fabric to the surrounding soil, likely indicates bioturbation. They are either partly infilled with loosely packed continuous aggregates, or completely filled by clayey material mixed with skeleton grains. These pedotubules probably result from termite and ant activity, which is prodigious at the site (Hart 1995).

Internally, the columnar peds are predominately massive. However, weakly developed intrapedal planar voids exist in a non-aggregated manner (Bullock et al. 1985). These voids display a zigzag to curved pattern that cluster along the edges of the ped. This results in weakly developed, fine to medium sub-angular blocky

Fig. 2 The study soil with characteristic columnar peds in the B horizon. Note the fractures penetrating into the thinly bedded lithic sandstone (Jurassic, Pilliga Sandstone)

structure with partly accommodated peds toward the outer edges, especially of the dome and less so along the vertical sides. Hubble et al. (1983) reported similar features and noted that whilst planar voids extended inward from the sides, they rarely joined up i.e. the columns are essentially massive. Clay mineralogy in the $<2\,\mu m$ fraction of the subsoil and bedrock is dominated by smectite (51–56%) and kaolinite (41–46%) (Slansky 1984, Walsh et al. in prep.). The clay has a linear shrinkage of 10%, which equates to a COLE of 7.7 (McKenzie et al. 1994).

2.3 Classification

Based on field features, this soil would normally be treated as a Solodized Solonetz in Australia (Stace et al. 1968, Hubble et al. 1983) or Solonetz elsewhere (e.g. Young 1976, Duchaufour 1982, WRB 1998), except that the subsoil pH is acid not alkaline, which makes it a Soloth. In the latest Australian soil classification (Isbell

1996) the soil is a "grey, magnesic-natric, bleached Kurosol". Based on available data the pedon is a "fine-loamy over clayey, mixed, semi-active, acid, thermic, torrertic Natrustalf" (Soil Survey Staff 1998).

3 Methods

3.1 Terminology

Following common geological practice the term fracture is used in a general sense to include joints or joint planes, planar voids, fissile cracks etc. (e.g. Hills 1972). Joints are a special class of fractures in which there is no apparent displacement normal or parallel to their surfaces (e.g. Twiss and Moores 1992). In a soil system under investigation the assumption of no displacement is problematic. We use the term joint only where it seems to satisfy four criteria of regularity, continuity, mode of development, and depth of penetration (Durney and Kisch 1994, Appendix B).

Though the analysis of planar voids is well understood in soil fabric studies (e.g. Brewer 1976, Bullock et al. 1985) we treat them here as fissile cracks which are largely a result of weathering exploiting anisotropic conditions in the soil (see Lafeber 1965). In comparison to planar voids, fractures appearing as joints are noted by their greater regularity and continuity, deeper penetration extending well into the underlying substrate as well as differing in origin. A fracture-plane refers to the planar surface bounding a fracture and a fracture-trace is the line that indicates the direction of the fracture-plane. The top of a columnar ped is the dome and each dome forms a polygon so that a set of columnar peds in plan-view displays a polygonal structure, which resembles columnar jointing in tabular igneous intrusives (e.g. Twiss and Moores 1992).

3.2 Trench Excavation

An integrated set of three trenches totaling 36 m long and up to 1.6 m deep, were excavated by backhoe. The largest trench was oriented 76° east of magnetic north, with the others normal to this with one dissecting the main trench and the other forming a T (Fig. 3). To assist in detecting fracture patterns the trenches were oriented at an oblique angle to the preferred orientation of the conjugate joint set in the Pilliga Sandstone. The trench walls were mapped and the spacing of fracture-planes in the bedrock and saprolite were measured from the walls and the trench floor.

To measure the size and geometry of the columnar peds, and the continuity of the fracture surfaces observed in the trench walls, the topsoil over a 50 m^2 area in the southeast and southwest quadrants of the trenches (Fig. 3) was removed

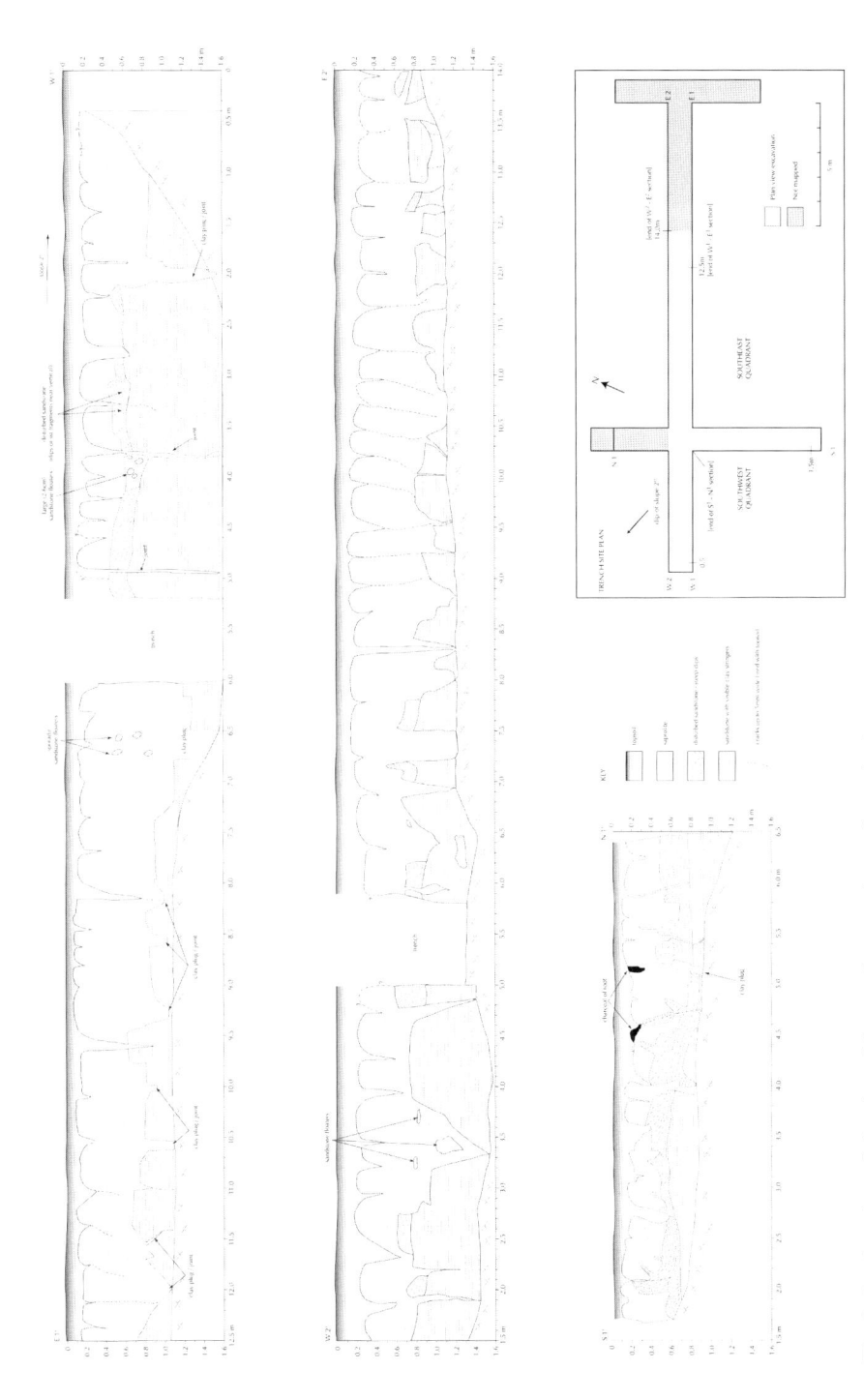

Fig. 3 Stratigraphy and site plan of the trench excavation at the study site

Fig. 4 Columnar peds exposed after removing the topsoil by air-blasting. The mallee, (multi-stem habit) of *Eucalyptus viridis,* is clearly evident. Southeast quadrant

using an air gun attached to a compressor with a capacity of $6.4\,m^3$/minute (Fig. 4). To quantify the geometry of the peds the following indices were recorded: orientation and length of the fracture-traces, angle of fracture intersections, size (average of the long and short axis) of each polygon, and the number of sides bounding each polygon. A calibrated Brunton compass was used to measure both fracture-plane orientations. Testing by repeated measurements in the field revealed an operator error in reading strike was $\pm4°$ and $\pm2°$ for the vertical and horizontal respectively.

3.3 Classification of Fractures

To assist in the collection and analysis of the fracture orientation and length data, each fracture- trace was assigned to one of four classes based on aperture (width); mega >5 cm, major 2–5 cm, minor 0.5–2 cm, and incipient <0.5 cm. The first three were continuous and defined the columnar peds. The latter represented a partial fracture across a dome. These classes were selected on the basis of perceived natural groupings and because the wider apertures appeared to be related to bedrock joints in the bedrock. The significance of this becomes apparent when their genesis is considered. If more than one aperture class was encountered along a trace (side of a polygon), the length was divided into the appropriate aperture class providing the secondary trace exceeded 10 cm.

When the length of a trace was less than 10 cm the dominant class was used. A 10 cm unit represented the average minimum length of the minor fracture system as determined from an initial measurement of a subsample. A new set of measurements was undertaken when the orientation of a trace in plan-view changed more than 10° along its length providing the length exceeded 10 cm. A line of best fit was applied when the change in direction of the trace was greater than 10° but less than 10 cm long.

Spray paint was applied along the trace immediately after measurement to avoid duplication. On completion of the measurements just described, the plan-view excavation was divided into 1 m² grids using string line, and overlapping digital images were taken from a step ladder. The images were combined in a graphics package and corrected for distortion to produce a composite image of the excavation (Fig. 5).

The scale of the study is similar to many studies on gilgai but these are often based on limited field-scale measurement sometimes with an analysis of planar voids at a microscopic scale (Knight 1980). An exception is the study by Florinsky and Arlashina (1998) who adopted a detailed microtopographic approach to calculate various hydrological parameters. Our approach is different but we try to captures the sense of geometry of the columnar peds and associated features.

3.4 Orientation Data Analysis

The cumulative length of all fracture-traces in the Pilliga Sandstone and overlying saprolite were plotted onto kite diagrams using 10° classes to determine visually preferred orientations. Kite diagrams avoid biases inherent in the

Fig. 5 Plan view excavation showing the dome tops of the columnar peds. The polygonal pattern and large fractures are clearly evident. Arrows show the location of material that has been extruded from the large fractures and draped over the tops of the columns (see Fig. 13). The dark stripes on some of the domes to the top left of the photograph are grooves 8 cm wide from the teeth of the backhoe bucket. The white oblongs label the metre square to assist in assembling the mosaic. Southwest quadrat to the left and southeast to the right. The full image is 9.5 m across

more commonly used version of rose diagrams where the areas of rectangles delineating class intervals are proportional to radius squared (Swan and Sandilands 1995). The method by Tanner (1955) was used to test whether the fracture-trace orientations follow a uniform (i.e. random) distribution in preference to statistics based on circular frequency distributions (e.g. Mardia 1972, Swan and Sandilands 1995). The latter is severely compromised where data are polymodal, as with fracture systems, and results in a loss of information (Williams 1972 and 1974, Gray 1974).

Tanner's (1955) method divides the circle into arbitrary uniform classes and totals are treated as linear data using measures of arithmetic mean and standard deviation for grouped data. Classes within one standard deviation of the arithmetic mean are considered to be not significantly different from a random distribution in that sector, whereas those classes more than one standard deviation away from the expected value, represent significant concentrations or voids at the 0.68 probability level. Classes that equal or exceed two standard deviations are considered highly significant at the 0.95 probability level. The 0.68 significance level is considered acceptable for detecting preferred direction since a higher level may be affected by overlapping tails especially where three or more broad modes are distributed over many class intervals (High and Picard 1971).

We report orientation data in the northern hemisphere (i.e. 270–90°) as per common convention even though kite diagrams are symmetrical about the north-south axis and reporting could be via the eastern hemisphere (0–180°).

4 Results

4.1 Size and Number of Sides of Polygons

In plan-view the polygons appear to be somewhat irregular in shape ranging from 19 to 87 cm across (mean 38, sd 12.2 cm, $n = 100$; Fig. 6a). However, they are mostly 5 (pentagonal) or 6 sided (hexagonal) polygons (mean 5.7, sd 0.97, $n = 130$; Fig. 6b). This result is identical to the number of sides in non-orthogonal sandstone polygons in Colorado and is similar to columnar jointed basalt which averaged 5.4 (Netoff 1971).

The results probably indicate a similar mechanistic response even though the processes differ (i.e., shrinkage on drying as opposed to cooling) as do the materials. The ratio of width to depth of columnar peds (= spacing of fractures to thickness of the clayey saprolite i.e. the S/T ratio of Bai and Pollard 2000) is mostly <0.8 ($n = 73$ from Fig. 3, mean 0.60, sd 0.34) with over half with a ratio of 0.5 or less. On this basis the columnar structured saprolite is fractured beyond an equilibrium state.

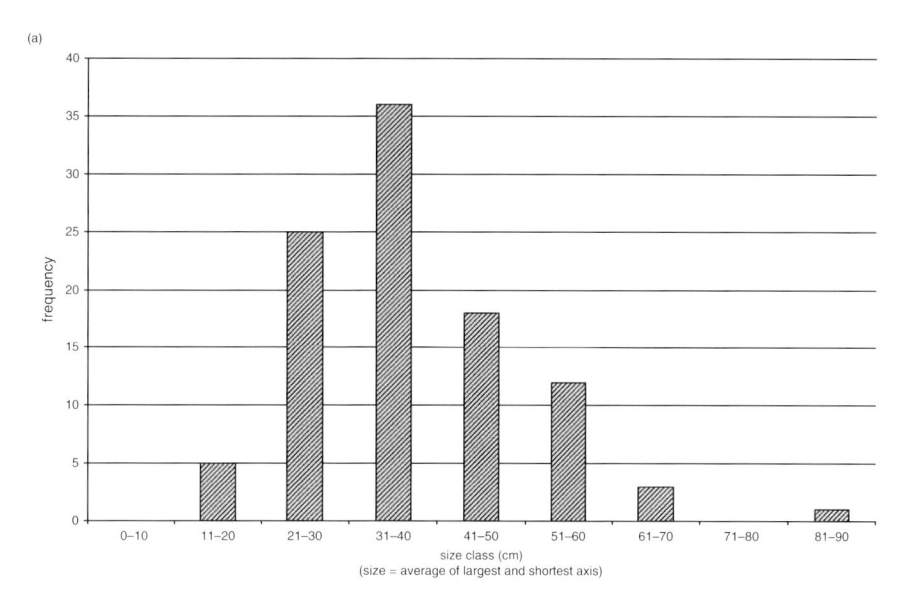

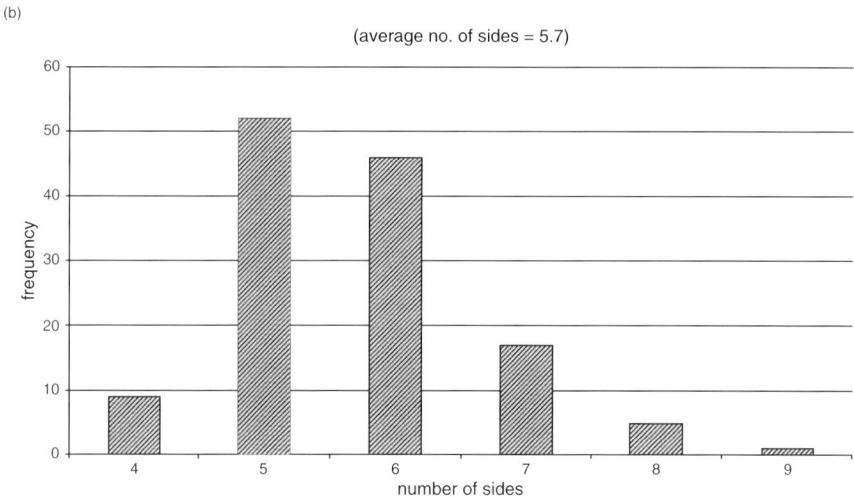

Fig. 6 Dimensions of the columnar peds. **(a)** size distribution of the average diameter of 100 polygons in the plan view excavation of Fig. 5. **(b)** Number of sides of 130 individual polygons in the plan view excavation of Fig. 5

4.2 Fracture size and Geometry

The aperture and length of fracture is quite variable and seems to correspond with the size of the columnar peds (Fig. 5). The pattern of fractures appeared to be more consistent on the eastern side of the trench, where two major fractures traverse this section in a north-south and another two in the east-west directions. Measurements of the angles of the fracture intersections for the two distinct sets show that the larger system is predominately orthogonal, and the smaller system is non-orthogonal (Fig. 7). Of the 46 fracture intersections measured in the large set (Fig. 7a), 48% were orthogonal and 30% were within 10° of being orthogonal. In the smaller set (Fig. 7b), 64% of the 50 fracture intersections measured were within 10° of 120°, whilst only 14 and 6% were orthogonal, or within 10° of being orthogonal respectively.

(a)

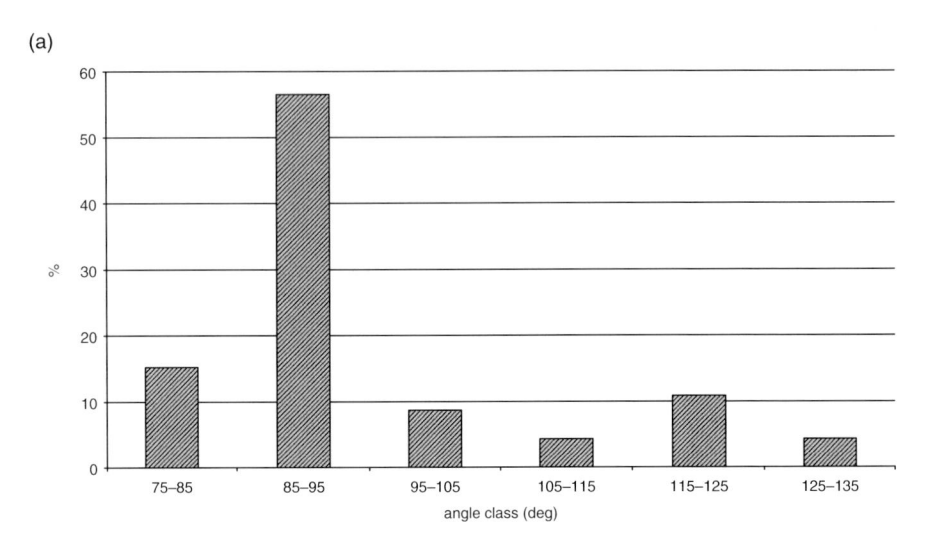

(b)

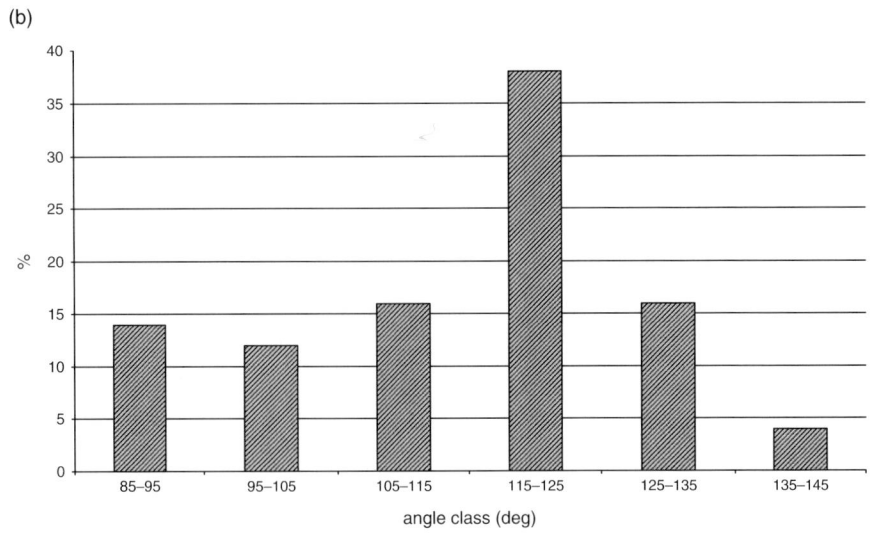

Fig. 7 Frequency of fracture intersection angles expressed as a percentage of the total. **(a)** Large fracture system. **(b)** Small fracture systems in the southeast quadrant of the plan view excavation

Netoff (1971) reported a similar pattern of fracturing in sandstone near Boulder, Colorado. He identified two distinctly different types of polygonal patterns, a larger orthogonal system on which was superimposed a smaller non-orthogonal system. He found that the former were characterised by a dominance of right-angle intersections forming large polygons (average size 3 m) that often had one or more curved sides. The smaller fracture system occurred within and subdivided the larger system, and was composed of small (average size of 40 cm) straight sided pentagonal and hexagonal polygons with a predominance of 120° intersections.

Of the 155 measurements on the angle of fracture intersection in the larger system, 55% were within 10° of being orthogonal. However, the presence of apparent non-orthogonal fracture system in saprolite needs to be treated with caution. Lachenbruch (1962) identified various problems associated with this type of fracture intersection in ice wedge polygons. In particular, when the radius of a bend in a primary crack is not large relative to the separation of the crack walls it proved to be difficult to determine whether the intersection at the bend was orthogonal or non-orthogonal.

Because of the width of the surface expression of ice wedges, the distinction between the two types of fracture intersections could not be made in many cases, and Lachenbruch (1962) noted that the same problem occurs when analysing shrinkage cracks in other materials such as concrete and mud. Hence the suggestion that a non-orthogonal fracture system is superimposed on the orthogonal system in the saprolite may be more apparent than real, despite the obvious appeal. Nevertheless, it was clearly evident in the walls of the trench that the larger fractures, which comprised the orthogonal system persisted well into the subsurface and into the bedrock without interruption or deflection (Figs. 2 and 3) and satisfies two of the four joint criteria presented in Appendix B. Based on these criteria, it is suggested that the orthogonal system at this site is a joint system inherited from the underlying sandstone.

4.3 Fracture Spacing

In vertical section it was also evident that the base of some columnar peds penetrated into the bedrock to form an elongated wedge or protuberances, which are referred to here as "clay-plugs". These generally coincide with the alignment of the fractures that delineated the sides of the columnar peds (Fig. 3). Some of these clay-plugs were up to 40 cm wide and encompassed two to three columns due to the complete alteration to saprolite between fractures (Fig. 3, section W2-E2, at 10.5 m).

In detail, a more complex picture of spacing emerges (Fig. 8). Spacing between fractures and clay-plugs (mean 35 cm, sd 11 cm) overlaps with the dominant mode of fracture/clay-plug spacing (mean 46, sd 15 cm). However, the fracture/clay-plug data also shows a weaker (37% of sample) but larger spacing (mean 140, sd 26 cm). Hence, the majority of fractures appear to be extensions of the clay-plugs/fractures in the sandstone, but that about a third of the columnar peds seem not to be directly associated with fracturing in the bedrock.

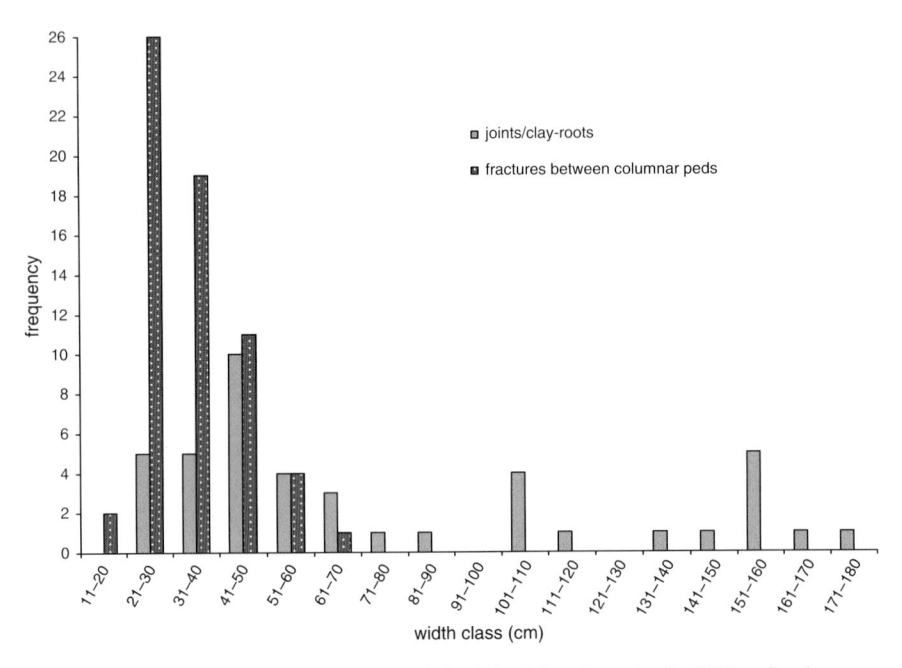

Fig. 8 Comparison between the spacing of the joints/clay-plugs in the Pilliga Sandstone to fractures delineating the sides of the columnar peds. Measurements taken from the walls of the trench. ($n = 43$ and 63 for the joints/clay-plugs and fractures, respectively)

4.4 Fracture-Trace Orientations in the Sandstone

Visual inspection of the kite diagram (Fig. 9a) for the cumulative lengths of fractures orientations in the Pilliga Sandstone shows four distinct peaks. Two significant fracture orientations occurred between 331–350° and 41–50° (Table 1), with the former and latter peaks striking parallel and roughly normal to the local slope (Fig. 3). In addition, two non-significant secondary peaks occur between 355–5°and 271–300° where the sandstone is broken into irregular polygonal blocks of varying size by clay-filled fractures.

A dominant north-south fracture set is exposed in the floor of the trench and is represented by the peak between 331 and 350°. A set of sub-parallel fractures, represented by the 355–5° peak, exists too though they do not always traverse the entire width of the trench. A west and northeast fracture set (271–300° and 41–50° peaks respectively) complete the network to form the irregular polygons on the surface of the sandstone. These sets are often truncated, or had their paths deflected by the dominant northwest set, which suggests that they formed later i.e. the truncation pattern provides a basis for assigning a relative age (Pollard and Aydin 1988).

Fault and fracture development in horizontally bedded, brittle and semi-brittle rock, normally consists of four major joint sets; two mutually perpendicular and two diagonal (Price 1966). The two perpendicular sets tended to conform to the slope and contour of the topography, whilst the diagonal sets are nearly symmetrical to

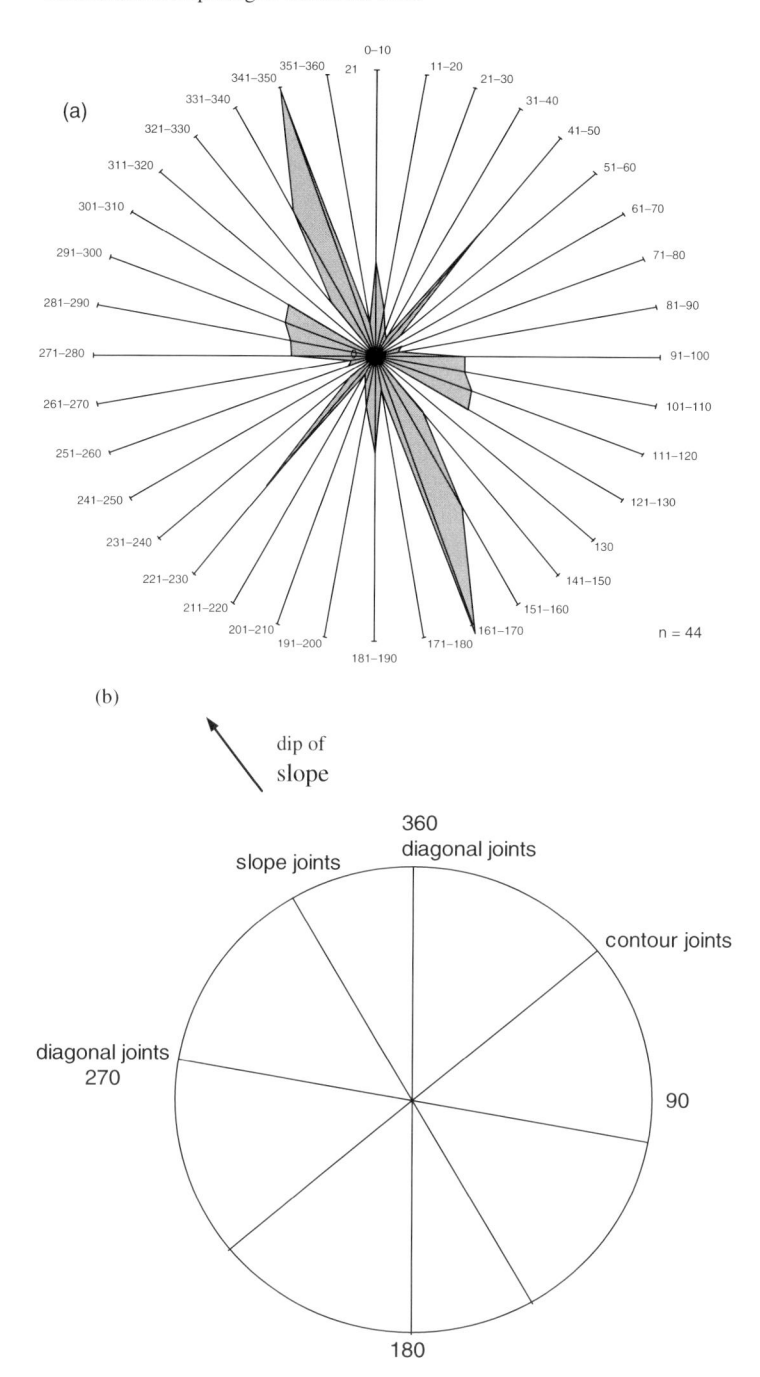

Fig. 9 Orientation of fractures in the Pilliga Sandstone: **(a)** Kite diagram of fractures (joints) exposed in the floor of the trench (Fig. 3). The scale refers to the cumulative length (m) of fractures in each interval class expressed as a percentage of the total length of all classes. **(b)** Stereogram of the orientations of the four fracture sets of Fig. 9a. The length of each joint trace is arbitrary

Table. 1 Orientation and lengths of the fractures in the saprolite of the study soil and joints in the underlying bedrock

Class interval	Mega		Major		Minor		Incipient	
	Cum length (m)	Tanner's (1955) test*	Cum length (m)	Tanner's (1955)	Cum length (m)	Tanner's (1955)	Cum length (m)	Tanner's (1955)
				Southeast quadrant				
1–10	0.61		0.60		1.48		**2.29**	**(p) 0.95**
11–20	0.67		0.77		0.61		**0**	**(v) 0.95**
21–30	**0**	**(v) 0.95**	0.78		0.76		**0.62**	**(v) 0.68**
31–40	0.25		2.36		**0.44**	**(v) 0.68**	0.87	
41–50	**3.02**	**(p) 0.68**	**2.61**	**(p) 0.68**	1.32		**0.91**	**(v) 0.68**
51–60	0.72		1.10		**2.0**	**(p) 0.68**	**0.61**	**(v) 0.68**
61–70	**0**	**(v) 0.95**	2.05		1.06		0.18	
71–80	**0**	**(v) 0.95**	1.38		**2.28**	**(p) 0.68**	**1.25**	**(p) 0.68**
81–90	1.30		**3.28**	**(p) 0.68**	1.18		1.06	
91–100	**0**	**(v) 0.95**	0.61		1.03		**0.29**	**(v) 0.68**
101–110	0.36		**3.30**	**(p) 0.68**	0.60		0.74	
111–120	0.62		1.51		1.56		0.73	
121–130	0.50		1.75		**2.17**	**(p) 0.68**	0.44	
131–140	**0**	**(v) 0.95**	1.99		0.81		0.55	
141–150	1.10		1.67		1.73		0.21	
151–160	**0**	**(v) 0.95**	**0**	**(v) 0.95**	**0.39**	**(v) 0.68**	0.53	
161–170	**5.38**	**(p) 0.95**	**0.57**	**(v) 0.68**	1.19		0.66	
171–180	1.89		0.93		0.60		0.50	
	û=0.91 (mean)	û±s=2.28, 0	û=1.51	û±s=2.46, 0.56	û=1.18	û±s=1.77, 0.59	û=0.69	û±s=1.20, 0.18
	s=1.37 (st.dev.)	û±2s=3.65, 0	s=0.95	û±2s=3.41, 0	s=0.59	û±2s=2.36, 0	s=0.51	û±2s=1.71, 0
			Southeast and Southwest Quadrants					
1–10	1.61		**1.44**	**(v) 0.68**	2.52		**2.99**	**(p) 0.95**
11–20	1.39		1.92		1.70		**0.23**	**(v) 0.68**
21–30	1.31		**1.36**	**(v) 0.68**	2.14		0.88	
31–40	2.0		**4.21**	**(p) 0.68**	**0.94**	**(v) 0.68**	0.99	
41–50	**3.23**	**(p) 0.68**	**3.76**	**(p) 0.68**	2.43		**2.58**	**(p) 0.68**
51–60	1.22		**1.44**	**(v) 0.68**	**2.79**	**(p) 0.68**	1.44	
61–70	1.51		2.84		1.93		**0.37**	**(v) 0.68**
71–80	0.40		3.51		2.37		**1.98**	**(p) 0.68**
81–90	1.11		**4.73**	**(p) 0.68**	2.38		0.76	
91–100	0.40		3.57		1.54		0.59	

P.G. Walsh and G.S. Humphreys

Table. 1 (Continued)

Class interval	Mega		Southeast quadrant Major		Minor		Incipient	
	Cum length (m)	Tanner's (1955) test*	Cum length (m)	Tanner's (1955)	Cum length (m)	Tanner's (1955)	Cum length (m)	Tanner's (1955)
101–110	0.80		3.30		**1.41**	**(v) 0.68**	0.86	
111–120	0.50		2.27		1.68		1.74	
121–130	2.42		2.37		**3.04**	**(p) 0.68**	1.69	
131–140	2.09		3.54		**2.81**	**(p) 0.68**	0.55	
141–150	1.10		2.52		**2.78**	**(v) 0.68**	0.45	
151–160	**0**	**(v) 0.95**	**0.80**	**(v) 0.68**	**1.26**		0.76	
161–170	**6.12**	**(p) 0.95**	1.61		2.24		0.89	
171–180	**3.87**	**(p) 0.68**	1.82		1.59		0.90	
	û=1.73 (mean)	û±s=3.21, 0.25	û=2.61	û±s=3.73, 1.49	û=2.09	û±s=2.70, 1.48	û=1.15	û±s=1.92, 0.38
	s=1.48 (st.dev.)	û±2s=4.69, 0	s=1.12	û±2s=4.85, 0.37	s=0.61	û±2s=3.31, 0.87	s=0.77	û±2s=2.69, 0

joints/Pilliga Sandstone

Class interval	Cum length (m)	Tanner's (1955)
1–10	0.60	
11–20	0.31	
21–30	0.18	
31–40	0.13	
41–50	**1.04**	**(p) 0.68**
51–60	0.20	
61–70	**0**	**(v) 0.68**
71–80	0.18	
81–90	0.15	
91–100	0.54	
101–110	0.55	
111–120	0.62	
121–130	0.65	
131–140	**0**	**(v) 0.68**
141–150	0.44	
151–160	**1.06**	**(p) 0.68**
161–170	**1.80**	**(p) 0.95**
171–180	0.21	
	û=0.48 (mean)	û±s=0.93, 0.03
	s=0.45 (st.dev.)	û±2s=1.38, 0

*significant orientation peaks (p) or voids (v) at given confidence level.

the other two sets, but not necessarily at right angles to each other. A stereogram of the fracture orientations in Fig. 9a, show that the fractures in the Pilliga Sandstone conform to Price's model (Fig. 9b). As previously noted the two significant peaks in the Pilliga Sandstone strike parallel and normal to the slope, whilst the two non-significant peaks (355–5°, 271–300°) occurred diagonally and are nearly symmetric to the two dominant sets.

4.5 Fracture-Trace Orientations in the Saprolite

The orientation measurements of fracture traces in the saprolite exposed in the plan-view excavation are presented below. The mega, major, minor, and incipient fractures are dealt with separately to assist in determining which, if any, of the fractures followed the same pattern in the underlying sandstone. To examine the influence of the mega fractures on secondary fracturing in the saprolite, the orientation data from the southeast quadrant (Fig. 3) were analysed separately since the mega set in this quadrant were more continuous and regular in their spacing, orientation, and size, compared with their counterparts in the southwest quadrant (Fig. 5).

Visual inspection of the kite diagrams for the fracture trace orientations in the southeast quadrant (Figs. 10a, b,c,d) show that all of the fracture systems display a

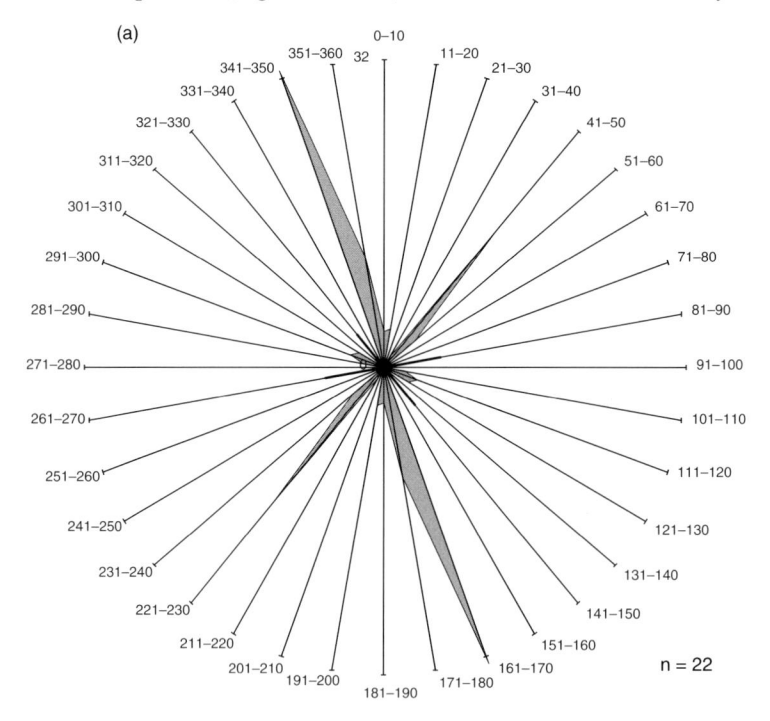

Fig. 10 Fracture trace orientations exposed in the subsoil of the plan view excavation for **(a)** mega, **(b)** major, **(c)** minor and **(d)** incipient fracture systems in the southeast quadrant of the plan view excavation. The scale refers to the cumulative length (m) of fractures in each interval class expressed as a percentage of the total length of all classes (= 30m to the edge of the circle in **(a)**; 12 in **(b)**; 10 in **(c)** and 16 in **(d)**

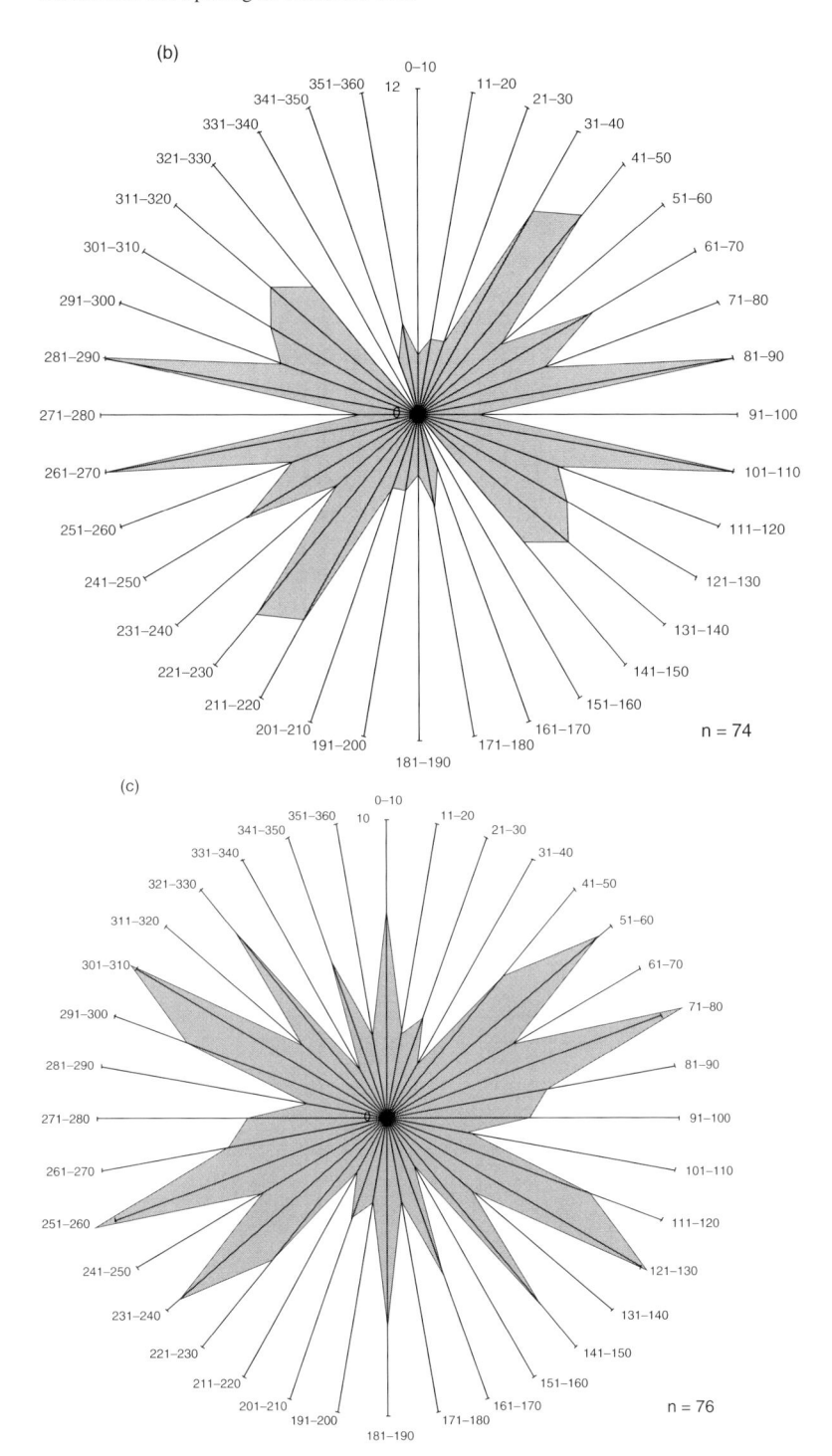

Fig. 10 (Continued)

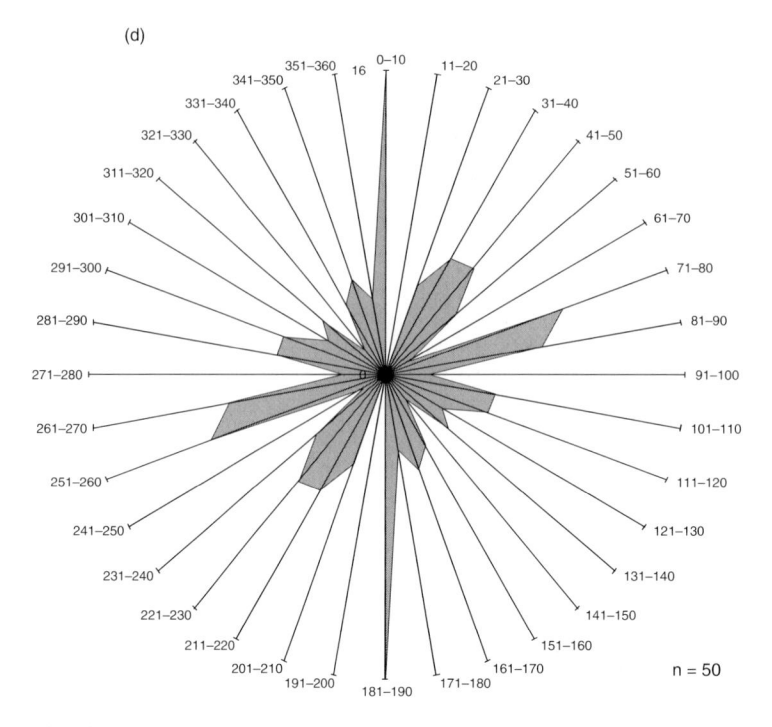

Fig. 10 (Continued)

number of distinct peaks in their orientation. The mega fracture system displayed two significant and three non-significant peaks (Fig. 10a). The significant peaks struck at 341–350° and at 41–50° (Table 1), which coincides with the two significant peaks identified in the Pilliga sandstone (Fig. 9a). The three non-significant peaks struck at 291–300°, 320° and 80°. The 291–300° peak coincides with the 271–300° non-significant diagonal set in the sandstone, whilst the 320° peak is subparallel to the significant slope fracture set shown in Fig. 9b.

The major fracture system shows three significant peaks in the 281–290°, 41–50° and 81–90° directions (Fig. 10b, Table 1). The 281–290° and 41–50° peaks coincide with the non-significant 271–300° diagonal fracture set and the significant 41–50° contour fracture set in the Pilliga sandstone respectively, whilst the 81–90° peak coincides with the small non-significant 80° peak in the mega fracture system. The minor fracture system also show three significant directional peaks (Fig. 10c), and as with the major fracture system, two peaks (301–310° and 51–60°) coincide with the non-significant 271–300° diagonal fracture set and the significant 41–50° contour fracture set in the sandstone.

The third peak at 71–80° is subparallel to the 41–50° peak. The incipient fracture system (Fig. 10d) shows two significant peaks, one at 355–0° (1–10° in Table 1) which coincides with the non-significant diagonal fracture set at 355–5° in the sandstone, and another at 71–80° which coincides with the same peak in the mega, major and minor fracture systems. Three other non-significant peaks are present too and coincide with the slope and contour fracture sets and the non-significant 271–300° diagonal fracture set in the sandstone.

When the orientation data for the respective fracture systems in the southeast quadrant are combined with the southwest quadrant, all of the fracture systems display significant preferred orientations (Fig. 11, Table 1). The two significant peaks in the mega fracture retain the same directions identified in the southeast quadrant (Fig. 11a), whilst the major fracture system retain two of the three significant peaks previously identified, both in the 0–90° portion of the kite diagram (Fig. 11b). The minor fracture system also retains two of the three significant peaks previously identified, one in each of the 90° portions of the kite diagram (Fig. 11c) whilst the incipient fracture system retains the two significant peaks identified in the southeast quadrant, but also gained a third significant peak in the 41–50° direction (Fig. 11d), which is concordant with the significant contour fracture set identified in the sandstone.

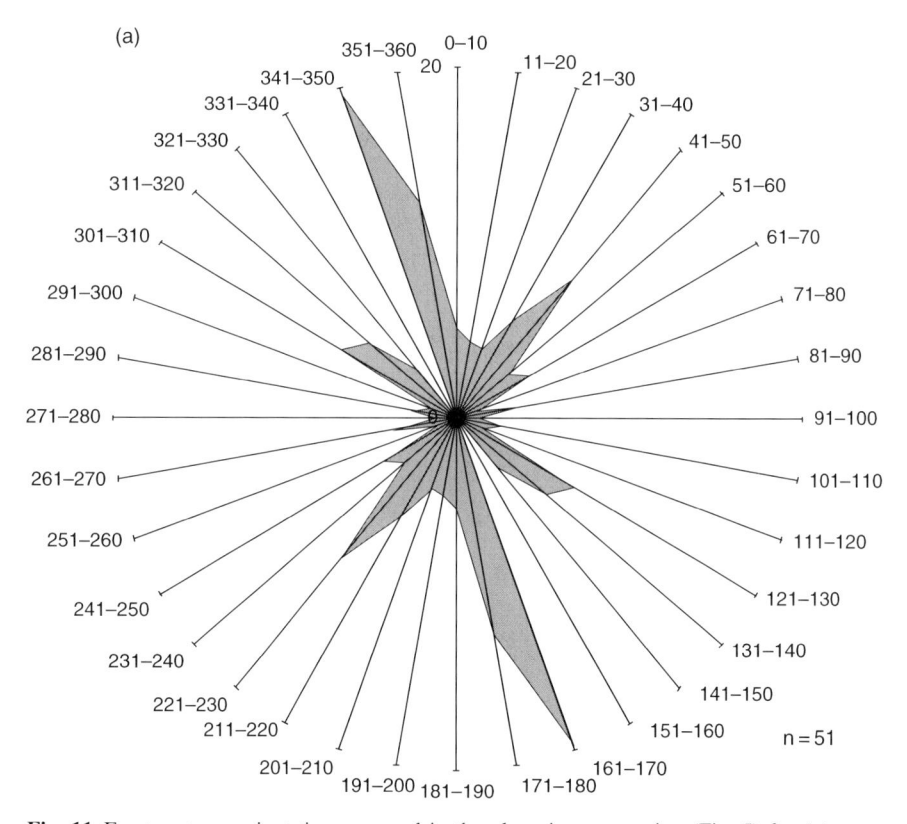

Fig. 11 Fracture trace orientations exposed in the plan view excavation (Fig. 5) for (**a**) mega, (**b**) major, (**c**) minor and (**d**) incipient fracture systems in the southeast and southwest quadrants. The scale refers to the cumulative length (m) of fractures in each interval class expressed as a percentage of the total length of all classes (=20 in (**a**); 10 in (**b**); 8 in (**c**); 14 in (**d**)

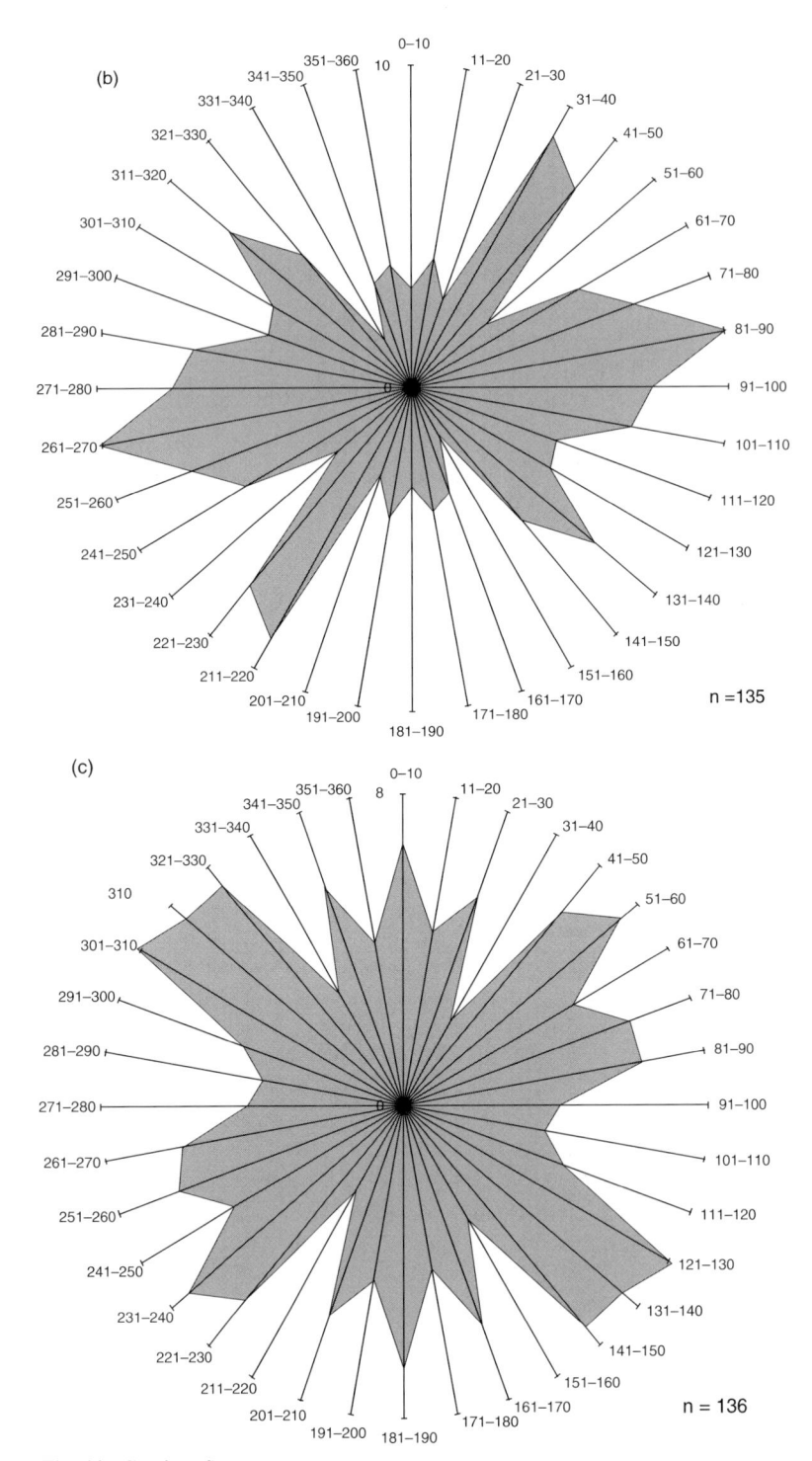

Fig. 11 (Continued)

(d)

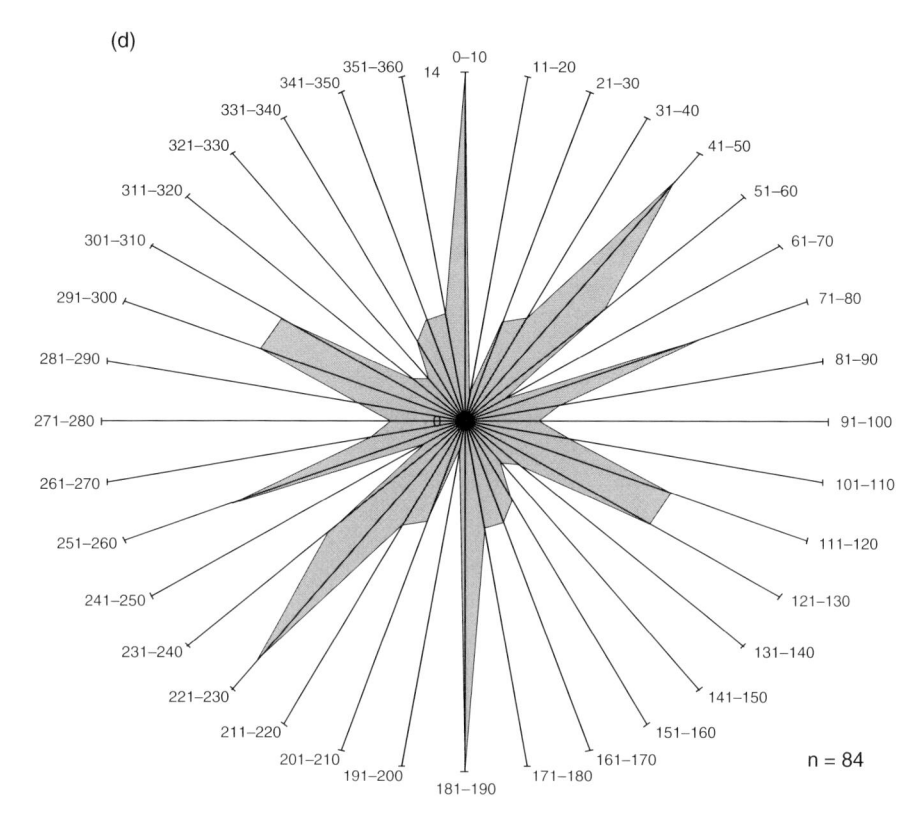

Fig. 11 (Continued)

To simplify the interpretation of the kite diagrams, all of the orientations that are judged significant for the respective fracture systems are plotted as straight lines of arbitrary unit length on single diagrams for the southeast and combined quadrants (Fig. 12a,b). Each line locates the direction of the centre of its respective class interval. The preferred orientations of the fracture systems (Fig. 12) display a marked similarity to typical stereograms of joint systems in rock as demonstrated by Price (1966). In the southeast quadrant the mega, major, minor, and incipient fractures displayed distribution peaks that either coincides with, or are subparallel to the four fracture sets identified in the sandstone (Fig. 9b). This relationship suggests that the anisotropic distribution displayed by these three fracture systems may be related to zones of prior weakness in the saprolite and thus are inherited from the underlying sandstone.

The major, minor and incipient fractures also have in common a significant orientation between 71–90°, which is not explicable in terms of inheritance from the underlying sandstone, as fractures in the trench floor do not register a significant group in that direction. Slight variations in the direction of the slope across the excavation site have likely caused those fractures to swing slightly toward the east. The combined dataset (Fig. 12b), retains the same marked similarity to the pattern of joint orientations in the sandstone (Fig. 9b).

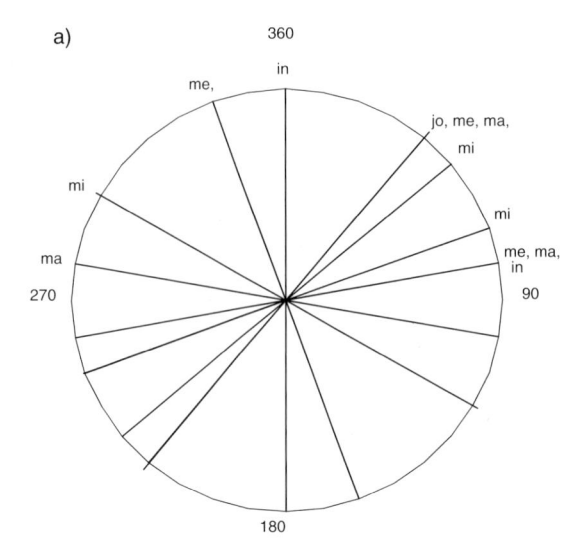

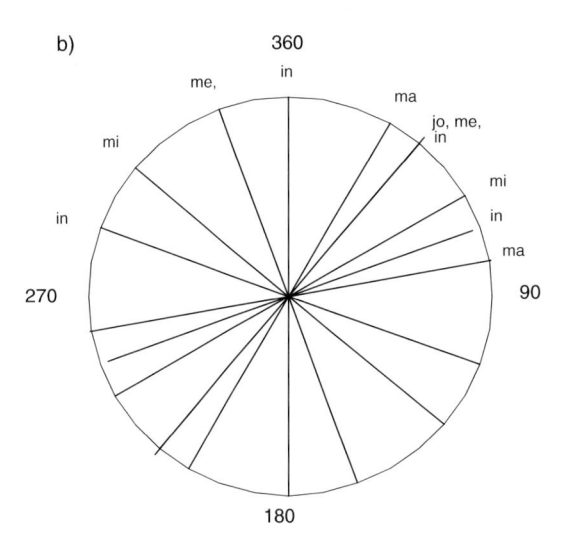

Fig. 12 Stereogram of significant orientation peaks simplified from (**a**) southeast quadrant and (**b**) southeast and southwest quadrants combined of the plan view excavation. jo = joints, me = mega, ma = major, mi = minor, in = incipient

4.6 Orientation of Fracture-Trace Voids in the Saprolite

The results presented in Table 1 also reveal the directions in which there is an absence of fractures. The mega fracture system in the southeast quadrant is quite regular (equivalent to a systematic joint system), with significant gaps in the distribution at

21–30°, 61–80°, 271–280°, 311–320°, and 331–340°. However, only the later gap is preserved when the two quadrants are combined. This finding confirms the initial interpretation that the mega fractures in the southeast quadrant are more continuous and regular in their spacing, orientation and size, compared to the southwest quadrant. These results also reveal considerable spatial variation that cannot be disentangled in the present study.

In contrast to the mega fractures, the number of significant gaps in the major fracture system increased when the two quadrants were combined (Table 1). In both quadrants a significant void is present between 331 and 350°, which coincides with the significant slope fracture set in the sandstone and mega fracture system. The minor fracture system also possessed a significant gap in the 331–340° direction in both quadrants and at 271–280° position. The same applies to the minor fracture system, except that it records a significant gap at 271–280°. However, the number of significant gaps in the incipient fracture system decreased when the two quadrants were combined even though the general orientation of the voids remained the same.

The degree of correspondence in the direction of significant gaps in the fracture orientation between both the quadrants and fracture aperture is poor. Only the 331–340° set is consistent between minor, major, and mega fractures in both quadrants. With rare exceptions of 341–350° between mega and major and 51–60° between minor and incipient fractures, the orientation of fractures at one class does not correspond to a gap at the next class level. This type of pattern is consistent with the presence of a stronger preferred orientation of the fractures.

4.7 Flow Features in the Soil

The analysis far emphasises brittle deformation, i.e. a fracture pattern, consistent with a material subjected to brittle failure (e.g. Twiss and Moores 1992). This is quite different from gilgai, which also has a pentagonal to hexagonal structure (Florinsky and Arlashina 1998) but is normally explained in terms of mechanisms consistent with ductile deformation (e.g. Paton 1974a). Despite the apparent prominence of brittle failure at the site there are features that are best viewed as examples of ductile behavior. In particular, the major plan-view excavation (Fig. 5) revealed two features that may indicate flowage and shearing. The first is light coloured material that partly in-fills many of the larger fractures and forms a convex-up cross-section the amplitude of which increases with width (Fig. 13a). This material is hard and riddled with vesicles when dry but slakes when wet indicating fragipan qualities.

The vesicles may have developed via the rapid expulsion of air when the material wets up. Viscous flow is also indicated by the squeezed toothpaste-like appearance. The second material drapes the top of some columnar peds (Fig. 13b). It consists of a mixture of topsoil, subsoil, and lithic sandstone fragments, and was in places up to 3 cm thick.

The size of the larger sandstone fragments varied from 2 cm to 10 cm (B_{max}). These features resemble miniature mud-volcanoes and may result from the rapid expulsion

(a) (b)

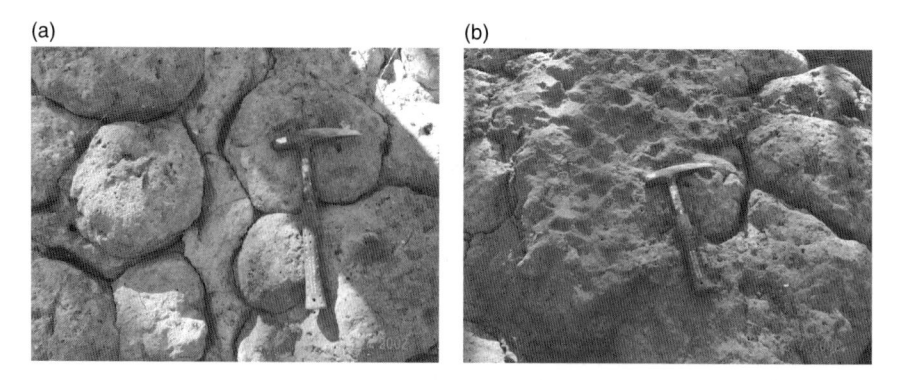

Fig. 13 Flow features. (**a**) A large fracture infilled with cemented E horizon material. Viscous flow appears to have occurred. (**b**) Extruded subsoil material drapes the side and top of some columnar peds. Fluidized soil appears to have been forced upwards along fractures

of fluidised soil along a fracture to erupt over adjacent domes. Collectively, these types of features attest to the stresses occurring between columnar peds and also assist in accounting for the re-oriented sandstone fragments as noted in Sect. 2.2. Similar phenomenon occur in ice wedge polygons, where seasonal freeze thaw cycles caused shearing and flow to occur in the wedges between the polygons forcing organic and mineral material to the surface (Black 1952).

5 Discussion

5.1 General Interpretation of Polygonal Structure

There is nothing unique in nature about a polygonal fracture pattern. It occurs in a range of materials that undergo volume change, including limestone, basalt, sandstone, concrete, frozen ground, ceramic glazes, mud, varnish, and paint. Their development in heterogeneous material, however, is complex and is attributed to several interacting factors including, the stress state that operated during formation, mechanical interactions between adjacent fractures, properties of the material and the local heterogeneities within it (Lachenbruch 1962, Tuckwell et al. 2003). However, these authors suggest that the detailed geometry of a fracture network in heterogeneous material is greatly affected by the early stress history during fracture nucleation whereby a small number of fracture seeds develop and dominate the resulting fracture network geometry. As well as representing the largest fractures in the system, these early structures significantly modify the stress field around them, preventing other seeds from developing and influence the propagation paths of nearby fractures.

The nature of the initial fracture system is also influenced by the degree of homogeneity of the material. Thus, fracture systems in basalt are established initially in a relatively thin surficial layer by cracks whose depth is generally of the order of the crack spacing or less. Subsequent deepening occurs incrementally, as the thermal stress penetrates to greater depth so that the final depth of fractures often exceeds spacing by two orders of magnitude (Lachenbruch 1961). In contrast, only some of the widely spaced cracks selectively deepen in heterogeneous materials such as mud (Tuckwell et al. 2003).

These influences are apparent at the study site where the depth of penetration increases with fracture aperture as follows: incipient <4 cm, minor <10 cm, major >20 cm and often extending to the saprolite/rock interface and occasionally beyond, and mega >1.6 m and always traversing the saprolite/rock interface. In some sections of the trench, there is a one to one relationship between the depth of saprolitic subsoil and the width of the columnar ped. This is most clearly evident at the scale of the mega fractures (Figs. 2 and 5) where the fractures most clearly conform to joints i.e. they are mega blocky peds $>1\,m^3$. Viewed this way the domes are simply the cobblestone or cauliflower surface or large blocks! This 1:1 relationship between fracture spacing (S) and bed thickness (T) is well documented in the geological literature (e.g., Price 1966, Narr and Suppe 1991) and indicates a net balance of stresses since critical spacing to unit thickness occurs (Bai and Pollard 2000).

When treated individually the columnar peds have an S/T of mostly much less than one (average of 0.60), This represents a state of saturated spacing i.e. additional fractures develop between the mega fractures as stress is transferred from adjacent layers. Presumably volume changes in response to wetting and drying may cause this.

5.2 Tentative Model for the Development of Polygonal Fractures and Columnar Peds

In considering how the polygonal pattern develops to define columnar peds at the study site there are at least three possibilities:

(i)   the fracture pattern is imposed from above where it is controlled by the weathering environment,
(ii) the fracture pattern is imposed from below via a pre-existing system i.e. it is inherited from the bedrock, or
(iii) a combination of both

The first is unlikely for two simple reasons. If the weathering environment such as via simple shrinking and swelling dominated, the fractures would have developed simultaneously and generated a non-orthogonal pattern (Lachenbruch 1962). The strong degree of orthogonality indicates otherwise. In addition one would expect the aperture of the fractures to be similar but they are not. In contrast, there is stronger support for the second explanation. It is clear from the preceding section that an initial stress field imposed on rock will lead to a fracture system that strongly influences subsequent development. Logically these influences could be translated to the development of peds. To be the sole explanation a strongly orthogonal pattern must

minor and incipient fracture systems relieve surficial stress created in the saprolite during rapid desiccation after rain

mega and major fracture systems represent seasonal or prolonged desiccation that relieve stress in the saprolite at greater depths

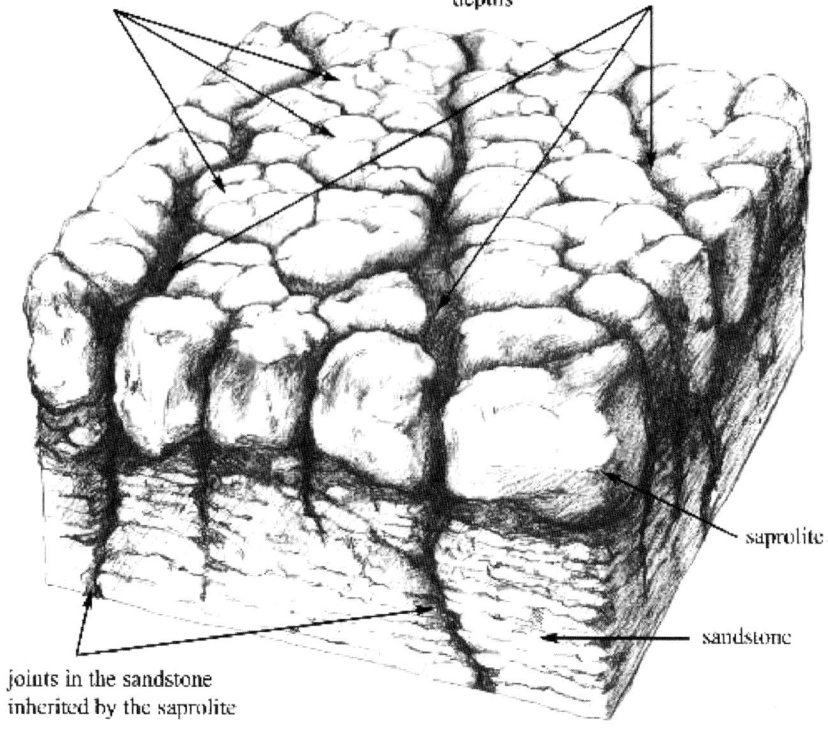

saprolite

sandstone

joints in the sandstone
inherited by the saprolite

Fig. 14 Illustration showing the role of the different fracture systems in relieving stress induced by contraction of the saprolite

prevail. In detail, orthogonality is best expressed in the larger aperture fractures but declines as the fractures narrow and is replaced by non-orthogonal trends. Clearly the evidence indicates that a combination is required with the non-orthogonal component superimposed but constrained by the inherited orthogonal fracture pattern.

A possible senario for the study site is as follows (Fig. 14). Tensile stress is imposed as the clayey saprolite begins to dry out. This leads to fracturing of the weakest (best established) relict joints (largest fractures) which follow loci of low strength (Koo 1982). Thus fracturing is along predetermined planes of weakness (Lachenbruch 1962) and they do not traverse the surface in a random sinuous mode. Continued desiccation, activates all of the relict joints or mega fractures divides the saprolite into mega blocks and at the same time increases the strength within them.

As desiccation continues, the blocks are subdivided along secondary (major) fractures which intersect the pre-existing primary cracks predominantly at right angles, and thus inhibit further fracturing by reducing stress primarily through superposition of zones of stress relief (Lachenbruch 1962). Thus the basic polygonal structure results from the application of tensile strength during desiccation of mega and major fractures.

The smallest or incipient fractures appear to be initiated in the centre of blocks in response to desiccation after rain and appear to relieve surficial stresses only. However, as desiccation of the saprolite moves beyond the surficial layer, some of these smaller fractures are activated and propagate to depths controlled by the stress relief zones of the mega and major fractures. These intermediate fractures are representative of the minor fracture system. Over time, the deepening of the respective fracture systems will be controlled by the depth of desiccation within the clayey saprolite, with the mega and major fractures operative over seasonal to decadal fluctuations in moisture, and the minor and incipient fractures operative over the interval between individual rainfall events.

5.3 Pedological Considerations

This study deliberately selected a site where bedrock was close to the surface and where a bedrock influence could be tested. The close correspondence between the macro fracture pattern and bedrock jointing attest to this influence. An obvious issue to consider is the application of this result to columnar peds in depositional materials. A partial answer to this is (Sect. 5.1) attributed to the pre-existing stress field will influence future patterns. It is interesting to note that inheritance of older stress fields may influence conditions in the future. A geological parallel exists where old tectonic structures may permeate through thousands of metres of more recent rock.

There is a need to consider the convex–up shape or dome that distinguishes columnar from prismatic peds. The current study has not considered this issue other than to note that fracturing (i.e. brittle deformation) is unlikely to be a direct cause. Likewise there is evidence for ductile deformation such as the re-oriented sandstone fragments in the basal zone (Sect. 2.2) and flow structures (Sect. 4.7). Alternatively and/or additionally the dispersive character of sodic soils may need to be considered though it is unclear how the balance between dispersion and swelling under changing soil moisture status (Sumner 1993) might contribute to the dome development. These are fruitful avenues for future research.

To date most explanations of ped development have emphasised aggregation (e.g. Tisdall and Oades 1982), which would seem to best apply to small (mm scale) crumb and granular peds and perhaps to polyhedral and sub-angular blocky peds. It is unlikely that aggregation explains larger peds and especially those with well defined faces such as blocky, platey and parallelopiped peds where splitting (i.e. brittle deformation) of a larger mass is more likely as might occur during volume change or shearing. But even these mechanisms may not adequately account for the very large peds (decimetre scale) as occurs in prismatic and columnar peds. The current study has indicated that pre-existing fractures especially joints and even bedding planes in the underlying bedrock may impart a dominating control on ped size and shape.

The soil studied is also of interest for other reasons. Firstly it has an acid reaction, texture contrast with a modest shrink-swell capacity and yet contains pronounced columnar peds. Nevertheless, such soils are widespread in sub-humid eastern Australia (Paton 1974b, Paton 1978, Hubble et al. 1983, Chartres 1993) but a full explanation of their genesis has yet to emerge. Traditionally, soils with this morphology have been treated as solonetz to soloth/solod with alkalinity also characterising the former and a degree of acidity the latter (see slightly different versions of this sequence in for example Young 1976, Paton 1978, Duchaufaur 1982). Such soils were often genetically linked to a primeval stage of solonchak.

The removal of free salts (solonization) lead to a solonetz and the subsequent leaching of exchangeable sodium and magnesium (solodization) led to solodized solonetz with further leaching and acidification to a soloth. However, this sequential model provided little explanation for the formation of columnar peds, though Duchaufaur (1982) viewed the columnar peds in the soloth as a degraded version of the solonetz stage (see his Fig. 13a). Very little of this seems to offer an explanation of how columnar peds develop and persist in the soil studied. Previously, Hallsworth and Waring (1964) also questioned the need for an initial solochak stage in the Pilliga region but they based this on an alternative explanation of texture contrast soil formation – an explanation that does not concur with the known Quaternary history of this region (Hesse and Humphreys 2001).

6 Conclusions

The results presented in this study indicate that the fracture system in the Pilliga Sandstone exerts a major influence on the size and orientation of the columnar peds developed in the clayey saprolitic B horizon of a texture contrast soil. This influence is most pronounced in the mega and major fractures (joints), which define orthogonally bounded blocks approximately $1\,\mathrm{m}^3$ in size. Superimposed on these blocks are smaller fractures that penetrate only into the upper $2\,\mathrm{cm}$ (incipient fractures) and $10\,\mathrm{cm}$ (minor fractures) of the tops of domes of the columnar peds. It is these smaller and non-orthogonally aligned fractures that provide the final definition of the mostly pentagonal or hexagonal columnar peds. Thus the origin of these peds relies on a combination of inherited joints and fractures superimposed by near surface derived mechanisms. In essence the columnar peds are a cauliflower-like upper surface of blocky to tabular mega-peds.

A tentative model to explain the development of the conspicuous polygonal fractures in the saprolite is proposed. Relict joints provided the flaws or zones of weakness along which initial fracturing of the saprolite takes place. Subsequent fracturing requires increasing levels of applied stress, since the fracturing process removes zones of weakness thereby increasing the strength of the bounded material.

With increasing magnitudes of desiccation, further fracturing subdivides the material into smaller blocks with a reduction in stress achieved primarily via the superposition of zones of stress relief of individual cracks. This model is based on the principles of brittle deformation and does not explain the rounded tops or domes of columnar peds. This awaits future research. In addition there is evidence of ductile features in this soil that indicates possible shearing and other responses to built up stresses that might occur during expansion as the soil wets up. But, these additional factors would appear to play a lesser role given dominating influence of the inherited joints.

Acknowledgments This chapter is dedicated to the memory of Geoffrey Steel Humphreys who passed away suddenly on the 12th August 2007. Geoff was a great teacher, mentor and friend. His contribution to pedology was enormous, and his influence is everywhere throughout these pages, in what is our contribution to an understanding of the genesis of columnar peds.

This study began on a student field trip where we camped on site for week at a time. We thank several groups of GEOS399 students for their assistance in this and other projects. We also express our gratitude to several colleagues who assisted in different ways: Liz Norris provided detailed vegetation lists, Dave Durney advised on fracture nomenclature, Pat Conaghan advised on the sedimentological features of the substrate Di Hart and Paul Hesse joined in many facets of the field work, and Peter Mitchell who provided valuable insights into the Pilliga Scrub. We also acknowledge the advice of Ron Paton who along with the late Cliff Thompson spent considerable time examining columnar pedal materials along the Moonie pipeline trench for CSIRO in the early 1960's. Funding was obtained from research grants at Macquarie University. Finally we acknowledge permission and logistic support from the NSW State Forest for allowing us to work within the Pilliga State Forest.

Appendix A: The Soil

Setting

Gentle (4% slope to the north west) lower footslope of upland terrain with a shallow, non-calcareous soil over sandstone (Pilliga Sandstone). The site is adjacent to an extensive alluvial plain (Hesse and Humphreys 2001).

Vegetation

An open woodland, 7–8 m high, dominated by Eucalyptus viridis (green mallee), and occasional Callitris glaucophylla (white cypress) and Brachychiton populneum (kurrajong) with a shrubby understorey of Dodonaea viscosa subsp.cuneata (hop bush), Olearia decurrens (daisy bush) and occasional Santalum acuminatum (quandong) and a sparse herbaceous layer of grasses (Stipa scabra subsp. scabra, S. setacea and Paspalidium constrictum) and several non-grasses (Einadia hastata, E. nutans and Oxalis corniculata).

Soil Morphology

Horizon	Depth (cm)	Description
Oi	surface	Crytogamic crust 2–5 mm thick underlying litter layer (mostly eucalypt leaves) up to 1 cm thick and covering 70% of the surface; exposed crust capped with a layer of single grain sand; sharp planar to:
A11	0–4	Brown, 10YR4/3d to very dark greyish brown 10YR3/2m; sandy loam to sandy clay loam; apedal; earthy fabric but layered with buried couplets of leaves and sediment (bowl structures*); common fine roots; clear, planar to:
A12	4–14 (range 10–18)	Dark yellowish brown, 10YR4/4d to 10YR3/4m; sandy clay loam, apedal, earthy fabric; common termites; clear, wavy to:
Ex	14–15 cm	Pinkish to light grey 7.5YR–10YR6–7/2d to brown 7.5–10YR4.5/3m; sandy loam to sandy clay loam; apedal with earthy fabric and fragipan behavior; occasionally penetrates (tongues) 30–40 cm along major fractures and forms a capping to the dome top; deeper material slightly darker (10YR5/3d,4/2m); occasional pebble of rounded quartz, jasper and chert on top of and between columnar peds; sharp, convoluted to:
Btn	15–50	Brown to dark greyish brown, 7.5YR-10YR4/2m; (sandy) medium clay; harsh; very coarse, columnar peds or domes (30 cm wide); few roots within the domes, large roots penetrate the major fractures; hard when dry; diffuse to:
Bn	50–65	Olive brown, 2.5Y4/3m; (sandy) medium clay; forms base of columns; many roots; distinctly weaker than upper part of columns (Btn) such that columns shear at this depth when a strong force is applied; clear, convoluted to:
B/C	65–95	Olive yellow, 2.5Y6/6m sandstone interbedded with many very dark greyish brown (2.5Y3/2m) sandy clay beds mostly 5–7 thick; occasional extensions of Bn material (clay-plug 20–30 cm wide) of olive brown (2.5Y4/3m) (sandy) medium clay penetrating into this layer; sandy medium clay from the base of domes also penetrates into the sandstone along major fractures; sandstone is more fragmented and often reoriented around major fractures and sometimes towards the base of the domes; clear, wavy to:

Horizon	Depth (cm)	Description
Cr/R	95–135+	Olive yellow, 2.5Y6/6m sandstone with occasional thin (1–4 mm) dark sandy clay bed; well bedded fine lithic sandstone; bedding intact and approximately parallel to the surface

*See Humphreys (1994)

Soil Chemical Properties

Property[1]	Sample depth (cm)					
	0–4 (A11)	5–10 (A12)	37–42 (Btn)	57–62 (Bn)	75–80 (B/C)	130 (S/st
pH	5.2	5.6	5.3	5.5	5.1	5.3
Olsen P (µg/ml)	6	3	3	7	4	11
Total N (%)	0.15	<0.04	<0.04	<0.04	<0.04	<0.04
Organic matter (%)	4.6	1.7	1.2	0.7	0.6	0.3
Avail K (cmol(+)/kg)	0.52	0.14	0.26	0.30	0.47	0.43
Avail Ca(cmol(+)/kg)	2.4	<0.5	<0.5	<0.5	<0.5	<0.5
Avail Mg (cmol(+)/kg)	3.9	2.0	9.0	14.9	16.0	14.2
Avail Na (cmol(+)/kg)	0.16	0.24	1.74	3.93	5.95	5.22
CEC (cmol(+)/kg)	11.5	5	16.4	21.8	26.3	22.0
Base Saturation (%)	59	53	71	88	83	90
EC (µS/cm)	1.1	<0.8	0.9	nd	4.4	nd
ESP (%)	1.4	4.8	10.6	18.0	22.6	23.7

[1]Samples were air dried at 35–40°C overnight and crushed to pass through a 2 mm sieve. Soil pH was obtained from a 1:2 (v/v) soil-water mix with a potentiometric determination. Exchangeable K, Ca, Mg and Na were obtained from a 1M neutral ammonium acetate extraction and measured by ICP-OES. CEC is the summation of extractable cations (K, Ca, Mg, Na) plus extractable acidity. Total carbon and total N were obtained by a Dumas combustion with $OM = 1.72 \times total\ C$. Available P was obtained using an Olsen extraction followed by Molybdenum Blue colorimetry. Total P was obtained from a nitric/hydrochloric digestion and measured by ICP-OES. Analyses were undertaken at Hill Laboratories, New Zealand, and internationally accredited facility.

Appendix B: Field Classification for Cleavage in Clastic Sedimentary Rocks (Based on Durney and Kisch, 1994)

The following four criteria were used to distinguish between relict joints and planar voids in the saprolite exposed in the excavations at the study site;

(i) Regularity. Fissile cracks are less regular than joints, and are characterised by curviplanar surfaces and seemingly unsystematic T-intersections in a 3D-network to produce a generally elongate or flat pattern of interlocking, irregular polygonoids. In contrast, joints are usually very planar and parallel, often appear in distinct orientational sets, and tend to produce flat sided angular blocks, parallelepipeds and prisms.

(ii) Continuity. Jointing tends to be more continuous in length than fissility, the former being commonly decimetric to kilometric in length, whereas the latter have lengths that are more commonly centimetric. Joints also tend to traverse minor lithological junctions without interruption or deflection.

(iii) Mode of development. Fissile cracks tend to develop parallel to tectonic and sedimentary fabric anisotropy planes, whereas joints do not enjoy the same exclusive relationship.

(iv) Depth of penetration. With the exception of sheeting joints and exfoliation structure, joints tend to persist well into the subsurface, in contrast to fissile cracks which decrease very rapidly with depth.

References

Arditto PA (1982) Deposition and diagenesis of the Jurassic Pilliga Sandstone in the southeastern Surat Basin, New South Wales. Journal of the Geological Society of Australia, 29: 191–203.

Bai T, Pollard DD (2000) Fracture spacing in layered rocks: a new explanation based on the stress transition. Journal of Structural Geology, 22: 43–57.

Beavis FC (1985) Engineering geology. Blackwell Scientific Publications, Melbourne.

Black RF (1952) Polygonal pattern and ground conditions from aerial photographs. Photogrammetric Engineering, 18: 123–133.

Brewer R (1976) Fabric and mineral analysis of soils. Reprinted by R.E. Kreiger Co. New York.

Bullock P, Fedoroff N, Jongerius A, Stoops G, Tursina T, Babel U (1985) Handbook for soil thin section description. Waine Research Publications, Wolverhampton.

Chartres CJ (1993) Sodic soils: an introduction to their formation and distribution in Australia. Australian Journal of Soil Research, 31: 751–760.

Duchaufaur P (1982) Pedology. Translated by T.R. Paton. George Allen and Unwin, London.

Durney DW, Kisch HJ (1994) A field classification and intensity scale for first-generation cleavages. AGSO Journal of Australian Geology and Geophysics, 15: 257–295.

Florinsky IV, Arlashina HA (1998) Quantitative topographic analysis of gilgai soil morphology. Geoderma, 82: 359–380.

Gray JM (1974) Use of chi-square on percentage orientation data: discussion. Geological Society of America Bulletin, 85: 833.

Hallsworth EG, Waring HD (1964) Studies in pedogenesis in New South Wales. VIII. An alternative hypothesis for the formation of the solodised solonetz of the Pilliga district. Journal of Soil Science, 15: 158–180.

Hart DM (1988) A fabric contrast soil on dolerite in the Sydney Basin, Australia. Catena, 15: 27–37.

Hart DM (1995) Litterfall and decomposition in the Pilliga State Forest, New South Wales, Australia. Australian Journal of Ecology, 20: 266–272.

Hart DM, Hesse PP, Mitchell PB (1985) The inheritance of soil fabric from joints in the parent rock. Journal of Soil Science, 36: 367–372.

Hesse PP, Humphreys GS (2001) Pilliga landscapes, Quaternary environment and geomorphology. In: J. Dargarvel, D. Hart and B. Libbis (Eds.), Perfumed Pineries, Environmental History of Australia's Callitris Forests. Centre for Resource and Environmental Studies, The Australian National University, Canberra, pp. 79–87.

High LR, Picard MD (1971) Mathematical treatment of orientation data. In: R.E. Carver (Ed.), Procedures in Sedimentary Petrology. Interscience Pubs. Inc., New York, pp. 21–45.

Hills ES (1972) Elements of structural geology. Chapman and Hall Ltd., London.

Hubble GD, Isbell RF, Northcote KH (1983) Features of Australian soils. In: Soils: An Australian Viewpoint. Division of Soils CSIRO & Academic Press, Melbourne, pp. 17–47.

Humphreys G, Norris E, Hesse P, Hart D, Mitchell P, Walsh P, Field R (2001) Soil, vegetation and landform in Pilliga East State Forest. In: J. Dargarvel, D. Hart and B. Libbis (Eds.), Perfumed Pineries, Environmental History of Australia's Callitris Forests. Centre for Resource and Environmental Studies, The Australian National University, Canberra, pp. 71–78.

Humphreys GS (1994) Bowl-structures: a composite depositional crust. In: A.J. Ringrose-Voase and G.S. Humphreys (Eds.), Soil Micromorphology: Studies in Management and Genesis. Developments in Soil Science 22, Elsevier, Amsterdam, pp. 787–798.

Hunt PA, Mitchell PB, Paton TR (1977) Laterite profiles and lateritic ironstones on the Hawkesbury Sandstone, Australia. Geoderma, 19: 105–121.

Isbell RF (1996) The Australian Soil Classification. CSIRO Publishing, Melbourne.

Knight MJ (1980) Structural analysis and mechanical origins of gilgai at Boorook, Victoria, Australia. Geoderma, 23: 245–283.

Koo YC (1982) Relict joints in completely decomposed volcanics in Hong Kong. Canadian Geotechnical Journal, 19: 117–123.

Lachenbruch AH (1961) Depth and spacing of tension cracks. Journal of Geophysical Research, 66: 4273–4292.

Lachenbruch AH (1962) Mechanics of thermal contraction cracks and ice wedge polygons in permafrost. Geological Society of America Special Paper 70.

Lafeber D (1965) The graphical representation of planar pore patterns in soils. Australian Journal Soil Research, 3: 143–164.

Mardia KVM (1972) Statistics of Directional Data. Academic Press, New York.

McKenzie NJ, Jacquier DJ, Ringrose-Voase AJ (1994) A rapid method for estimating soil shrinkage. Australian Journal of Soil Research, 32: 931–938.

Narr W, Suppe J (1991) Joint spacing in sedimentary rocks. Journal of Structural Geology, 13: 1037–1048.

Netoff DI (1971) Polygonal jointing in sandstone near Boulder, Colorado. Mountain Geologist, 8: 17–24.

Ollier C, Pain C (1996) Regolith, Soils and Landforms. John Wiley and Sons, London.

Paton TR (1974a) Origin and terminology of gilgai in Australia. Geoderma, 1: 221–242.

Paton TR (1974b) The Moonie pipe line trench. Unpublished report. Mimeo held at Macquarie University, Sydney.

Paton TR (1978) The Formation of Soil Material. George Allen and Unwin, London.

Paton TR, Humphreys GS, Mitchell PB (1995) Soils: A New Global View. UCL Press, London.

Pollard DD, Aydin A (1988) Progress in understanding jointing over the past century. Geological Society America Bulletin. 100: 1181–1204.

Powell B, Ahern CR, Baker DE (1994) Soil dispersion in Queensland sodic soils and their classification. In: R. Naidu, M.E. Sumner and P. Rengasamy (Eds.) Australian Sodic Soils. Distribution, Properties and Management. CSIRO, Melbourne, pp. 81–87.

Price NJ (1966) Fault and Joint Development in Brittle and Semi-Brittle Rock. Permagon Press, Oxford.

Slansky E (1984) Clay mineralogy. In: Hawke JM and Cramsie JN (Eds.), Contributions to the Geology of the Great Australian in New South Wales. Geological Survey of NSW. Department of Mineral Resources, Sydney, Bulletin 31: 179–203.

Smith RM, Cernuda CF (1951) Some characteristics of the macrostructure of tropical soils in Puerto Rico. Soil Science, 73: 183–192.

Soil Survey Staff (1998) Keys to Soil Taxonomy. United States Department of Agriculture Natural Resources Conservation Service, Washington, D.C.

Stace HCT, Hubble GD, Brewer R, Northcote KH, Sleeman JR, Mulcahy MJ, Hallsworth EG (1968) A handbook of Australian soils. Rellim Tech. Publ., Glenside, South Australia.

Sumner ME (1993) Sodic soils: new perspectives. Australian Journal of Soil Research, 31: 683–750.

Swan ARH, Sandilands M (1995) Introduction to Geological Data Analysis. Blackwell Science, Oxford.

Tanner WF (1955) Paleogeographic reconstruction from cross-bedding studies. American Association of Petroleum Geologists Bulletin, 50: 2457–2465.

Tisdall JM, Oades JM (1982) Organic matter and water stable aggregates. Journal Soil Science, 33: 141–163.

Tuckwell GW, Lonergan L, Jolly RJH (2003) The control of stress history and flaw distribution on the evolution of polygonal fracture networks. Journal of Structural Geology, 25: 1241–1250.

Twiss RJ, Moores EM (1992) Structural Geology. W.H. Freeman Co., New York.

Walsh PG, Conaghan PJ, Humphreys GS (in prep) The inheritance and formation of smectite in a texture contrast soil in the Pilliga State Forests, NSW.

Williams PW (1972) Morphometric analysis of polygonal karst in New Guinea. Geological Society of America Bulletin, 83: 761–796.

Williams PW (1974) Use of chi-square on percentage orientation data: reply. Geological Society of America Bulletin, 85: 833–834.

WRB (1998) World Reference Base for Soil Resources. FAO, Rome and ISRIC, Wageningen.

Young A (1976) Tropical Soils and Soil Survey. Cambridge University Press, London.

Physical Fractionation and Cryo-Coupe Analysis of Mormoder Humus

J. van Mourik and S. Blok

Abstract Soil micromorphology plays a key role in the research of the decomposition, storage, and mineralisation of soil organic matter. It is often difficult to study the internal fabrics and stability of soil aggregates, based on soil micromorphological properties. Similar to physical fractionation of soil inorganic particles organic matter from ectorganic and endorganic horizons can also be fractionated. Quantification of aggregate stability can be achieved, by the weight percentage of such fractions. Also the micromorphological properties of soil particles can be studied using cryo-coupes from gelatin suspensions prepared in a cryostat. The results of physical fractionation and cryo-coupe analysis improve our knowledge on the development of the humus form in the soil.

Keywords Soil micromorphology · cryo-coupe · physical fractionation · humus form · soil aggregate

1 Introduction

Green et al. (1993) developed a new global classification system for humus forms. In forest soil research, determination of humus forms adds information to the regular soil classification. It has been shown that variations in forest soil ecology might occur within the same humus form class (Van Mourik 2003). Refinement of the humus form classification can be based on soil micromorphological studies. In

J. van Mourik
Institute for Biodiversity and Ecosystem Dynamics (IBED), University of Amsterdam, Nieuwe Achtergracht 166, 1024 AD Amsterdam, The Netherlands, e-mail: jmourik@science.uva.nl

S. Blok
Institute for Biodiversity and Ecosystem Dynamics (IBED), University of Amsterdam, Nieuwe Achtergracht 166, 1024 AD Amsterdam, The Netherlands

S. Kapur et al. (eds.), *New Trends in Soil Micromorphology*,
© Springer-Verlag Berlin Heidelberg 2008

preliminary studies (Dijkstra and Van Mourik 1995, 1996) the micromorphological description of mormoder humus forms was based on field characteristics and observations in 'normal' thin sections. This article presents a method for physical fractionation of soil samples, based on dissolved air flotation (Edzwald 1995). Based on the weights of the obtained fractions better impression is required of the total amount of mineral grains, decaying organic skeleton grains, amount and stability of organic aggregates and fine materials.

Cryo-coups have been prepared from the organic fractions of the soils studied. Micropedological observations in cryo-coups, additional to observations in conventional thin sections, contribute to understand the processes of aggregation and disintegrating in humus forms.

During the last decades soil development under forest plantations was an important theme in soil chemistry (Emmer 1995) and pollen analysis (Dijkstra and Van Mourik 1995, 1996, Van Mourik 2003) in The Netherlands. Soil micromorphology was an important tool to understand soil development and palynological fingerprints from organic soil horizons. However, based on observations from conventional thin sections some questions could not be answered satisfactorily. Due to the dense setting of various compounds, it is often difficult to understand the internal fabric and stability of individual soil aggregates in ectorganic and endorganic horizons.

Physical fractionation of soil material by dissolved air flotation appeared to be a promising method to separate fresh, not consumed organic skeleton grains from mineral grains, soil aggregates and fine materials. Fractionated soil material can be suspended in gelatin. In a cryostat a number of cryo-coupes with a thickness of 14 μm can be produced. Soil micromorphological observations in cryo-coups provide additional information about aggregation and pollen conservation.

The objective of our study was to present the result of a pilot study of physical fractionation and cryo-coupe analyses, applied on the horizons with mormoder humus formed under oak forest in the coastal dune region in The Netherlands.

2 Materials and methods

2.1 Physical fractionation

The applied method of physical fractonating is based on Edzwald (1995). The main target of physical fractionation is to separate individual of inorganic and organic particles and 'stable' aggregates. The liberation of particles takes place in three stages after pretreatment of the samples. The air-dried samples were crushed over a sieve (opening 4 mm). With a two-way splitter sub samples of approximately 25 cm^3 were taken. These were placed in glass tubes and immersed in 250 mL water (Fig. 2). The tubes were sealed in stainless steel containers. After lowering air pressure to near vacuum, the containers were positioned horizontally on a shaking machine. They were then shaken transversally at a rate of 200 cycles per minute for one hour.

Throughout all stages of the liberation and fractionation dematerialized water conditioned with isopropanol (0.02% V) was used.

The samples were fractionated in three successive stages, yielding finally four fractions. The Fractionation procedure is summarized in Fig. 1. In the first stage of fractioning, sedimentation is used to isolate the suspended (SUS) fraction, composed of light and small particles. The sample holders were filled with water, sealed and held upside down for 30 s. Particles were then allowed to settle for 60 s

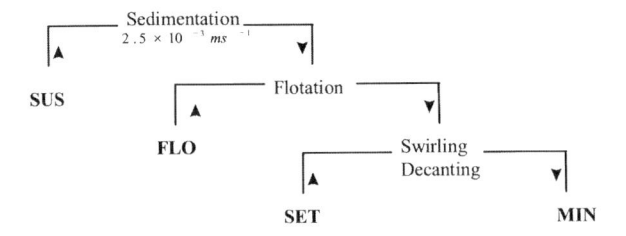

Fig. 1 Schematic overview of the physical fractionation procedure

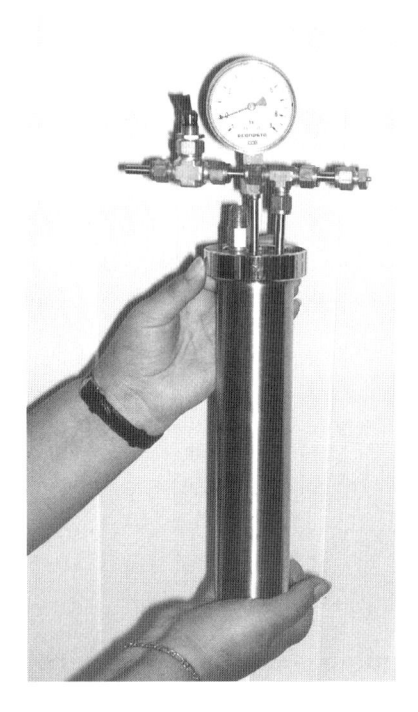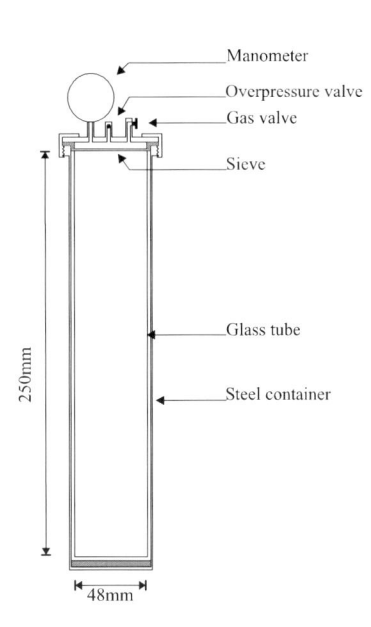

Fig. 2 Picture and diagram of the fractionation unit

before the top 150 mm were siphoned off, with the exception of floating particles. This procedure was repeated two or three times depending on the concentration of fine particles in the sample. The resultant suspension was centrifuged at 2000 rpm for 30 min and the supernatant discarded.

In the second stage, flotation is applied to the residue from the first stage to isolate the floated (FLO) fraction, with hydrophobic surfaces or sites. The level of the suspension in the glass tube was raised to 190 mm. The glass tubes were again sealed in the sample container. The samples were then shaken for 30 min at elevated pressure (3.5 bar N_2). On release of the pressure, gas bubbles nucleate in the suspension and form agglomerates composed of hydrophobic particles. All the particles that were collected at the surface were decanted. This procedure was carried out twice on the residue and once on the combined proceeds.

In the third stage, the residue of the second stage is raised by swirling and decanting, yielding the predominantly organic settled (SET) fraction, the organic part of the residue and the mineral (MIN) fraction, the mineral part of the residue (sand sized mineral grains). The FLO and SET fractions were drained by vacuum filtration using a hard paper filter (Schleicher and Schuell ref. No. 1575). The liquid phase was discarded. They were split into representative sub-samples by cutting out a section of the filter and collecting the precipitate. Taking a representative sub-sample of SUS fraction posed no problems as it could easily be homogenized and decanted. All organic fractions were stored in a 2% para-formaldehyde citrate/phosphate buffer solution (pH 5).

2.2 Cryo-coupe preparation

Cryo-coups are applied in biomedical studies of all types of tissues. Cryo-coupes can also be prepared from gelatin suspensions of various soil particles for micromorphological analysis.

The cryo-coupes, used in this study, were prepared with a Leica Jung Frigocut 2800 N cryostat. To prepare cryo-coupes, the physical fractions must be washed out with distilled water and transferred to aluminum foil tubes. They must be embedded in 10% gelatin, fixated with 2% para- formaldehyde and preserved in a cold room for several days. The gelatin blocks were then trimmed and frozen onto a cryostat-mounting stub with solid CO_2. Sections were cut at a temperature of −35°C. Section thickness was 14 μm. Sections were transferred to glass slides and embedded in gelatin glycerin 1:1 with water on a hot plate to prevent the formation of air bubbles. The cryo-coupes can be studied under microscope, but must be stored in the cooler temperatures.

2.3 Sampling and Pollen Extraction

For the application of physical fractionation and cryo-coup micromorphology, a soil profile under an oak stand was selected in the coastal dune region (Fig. 3a). Van Til and Mourik (1999) have described the development of landscape and forest

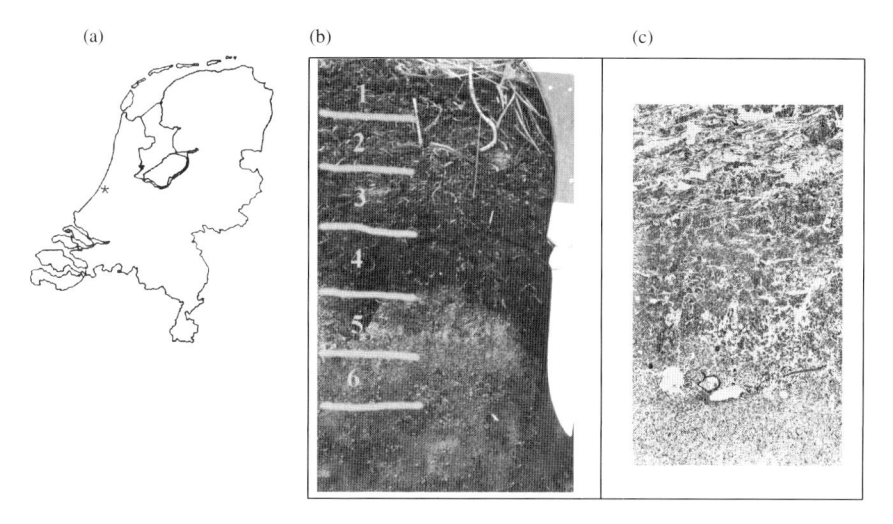

Fig. 3 (a) Location of the site Paardenkerkhof in The Netherlands **(b)** Soil column of the sampled mormoder. The boundaries between the six test samples (thickness 2.5 cm) are indicated **(c)** Thin section of the sampled mormoder (same scale as 3b). Sampled horizons: LF (T1), F (T2, T3), H (T4) and AE (T5, T6)

of this site, locally named 'Paardenkerkhof'. The parent material consists of acid (pH ≈ 4) moderately fine eolian sand. The soil was classified a Cambic Arenosol with a mormoder humus form.

Additionally a soil column was taken with a humus sampler and divided in six test samples, each one 12 cm long, 4 cm width, and 2.5 cm thick (Fig. 3b). An undisturbed sample was taken for the production of a conventional thin section (Fig. 3c). The test samples were divided in sub samples for soil pollen analysis and physical fractionation. Pollen extraction and analysis is based on the extraction technique and determination using Moore et al. (1991). Extractions, all in duplicate, were made from bulk samples (1 mL) as well as from the organic fractions (1 mL) to control pollen distribution in the soil and over the fractions. The cryo-coupes offer a unique opportunity to improve knowledge of pollen infiltration and preservation in terrestrial soils (Van Mourik 1999, 2001, 2003).

3 Results and Discussion

3.1 Distribution of physical fractions in the soil profile

Figure 4a,b shows the result of physical fractionation of the six test samples of the pilot profile. Since the particles were liberated by mechanical separation, the mechanical stability of fabric units is reflected by the degree of particle interaction.

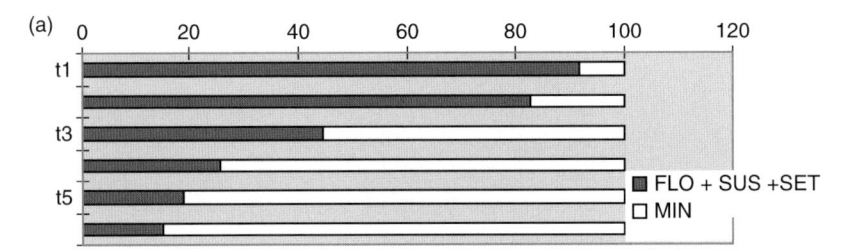

Fig. 4a Weight percentages of mineral (MIN) and organic (FLO + SUS + SET) fractions

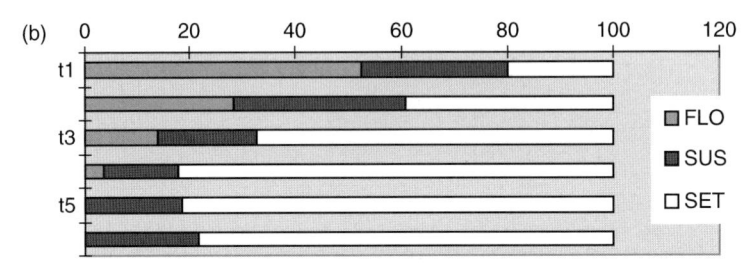

Fig. 4b Weight percentages of Floated (FLO), suspended (SUS) and settled (SET) organic fraction

Figure 4a shows the proportions of the mineral and organic fractions. The boundaries between the ectorganic F and H horizons and the endorganic Ah horizon, abrupt in the field description, are more gradual in this diagram.

Figure 4b shows the distribution of the FLO, SUS and SET fractions. As suggested in previous studies (Dijkstra and Van Mourik 1995, 1996), various microbes and soil animals are altering fresh litter into excremental and plasmatic humus. The distribution of the various fractions is characteristic for a mormoder humus form. The decrease of FLO from t1 to t4 and the increase of SET from t1 to t4, followed by a slight decrease in t5 and t6, compensated by an increase of SUS is obvious in this context. In the upper part of the humus form, the main process is decomposition of litter and development of stable aggregates. In the lower part loss of aggregate stability takes place.

3.2 Distribution of Pollen Grains in the Soil Profile

The pollen diagram (Fig. 4) shows a palyno-relict of dune grassland vegetation (preserved in the Ah-horizon) preceding the rise of the oak forest (preserved in the F horizons). The main indicators for the dune-grassland relict are *Hippophaë, Ligustrum, Plantago, Ranunculus* and *Rumex*. The pollen content of the H horizon shows the transition from grassland to planted-forest, the F horizon is the clear palynological expression of the oak forest stand. The age of the forest stand is estimated to be around 200 years (Van Til and Mourik 1999). Pollen grains are

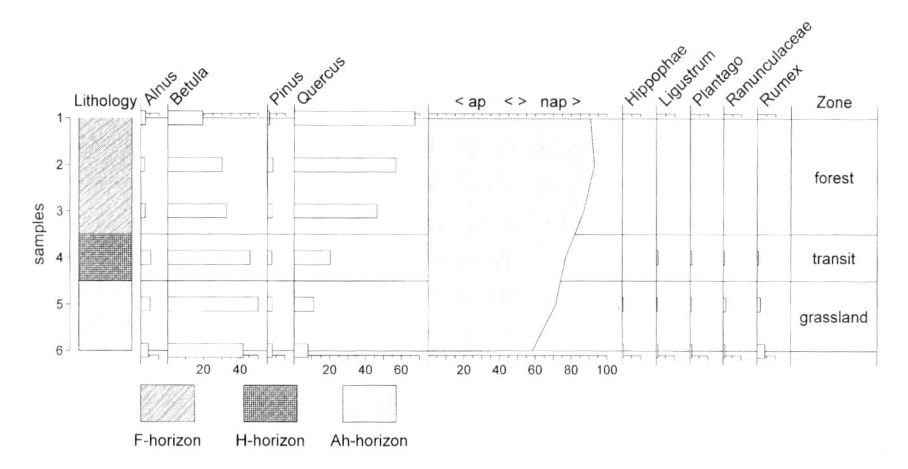

Fig. 5 Pollen diagram Paardenkerkhof

part of the annual eolian deposition at the soil surfaces. Various soil processes are responsible for the incorporation and infiltration of the pollen grains into soils. Microbes, litter consuming micro arthropods and earthworms are the important actors of this environment (Van Mourik 1999, 2001, 2003, Davidson et al. 1999). In weak acid sandy soils, earthworms can include pollen grains directly in their large sized excrements (diameter >200 μm). The internal fabric of these excrements is a perfect microenvironment for pollen preservation. In more acid soils, earthworms are absent. Small sized arthropod excrements (diameter 50–100 μm) and fine residual particles embed free pollen grains. The rate of protection of pollen grains in such aggregates seems to be similar to the earthworm excrements untill its disintegration.

The total pollen concentrations in extractions from bulk samples increased from t1 to t4 and decrease in the mineral soil. The concentrations of pollen grains, attached on particles of the FLO fraction decrease rapidly from t1 to t4 and are zero in t5 and t6.

Table 1 Pollen concentrations (kilo grains/mL) and weight percentages (%) of the fractions SET, SUS and FLO of the test samples of profile Paardenkerkhof (data based on counting's in duplicate slides of fractions and bulk samples, 4000 pollen in each slide

Sample	Organic fractions						Bulk sample	
	SET		SUS		FLO			
t1	39.1	20%	73.1	25%	34.9	55%	45.3	100%
t2	55.0	40%	60.9	30%	22.9	30%	47.1	100%
t3	64.9	50%	35.1	35%	19.3	15%	47.6	100%
t4	58.6	80%	31.9	15%	11.2	05%	52.2	100%
t5	29.0	80%	29.7	20%	–	–	29.1	100%
t6	22.5	75%	24.6	25%	–	–	22.9	100%

The pollen concentrations of SUS fraction decrease from t1 till t4 in contrast to the concentrations of SET. This reflects probably the process of embedding of the free pollen grains in finally stable aggregates, surviving the physical fractionation process.

In t5 and t6 there is an increase of the percentage of SUS and decrease of SET fractions, indicating loss of aggregate stability, resulting in some dispersion and liberation of pollen grains (Van Mourik 2003).

3.3 Cryo-Coupe Analyses

Particles in cryo-coupes are slightly different from the original fabric of a soil horizon as evident in a conventional thin section. A first consideration is to assess the aggregates impact of the liberation procedure. Since aggregates were liberated by mechanical means, the stability of internal fabric is reflected by the degree to which they are preserved in the SET fraction.

A second consideration was the reliability of particle analyses, as the separation was never completed to 100%. Many cases some fine particles, belonging to the SUS fraction, were attached on FLO or even SET particles. Despite this inaccuracy cryo-coupes enable observations on relatively clean particles in the various fractions. Based on cryo-coupe analyses, several typical situations can be described.

1. T1, T2: The original fabric consists of loosely stacked plant residues with some products of primary consumption in the form of intact small faecal pellets (Fig. 6a,c). During liberation, the plant residues are completely separated, whereas during the fractionation these were found almost exclusively in the FLO (Fig. 6b). Faecal pellets of the micro fauna are preserved, whereas macro faunal droppings disintegrate into constituents of plant residues. During fractionation, these end up in mainly SUS and also SET fractions (Fig. 6g,h,i,j). In this part of the profile, pollen grains occur attached on litter fragments (Fig. 6d), and in aggregates (Fig. 6c,j and Table 1).
2. T3, T4: The original fabric consists of plant residues in a more advanced stage of degradation embedded in a matrix of fine materials (Fig. 6e). The matrix is composed of small aggregates up to hundreds of micrometers, mainly (clusters of) faecal pellets and tissue. Part of the latter materials is released as particles during liberation and end up in SUS (Fig. 6k), during fractionation. The remainder stays attached to decayed plant residues. In general, only relatively fresh plant residues with little or no fine materials attached are found in FLO, where the root litter dominates the recognizable plant residues. (Fig. 6f). As the cohesion of the aggregates increases, more plant residues remain enclosed by fine materials and end up in SET (Fig. 6m).
3. T5, T6: The original fabric consists of quartz grains and fine organic materials subjected to secondary (microbial) consumption. Of the aerial litter, only recalcitrant tissues such as nutshells and bud scales remain. Brittleness and increased porosity due to ageing appear to determine the release of fine particles found in SUS during fractionation. With very few floatable materials remaining, the bulk of the particles are still found in SET.

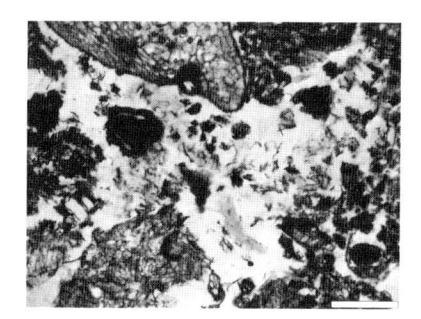

Fig. 6a F-horizon (level t1) in thin section, a mix of decomposing litter fragments and fine organic particles. Rate of decomposition is hardly observable. Scale bar 200 µm

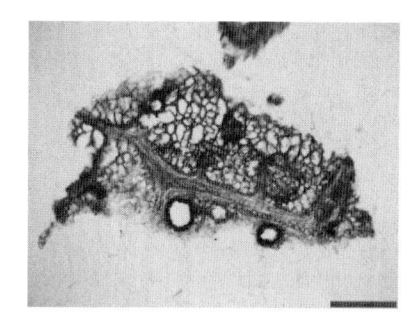

Fig. 6b F-horizon (t1) FLO-fraction: Clean litter fragment with low rate of decomposition and clear tissue structure. Scale bar 200 µm

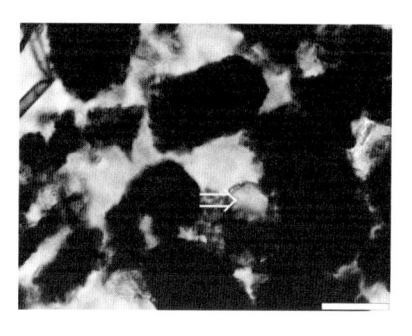

Fig. 6c F-horizon (level t2) in thin section; small excrement's (intern fabric hardly visible) and fine organic particles are embedding a pollen grain (arrow). Scale bar 50 µm

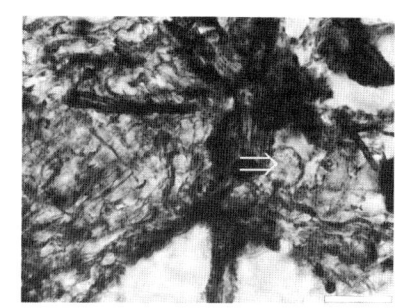

Fig. 6d F-horizon (t2) FLO-fraction: pollen grain (arrow) attached on clean litter fragment. Scale bar 50 µm

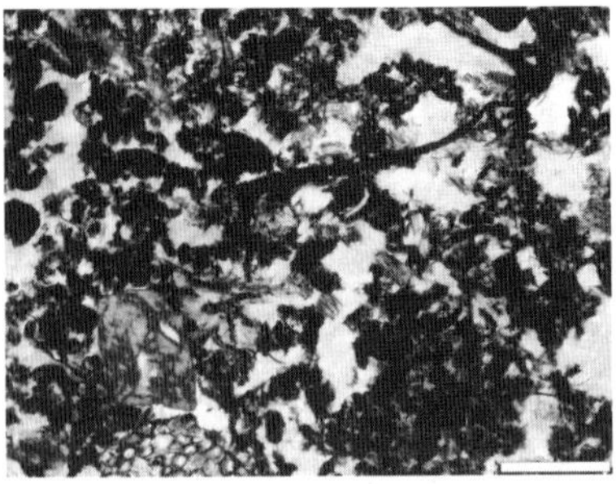

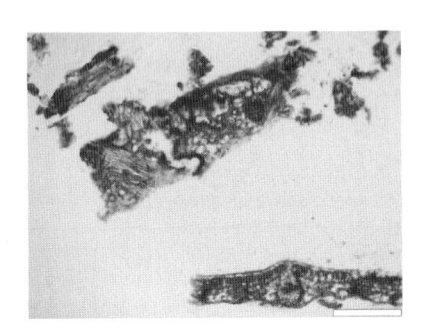

Fig. 6e H-horizon (level t4) in thin section; mix of organic aggregates, organic plasma and some mineral grains. Scale bar 200 μm

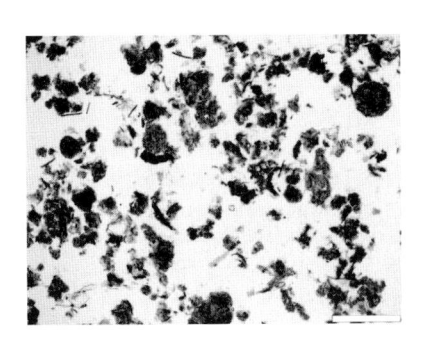

Fig. 6f H-horizon (t4) FLO fraction: Clean root fragments with decaying tissue structure. Scale bar 200 μm

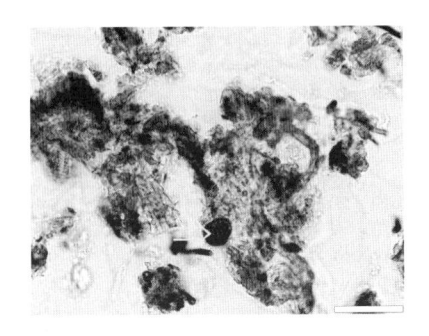

Fig. 6g F-horizon (t2) SUS fraction, fragments of hypha, small organic skeleton grains and plasma are visible. Scale bar 200 μm

Fig. 6h F-horizon (t2) SET fraction, organic aggregate; internal fabric is a dense mass, composed of hypha fragments, organic skeleton grains and organic plasma. Scale bar Scale bar 50 μm

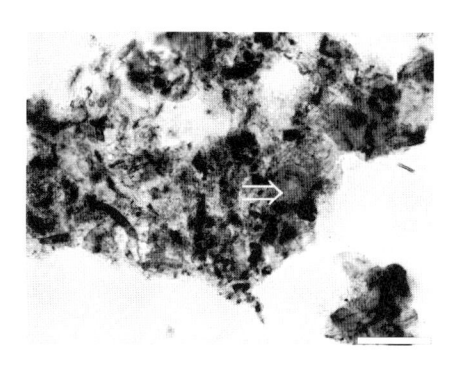

Fig. 6i F-horizon (t2) SUS fraction, organic particles, embedding pollen grain (arrow). The internal fabric is similar as in picture 6j, but disintegrating. Scale bar 50 μm

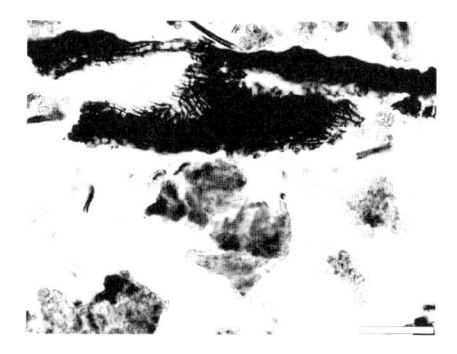

Fig. 6j F-horizon (t2) SET fraction, organic; internal fabric is a dense mass, composed of hypha fragments, organic skeleton grains and organic plasma. A pollen grain is visible (arrow). Scale bar 50 μm

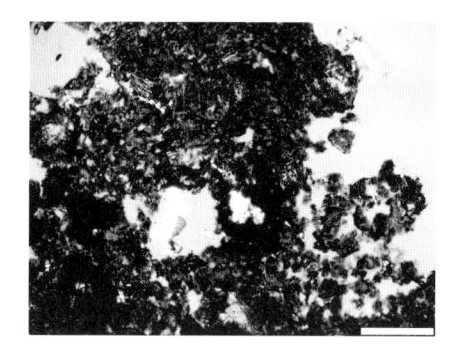

Fig. 6k H-horizon (t4) SUS fraction, a mix of organic skeleton grains, hypha and aggregated fine organic particles. Scale bar 50 μm

Fig. 6l H-horizon (t4) SET fraction, organic aggregate with embedded mineral grains; intern fabric is visible. Scale bar 200 μm

4 Conclusions

Physical fractionating is a promising tool for the quantification of organic compounds as tissue fragments, aggregates and fine material. The weights of these fractions offer additional quantitative information to the soil micromorphological observations in conventional thin sections.

Physical fractionation applied on organic and mineral horizons, provides information about aggregate stability. The ratio between settled and suspended fractions is especially an indicative parameter.

Micromorphological observations in cryo-coups provide more detailed information on the internal fabric of compounds of the soil matrix. The particles in the cryo-coupes are relatively 'clean' and 'thin'. The particles in the slides have an optimal dispersion compared with the dense packing of compounds in conventional thin sections.

The results of physical fractionation and cryo-coupe analysis improve our knowledge on the development of the humus form in the soil.

References

Davidson DA, Carter S, Boag B, Long D, Tipping R, Tyler A (1999) Analysis of pollen in soils: processes of incorporation and redistribution of pollen in five soil profile types. Soil Biology & Biochemistry 31, 643–653.

Dijkstra EF, Van Mourik JM (1995) Palynology of young acid forest soils in the Netherlands. Mededelingen Rijks Geologische Dienst 52, 283–296.

Dijkstra EF, Van Mourik JM (1996) Reconstruction of recent forest dynamics based on pollen analysis and micro morphological studies of young acid forest soils under Scots pine plantations. Acta Botanica Neerlandica 45, pp. 393–410.

Edzwald JK (1995) Principles and applications of dissolved air flotation. Water Science Technology 31, pp. 1–24.

Emmer IM (1995) Humus form and soil development during a primary succession of monoculture Pinus sylvestris foresta on poor sandy substrates. Thesis Universiteit van Amsterdam, p. 134.

Green RN, Towbridge RL, Klinka K (1993) Towards a Taxonomic Classification of Humus Forms. Forest Science, Monograph 29, pp. 1–49.

Moore PD, Webb JA, Collinson ME (1991) Pollen Analysis. Blackwell Scientific Publications, Oxford. 216pp.

Van Mourik JM (1999) The use of micromorphology in soil pollen analysis. Catena 35, 239–257.

Van Mourik JM (2001) Pollen and spores, preservation in ecological settings. In: Briggs, E.G. & Crowther, P.R. (eds.). Palaeobiology II. Blackwell Science, Oxford, pp. 315–318.

Van Mourik JM (2003) Life cycle of a pollen grain in mormoder humus forms of young acid forest soils; a micromorphological approach. Catena 54, pp. 651–663.

Van Til M, Mourik J (1999) Hieroglyfen van het zand, vegetatie en landschap van de Amsterdamse Waterleidingduinen. Architectura & Natura, Amsterdam. 272p.

Regional Manifestation of the Widespread Disruption of Soil-Landscapes by the 4 kyr BP Impact-Linked Dust Event Using Pedo-Sedimentary Micro-Fabrics

Marie-Agnès Courty, Alex Crisci, Michel Fedoroff, Kliti Grice, Paul Greenwood, Michel Mermoux, David Smith, and Mark Thiemens

Abstract The co-occurrence of a sharp dust peak, low lake levels, forest reduction, and ice retreat at ca. 4-kyr BP throughout tropical Africa and West Asia have been widely explained as the effect of an abrupt climate change. The detailed study of soils and archaeological records provided evidence to re-interpret the 4 kyr BP dust event linked rather to the fallback of an impact-ejecta, but not climate change. Here we aim to further investigate the exceptional perturbation of the soil-landscapes widely initiated by the 4 kyr BP dust event. Results are based on soil data from the

Marie-Agnès Courty
UMR 5198, CNRS-IPH, Centre Européen de Recherches Préhistoriques,
Avenue Léon-Jean Grégory, 66720 Tautavel, France, e-mail: courty@tautavel.univ-perp.fr

Alex Crisci
CMTC INP, 1260 rue de la Piscine, BP 75, 38402 Saint Martin d'Hères, France

Michel Fedoroff
Laboratoriy of Electrochimie & Chimie analytique, ENSCP,
11 rue Pierre et Marie Curie, 75231 Paris, France

Kliti Grice
Department of Applied Chemistry, Curtin University of Technology,
GPO Box 1987, 6845 Perth, Australia

Paul Greenwood
Department of Applied Chemistry, Curtin University of Technology,
GPO Box 1987, 6845 Perth, Australia

Michel Mermoux
LEPMI-ENSEEG, Domaine Universitaire, BP75, 38042 Saint-Martin d'Hères, France

David Smith
Labority of LEME/Nanoanalysis, MNHN, Laboratoire de Minéralogie,
61 rue Buffon, 75005 Paris, France

Mark Thiemens
Department of Chemistry and Biochemistry, University of California,
San Diego 92093-0352, USA

S. Kapur et al. (eds.), *New Trends in Soil Micromorphology*,
© Springer-Verlag Berlin Heidelberg 2008

eastern Khabur basin (North-East Syria), the Vera Basin (Spain), and the lower Moche Valley (West Peru) compared with a new study at the reference site of Ebeon (West France). The quality of the 4 kyr BP dust signal and the related environmental records are investigated through a micromorphological study of pedo-sedimentary micro-fabrics combined with SEM-microprobe, mineralogical, and geochemical analyses.

In the four regions studied, the intact 4 kyr BP signal is identified as a discontinuous burnt soil surface with an exotic dust assemblage assigned to the distal fallout of an impact-ejecta. Its unusual two-fold micro-facies is interpreted as (1) flash heating due to pulverization of the hot ejecta cloud at the soil surface, and (2) high energy deflation caused by the impact-related air blast. Disruption of the soil surface is shown to have been rapidly followed by a major de-stabilisation of the soil cover. Local factors and regional settings have exerted a major control on the timing, duration, and magnitude of landscape disturbances. Studies showed how a high quality signal allows to discriminate the short-term severe landscape disturbances linked to the exceptional 4 kyr BP dust event from more gradual environmental changes triggered by climate shift at the same time.

Keywords Impact-ejecta · soil surface · dust · flash heating · air blast · soil destabilization

1 Introduction

High resolution records in ice, marine and lake cores show recurrent climate shifts of global extent, possibly with millennial scale cyclicity throughout the Holocene (Alley et al. 1993, Bond et al. 1999, Mayewski et al. 2004). The 4000 yr BP dust event first identified in soils of northern Mesopotamia (Weiss et al. 1993), later traced as a sharp dust peak in the Arabian Gulf (Cullen et al. 2000), in Huascaran ice cores from the Andes of northern Peru, and a thick dust layer in the Kilimanjaro ice (Thompson et al. 2002) is commonly assigned to the millennial cyclicity (de Menocal 2001). The inferred 300-year long drought was concluded to have forced the abandonment of agricultural settlements in northern Mesopotamia (de Menocal 2001, Weiss and Bradley 2001). Major societal disturbances apparently synchronous around this period of time throughout the Mediterranean basin and Asia were attributed to the severe decrease of annual precipitation, widespread cooling, forest removal, and drastic flow reduction of major rivers broadly coincident with the radical increase in airborne dust (Weiss and Bradley 2001, Wang et al. 2004). This short-lived shift would have been triggered by large scale changes in the ocean-atmosphere-vegetation boundary conditions, comparable in amplitude to the global aridification of the Younger Dryas (12 900-11 500 yr BP), (Cullen et al. 2000, Weiss 2001).

Recently, the intriguing dust peaks were suggested to represent regional manifestation of the 4 kyr BP pronounced environmental shift (Marchant and Hooghiemstra 2004).

Combined microscopic and geochemical investigations from a wide range of soil-sedimentary, and archaeological sequences in the Near East have allowed us to interpret the 4 kyr BP dust-event as the fallout of a distal impact-ejecta rather than a sudden drought (Courty 1998, 2001, 2003). Its exotic petrography with geochemical anomalies, and the related unusual pedo-sedimentary micro-fabrics helped us to link the fallout of the far-traveled dust with high temperature effects at the soil surface and violent deflation of surface horizons by high speed winds. We suggested the micro-debris fallout and the related manifestations to possibly trace effects of the distal dispersion of impact-ejecta.

Our objectives were to: (1) consolidate the widespread extent of the exceptional 4 kyr BP dust event based on soil sequences from contrasting geomorphic, geologic, and climatic settings; (2) compare the regional manifestations of the timing, spatial variability and controlling factors of soil disturbances, which started with the 4 kyr BP dust event; and (3) further elucidate the confusing resemblance of this exceptional dust event with a climate-triggered drought.

2 Materials and Methods

2.1 Site Selection

Soil anomalies at about 4 kyr BP were searched on soil sequences of the eastern Khabur basin (north-east Syria, Fig. 1a), the Vera basin (Spain, Fig. 1b) and the lower Moche valley (Peru, Fig. 1c). This focus is part of the interdisciplinary projects studying the evolution of the Holocene soil-landscapes with respect to anthropogenic forcing and climate changes (Courty 1994, Courty et al. 1995). A unique perturbation of soil-landscapes throughout the Holocene was identified from a stratigraphic discontinuity showing similar unusual burnt facies in all three regions.

Absolute radiometric dating and archaeological data were used to independently confirm attribution of the burnt facies to the 4 kyr BP event. At Ebeon (western France, Fig. 1d) the unique section exposed during the excavation of a Middle Neolithic ditch displays an unusual fired-strata dated at ca. 4 kyr BP, incompatible with a human activity but strongly resembling the 4 kyr BP burnt soil surface signal.

2.2 The Eastern Khabur Basin (North-East Syria)

The eastern Khabur basin (north-east Syria) forms a gently undulated flood plain within an endoreic alluvial basin under a semi-arid Mediterranean climate. Soil

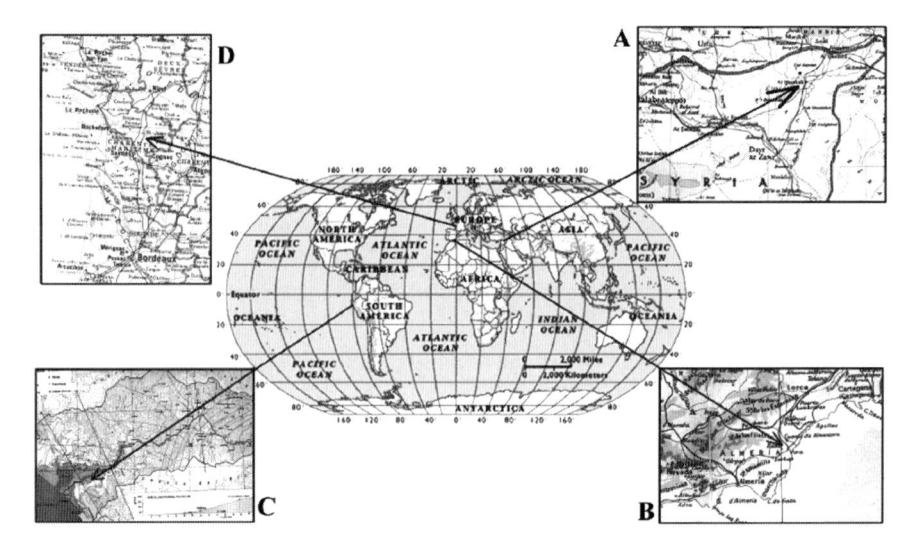

Fig. 1 Location of the studied regions: **(a)** Eastern Khabur basin (northeast Syria); **(b)** Vera basin (Spain); **(c)** lower Moche valley (Peru); **(d)** site of Chemin Saint-Jean at Authon-Ebeon (Charente Maritime, France)

landscapes have progressively aggraded since the last phase of channel incision at the end of the Pleistocene (Courty 1994) and supply of airborne dust from distant sources and nearby regions. The cumulative soil sequence offers a nearly continuous record of the Holocene climate fluctuations and a well-preserved signal of the 4 kyr BP burnt surface that can be spatially traced (Fig. 2a).

The related sequences display successive pedo-sedimentary units with characteristics of Calciorthids and Camborthids (Soil Survey Staff 1993). The weakly expressed horizon boundaries mostly result from gradual structure changes and abundance of pedogenic carbonates (Gaffie et al. 2001). Clayey to loamy-clay, homogeneous parent materials with 30–40% total carbonate content, have a clay mineral assemblage dominated by palygorskite and smectite.

2.3 The Vera Basin (Spain)

The Vera basin located in the most arid zones of southeast Spain is part of the transcurrent shear zone of Palomares. The combined effects of its tectonic instability, climate fluctuations and sea level changes have maintained a recurrent erosional crisis (Courty et al. 1995). Since the last phase of active tectonic uplift during the Middle Pleistocene (Ott d'Estevou et al. 1990, Zazo et al. 1993), the Vera basin has suffered from recurrent erosion. Devonian/Permian limestones and Triassic marls with deep gullied narrow basins (Fig. 3) are common. Soils are also highly eroded in the mid-altitude zones with extensive bad-lands developed from the Neogene marls and turbidites, and patches of Plio-Pleistocene conglomerates.

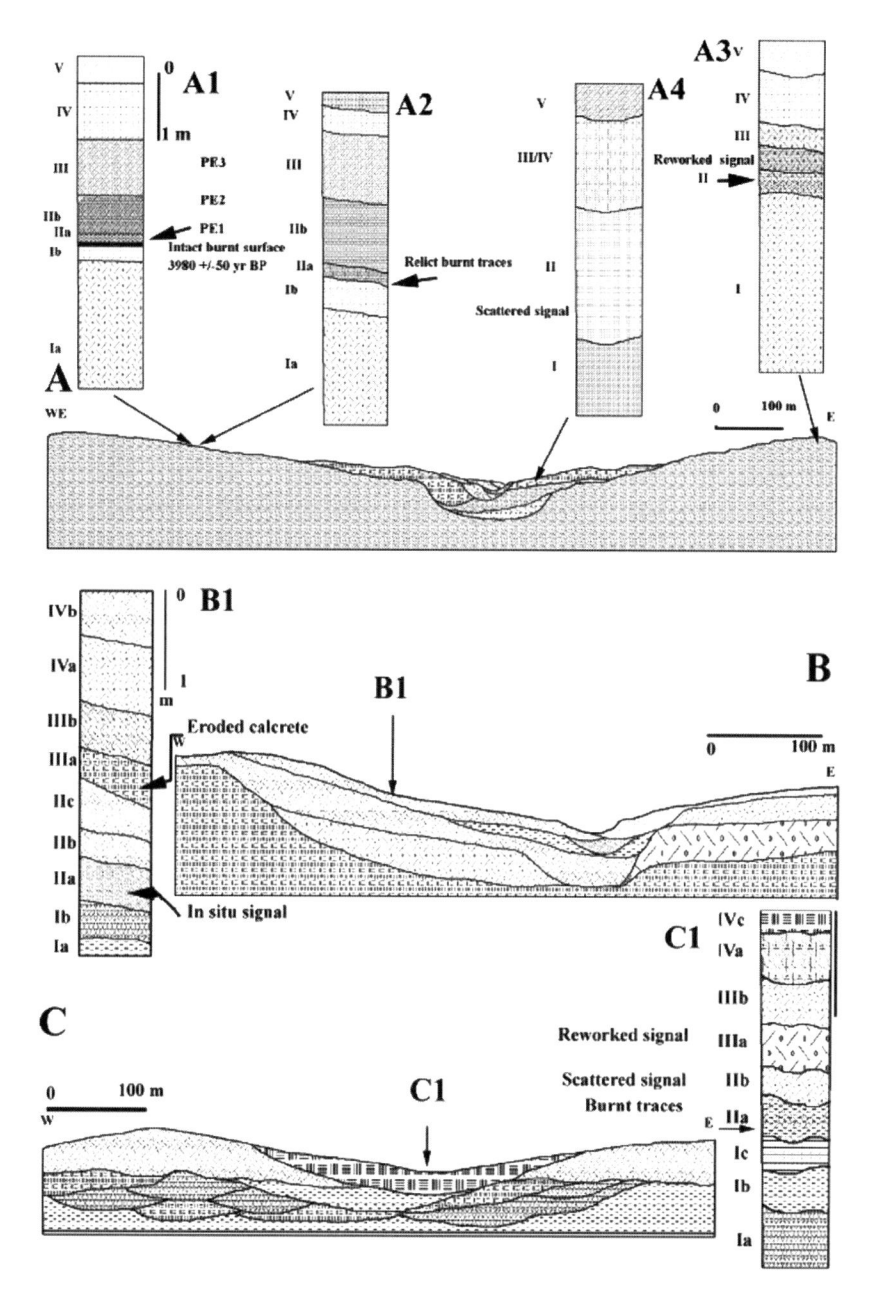

Fig. 2 Soil sequences in the eastern Khabur basin (north-east Syria). **(a)** undulated flood plain of Wadi Jarrah: **(a1)** intact 4 kyr BP signal **(a2)**, well preserved primary signal (IIa), fossilised by rapid burial (IIb); **(a3)** reworked primary signal (II); **(a4)** secondary signal (II) (II) in the lower flood plain. **(b)** Flood plain of wadi Rijlet Aaoueji: **(b1)** thick 4 kyr BP signal along the slope (II); post-event erosion of the calcrete (IIIa). **(c)** southern Radd basin: **(c1)** shift at 4 kyr BP (IIa) from alluvial discharge in a wide endoreic basin (Ia to Ic) to dust deposition in a closed depression (IIIa to IVc)

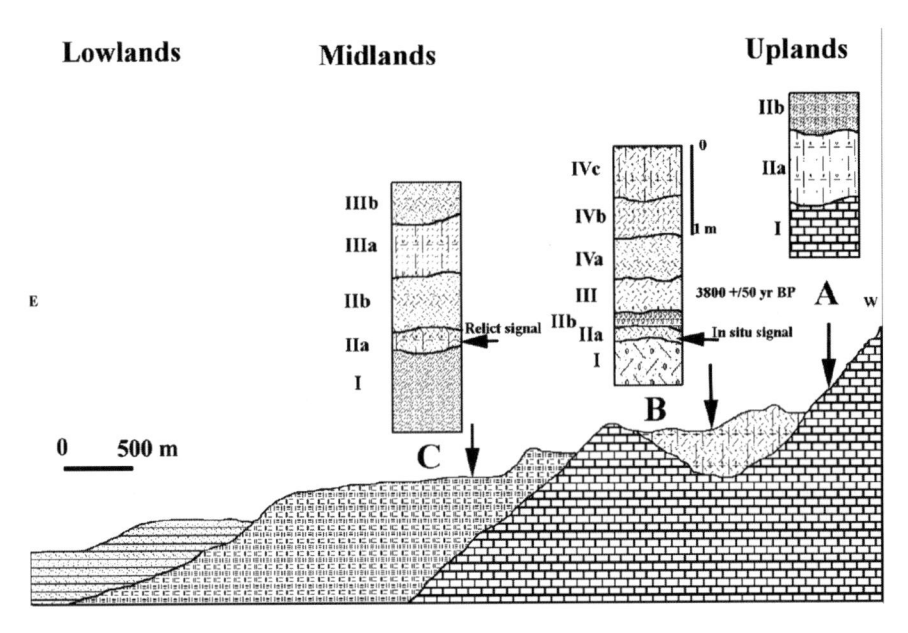

Fig. 3 Sections in the lower Aguas valley, Vera basin (Spain): **(a)** Highly eroded soil sequence in a dolina of the Permian limestone; **(b)** Rambla Añofli soil sequence with an intact 4 kyr BP burnt surface (IIa & IIb) sealed by the schistaceous mudflow (III); **(c)** Soil sequence on Miocene gypsiferous marls (I) showing a relict 4 kyr BP signal (IIa) preserved by erosion of the Gysiorthids (IIb)

The lowlands delineated by relicts of Pleistocene coarse-textured terraces, are presently flowed by three meandering channels, the Aguas, the Antas and the Almanzora. Tracing the 4 kyr BP fired strata in the lower alluvial plain was a delicate task due to the scarcity of natural exposures, the lack of reliable dating and the high amount of silting caused by re-activated erosion of the Neogene marls with modern agriculture.

2.4 The Lower Moche Valley (Peru)

The study area along the lower Moche valley belongs to the hyper-arid North Peruvian coast. In the flat littoral fringe, the Moche river is widely braided and weakly incised. The upper flood plain is blanketed by undulating sand dunes formed from deflation of the continental plateau during the late Pleistocene marine transgression,

The leveled pampas formed of complex sedimentary deposits of marine, eolian, colluvial-alluvial, and alluvial origin are delineated by Cretaceous granodiorite outcrops (Fig. 4). The Moche hydrographic regime is controlled by seasonal variations of heavy rainfall over the upper catchment basin in the Occidental Cordillera of the Andes at 4000 m altitude. Turbulent flooding during the recurrent El Niño episodes

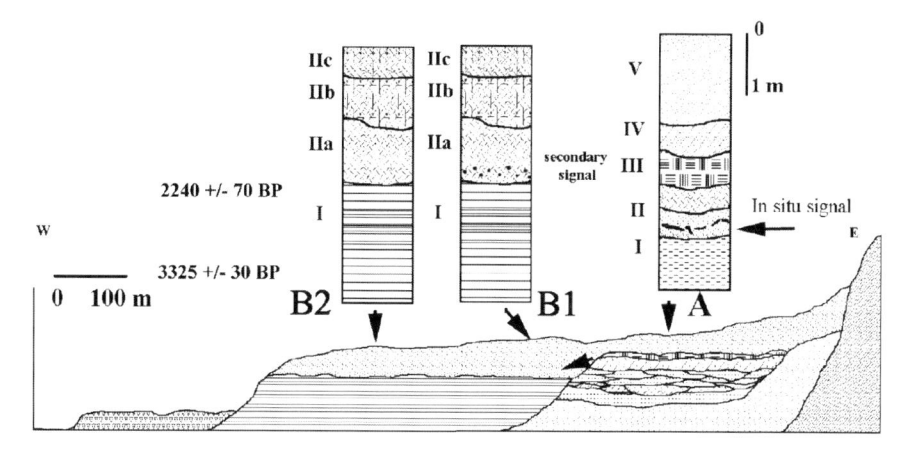

Fig. 4 (a) Soil sequences in the lower Moche valley (Peru). (I) Fluvisols on braided channel deposits, (II) Entisol on aeolian sand associated to the 4 kyr BP burnt soil surface, (III) the mudflow deposit. **(b)** Micro-stratified flood deposits with cyclical wild fires (I); B1: secondary 4 kyr BP signal eroded from the former river bank (IIa), not present in B2

leads to severe devastation of the lower flood plain, as illustrated by the major soil loss, following the last 1998 El Niño event.

2.5 The Neolithic Ditch Fill at Chemin Saint-Jean, Authon-Ebeon (Charente-Maritime, France)

The ditch dug into late Jurassic clay-rich limestone occupies a high position in the weakly undulated landscape. The Regosol (IUSS-WRB 2006) of about 50 cm thick shows only weak horizonation due to deep ploughing up to the underlying limestone. The distinct burnt layers contrast from the underlying and overlying filling units by their fine texture and low amount of coarse limestone fragments derived from the collapse of the unstable ditch sides (Fig. 5).

2.6 Analytical Procedure

The stratigraphical discontinuity of the 4 kyr BP event as previously defined in north-east Syria (Courty 1998, 2001) was detected in the field based on anomalies of colour, structure and texture, and patterns of burnt traces (Courty et al. 2005, 2006). The quality of the 4 kyr BP signal was then established from the compositional range, morphology and particle size distribution of the diagnostic micro-debris as given by successive test sampling on large amounts of bulk sediments collected along the stratigraphic discontinuities.

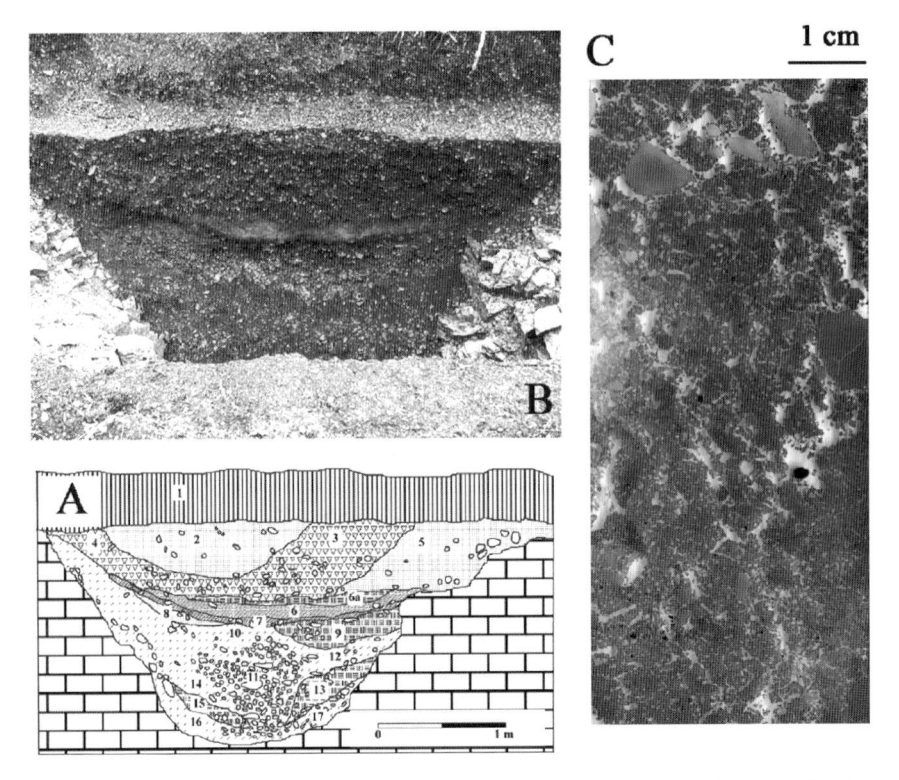

Fig. 5 Section in the Neolithic ditch fill at Chemin Saint-Jean, Authon-Ebeon (Charente-Maritime, France). (**a**) strata 17 to 8: collapse of the ditch talus; strata 6 & 7: anomalous units bearing the 4 kyr BP signal; strata 5 to 2: upper fill; strata 1: ploughed soil horizon. (**b**) dark brown strata in the field (7: FIa) overlaid by the reddish strata (6a and 6: FIb). (**c**) View in thin section of the homogeneous red strata (6) (see Fig. 8g for detailed view)

The spatial extent and micro-stratigraphical pattern of the intact 4 kyr BP signal were controlled through careful exposure of the host burnt surface. Undisturbed blocks and a series of refined bulk samples extracted from each micro-strata were collected throughout the exposed burnt surface. Continuous vertical columns of undisturbed samples were also collected at locations showing the most complete micro-stratigraphical succession. Thin sections were prepared from the blocks and studied under the petrographic microscope following the principles and terminology adapted from sedimentary petrography (Humbert 1972) and soil micromorphology (Bullock et al. 1985, Courty et al. 1989). A preliminary petrographical classification of the 4 kyr BP coarse fraction assemblage was established under the binocular microscope. XRD, isotope geochemistry, SEM/EDS, EPMA, TEM, GC-IR-MS and Raman spectrometry were used to characterize components on the 4 kyr BP signal. Thin sections were prepared from both selected areas of the large impregnated blocks and from a selection of extracted coarse grains for SEM and microprobe analyses.

3 Characterization of the 4 kyr BP Signal

3.1 Diagnostic Tracers of the 4 kyr BP by Petrographic Assemblage

In the four studied regions, a spatially irregular stratigraphic discontinuity with similar unusual burnt traces, in particular patches showing a completely incinerated surface horizon, is identified associated to a similar assemblage of exotic micro-debris (see Figs. 2a1, 3b, 4a, 5b), (Courty 2003, Courty et al. 2006). For the four situations encountered, attribution of these anomalies to the 4 kyr BP event was based on the common petrographic dust assemblage and age estimate (3800–4200 yr BP uncal.) depending upon the absolute radiometric dating and locally available archaeological data (Burnez et Fouéré 1999, Courty 2003, Courty et al. 1995, 2005).

In contrast from the local sediments, the exotic micro-debris predominantly consist of particles between 100 and 400 µm, with rare occurrence of larger grains up to 7 cm. Finer particles are also abundant but more difficult to isolate because of the dilution in the local background. Based on repeated sampling, the dust components are shown not to form a distinct stratum but to occur as discrete concentrations within a few cubic centimetres, with rapid variations at short spatial scales in the nature and abundance of the exotic particles (Courty et al. 2005). A similar assemblage from the different localities (Fig. 6) was obtained from water sieving of some ten kilograms of bulk sediments from the distinctive burnt surface.

It comprises a full range of spherules, droplets, teardrops and dumbbells that are often welded, glazed vesicular beads with an irregular shiny, sub-metallic, greyish yellow surface spread with spherules, and grey vesicular fragmented beads (Figs. 6a,b,c, and 7d). Angular fragments of white, whitish blue, dark brown and black, vesicular glass are common together with carbonate beads, often with metallic mounds lining the vesicles (Fig. 6e,g,h).

Vesicular beads showing partial devitrification and igneous-like coarse-textured re-crystallization are frequent (Fig. 6d). Black vesicular carbonaceous grains are common, and green carbonaceous materials associated to metallic deposits always occur as individual fibres, intricate vesicular fibres and angular fragments, together with metal-associated bright yellow and blue grains, translucent grains both as flakes and as blocky grains, and metallic particles present as platy ribbon, flakes, thin films, angular grains and spherules (Fig. 6f).

Various types of sulphates (Ca, Mg, Sr, Ba), and chlorides (K, Na) occur together with iron sulphide, phosphide, phosphate, and silicon-phosphate either as inclusions within the glassy debris (Fig. 6c) or as individual grains (Courty et al. 2005, 2006). The debris assemblage also comprises dumbbell millimetre-sized reddish brown to dark brown, carbonaceous aggregates with highly resistant, charred fine roots (Fig. 7a,b,c).

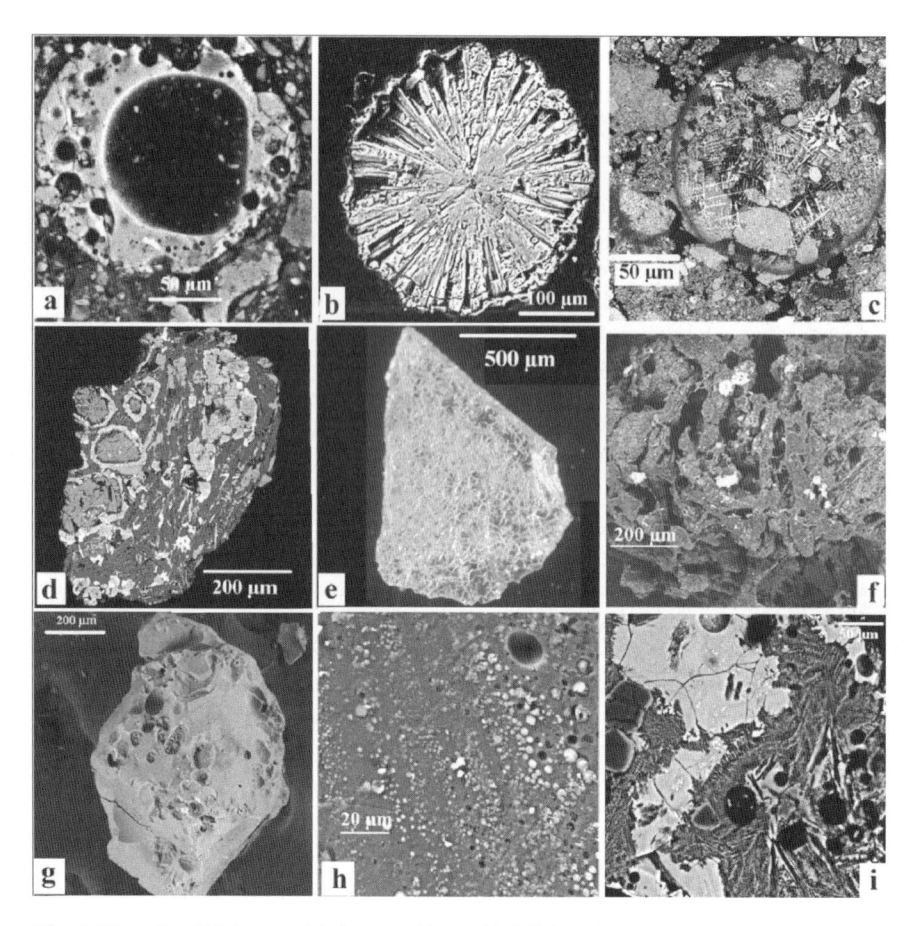

Fig. 6 The 4 kyr BP impact-debris assemblage: (**a**) Silicic spherical droplet. NE Syria, Tr.A1, IIa. SEM-BSE. (**b**) Heated re-crystallized foraminifera. NE Syria, Tr.A1, IIa. SEM-BSE. (**c**) KCl carbonaceous spherule. Moche valley, Peru, Tr. A, II. SEM-BSE. (**d**) Recrystallized igneous clast. NE Syria, Tr.A1, IIa. SEM-BSE. (**e**) C-rich silicic glass. Vera basin, Tr. B, IIa. (**f**) Metal-rich green vesicular fibrous carbonaceous glass (Ag and Fe-Cr-Ni). Ebeon, France, unit 7. SEM-BSE. (**g**) Whitish blue silicic glass. Ebeon, France, unit 7. SEM-BSE. (**h**) Enlarged view of (**g**) showing carbon-rich nickel phosphide splash. SEM-BSE. (**i**) Finely imbricated glass phases from rapid quenching of incompletely melted silicic and carbonate precursors. SEM-BSE

3.2 Anomalous Micro-Facies of the Buried Burnt Soil Surface

The most striking anomaly of the burnt soil surface associated to the exotic micro-debris is the succession of a dark stratum at the bottom overlain by a reddish layer (Fig. 5b). This reversed microstratification in contrast to the thermal record of wild fires or anthropogenic hearths (Courty et al. 1989) reveals the occurrence of two different micro-strata under the petrographic microscope with distinctive anomalous micro-facies (FIa and FIb) and related micro-debris (Courty et al. 2006).

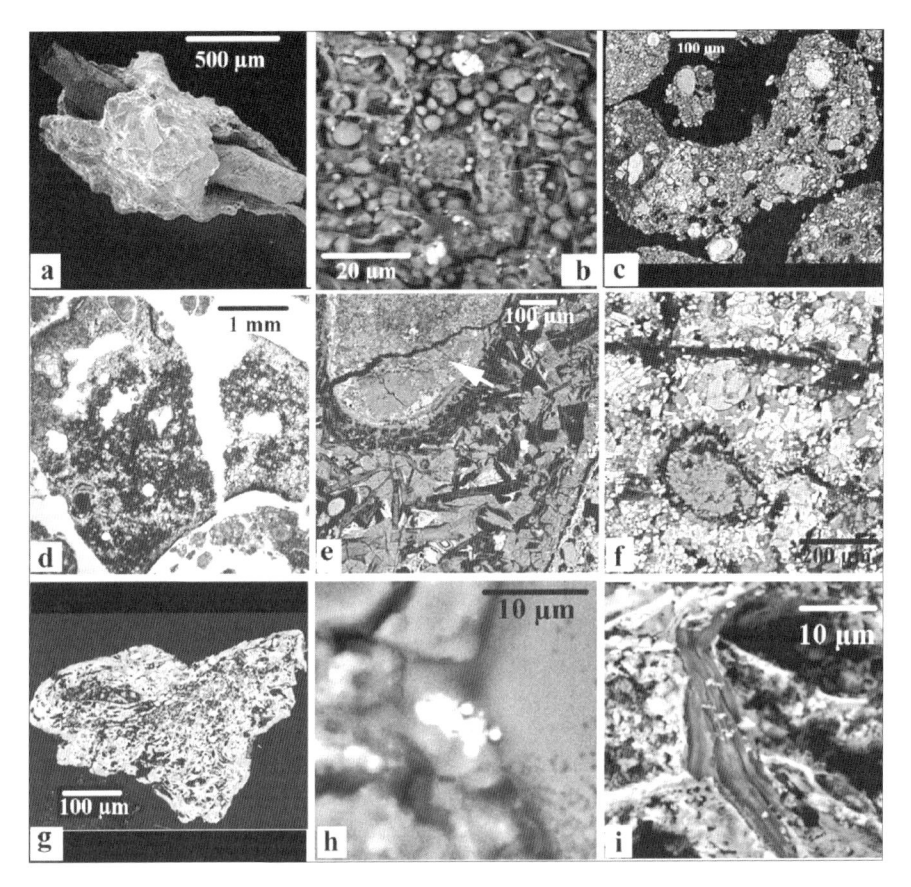

Fig. 7 Pulverisation of the hot ejecta cloud at the soil surface. (**a**) Charred fine root SEM-SE. Ebeon, France, unit 7. (**b**) Enlarged view showing pulverisation of carbonaceous micro-spherules, metal (bright particles) and glass micro-debris. SEM-BSE. Ebeon, France, unit 7. (**c**) Dumbbell carbonaceous aggregate. NE Syria, Tr.A1, IIa. SEM-BSE. (**d**) Vesicular glass bead. NE Syria, Tr.A1, IIa. SEM-BSE. (**e**) Enlarged view of (**d**): complex C-rich glass phases with a recrystallized CaCO3 clast (arrow). (**f**) finely cracked quartz and feldspar grains and the heat-transformed fine mass. (**g**) melted bone fragment showing injection of the metal-rich carbonaceous impact glass within the osteocytes. SEM-BSE. Ebeon, France, unit 7. (**h**) Silt-sized zircons-ZrSiO4– transformed into baddelyite-ZrO2 in the impact glass. SEM-BSE. Ebeon, France, unit 7. (**i**) Flow glass injected within the host soil materials. SEM-BSE. Ebeon, France, unit 7

The lower micro-facies (FIa) consists of a few millimetre to a few centimetre thick ashy unit formed of the incinerated original soil surface (Fig. 8b). The FIa on the undisturbed underlying horizon (Fig. 8a) displays an open packing of finely disrupted micro-aggregates with a darkened fine mass, associated to carbonised micro-rootlets, partially replaced by reddish iron oxides (Fig. 8c,d,h). In general, the birefringence of the fine mass in the burnt aggregates does not appear significantly modified as compared to the underlying weakly burnt aggregates.

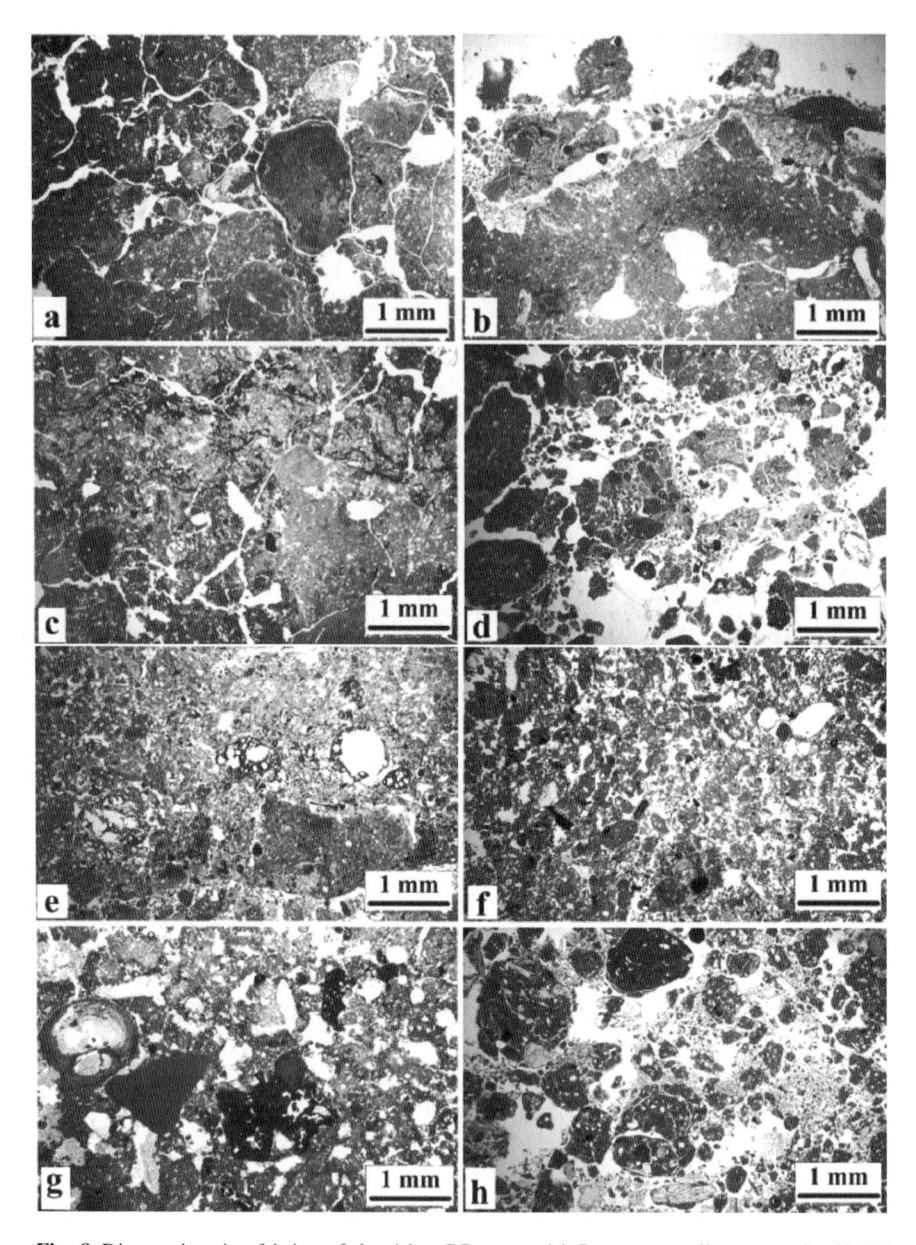

Fig. 8 Diagnostic microfabrics of the 4-kyr BP event. (**a**) Pre-event well structured soil. NE Syria, Tr.A1, Ia/Ib, PolM, PPL. (**b**) Calcinated vegetation, FIa microfacies. NE Syria, Tr.A1, IIa, PolM, PPL. (**c**) FIa microfacies: in situ carbonised rootlets. NE Syria, Tr.A1, IIa, MPol, PPL. (**d**) FIa microfacies: open packing of finely-disrupted heated micro-aggregates. NE Syria, Tr.A1, IIa, MPol, PPL. (**e**) FIa microfacies: impact glass beads within the host soil surface. NE Syria, Tr.A1, IIa, MPOl, PPL. (**f**) FIb microfacies: loose air-transported biogenic aggregates. NE Syria, Tr.A2, IIa, MPol, PPL. (**g**) FIb microfacies: air-transported biogenic aggregates, limestone clasts and burnt soil aggregates. Ebeon, France, unit 6 bottom, MPol, PPL. (**h**) FIa microfacies: loose heated micro-aggregates. Vera basin (Spain), Tr.B, IIa, MPol, PPL

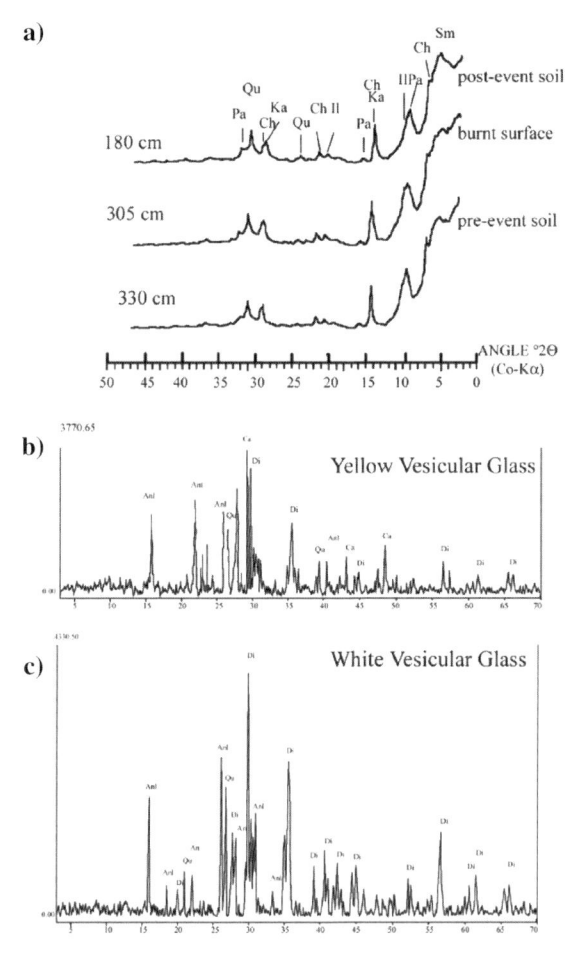

Fig. 9 **(a)** X-ray diffraction analysis, Leilan Tr. A1, NE Syria: moderate heating in the 4 kyr BP burnt surface (305 cm) as compared to the underlying soil horizon (330 cm) and the overlying soil unit (180 cm). **(b)** & **(c)** Mixing of high temperature recrystallized minerals and exogenous detrital minerals in the impact glass: diopside (Di), analcime (Anl), anorthite (Ant), calcite (Ca) and quartz (Qu)

Evidence for an overall moderate heating is confirmed by X-ray diffraction that does not show the irreversible change of the clay fraction (Fig. 9a). The most exotic micro-debris occur either as randomly distributed individual grains in the packing voids, as minute particles stuck to the micro-aggregates and within the carbonised roots (Figs. 7a,i, and 8d,h). Local concentrations of vesicular glass bodies are associated with high temperature transformation of the embedding matrix (Figs. 7d, 8e, 12c), as indicated by the development of a poorly crystalline phase of the clay-rich and carbonate-rich fine mass. This is confirmed by the transformation into diopside of the original alumino-silicate as shown by X-ray diffraction (Fig. 9b,c), of heated sample at around 700–850°C temperatures (Heimann 1982).

Fine vesicularity in the host materials attached to the glazed beads associated to the destruction of the poorly crystalline carbonate attests to the in situ carbonate decomposition by the loss of CO_2, due to local heating at temperatures around 700°C (Fig. 7e). Morphological and geochemical evidence for the thermal alteration of the host materials at the contact with the exotic components indicate temperature elevation up to ca. 1200°C (Fig. 7e,f). In contrast, the presence of heat-decomposed silt-sized zircons within the glassy phases of the glazed beads (Fig. 7h) attests for their formation at temperatures from 1400 to 2000°C (Courty et al. 2006) based on experiments (Pavlik et al. 2001). The similar compositional range of the high temperature phases between the vesicular glass bodies from distant sites suggests an exotic origin from a common source (Table 1).

The Ca-rich flow glass pattern with micro-scale compositional heterogeneities (Table 1) resembles a rare type of impactite glass (See et al. 1998, Graup 1999). The coarse glass bodies would have formed within an impact ejecta cloud possibly by rapid quenching of incompletely melted source materials (Kieffer 1977, Kring and Durda 2002). Penetration within fine pores of the host materials with interaction limited to an interfacial layer suggests final pulverisation of highly viscous hot liquid materials at the soil surface.

The darkened micro-aggregates locally display heat-induced amorphisation and abundant glassy micron-sized exogenous minerals similar to the coarse vesicular bodies (Fig. 7i). Glassy particles associated to clusters of metal-rich carbonaceous micro-spherules also occur around the charred fine roots, and within the calcinated layer (Fig. 7a,b,c). These suggest penetration into the surface soils of a carbonaceous volatile-rich phase associated to the hot mineral debris, possibly in the form of condensed droplets (Fig. 7g) through rapid dehydration (Hubbert et al. 2006). This would have resulted in a decreased wetting-ability of micro-aggregates, thus explaining their resistance to water treatment (Fig. 7a).

The upper microfacies (FIb) consists of loosely-packed reddish brown to dark brown dumbbell-shaped micro-aggregates. They are morphologically similar to the darkened ones present in the underlying facies, and also contain micro-particles diagnostic of the 4 kyr BP assemblage. Although their shape is typical of a biogenic origin, their random packing and lack of linkage to biological channels indicate a transport from their primary position (Figs. 7c and 8f). The direct juxtaposition of domains showing various intensities of thermal alteration indicates mixing of the burnt surface materials.

The common coarse-sized vesicular exotic debris typical of the 4 kyr BP assemblage occur as distinct particles without signs of interaction with the host materials, whereas the other categories of exotic debris show evidence of surface abrasion by aeolian transport (Fig. 8g). Rock fragments of millimetre to centimetre size derived from the surroundings and also from more distant regions are common in all the dust assemblages studied. The upper microfacies (FIb) is, thus, interpreted to result from deflation by high energy hot air flow of the burnt soil surface bearing the pulverized hot debris. The direct stratigraphic continuity between the two microfacies and the lack of turbation of the incinerated burnt layer suggests the flash heating to have immediately been followed by violent wind swirls. The explosive

Table 1 Major element comparison.XRF data in wt% on bulk devritrified glass, bulk glasses and host soils

	Bulk devritrified glasses			Bulk glasses		Glass phases						Host soils		
	RGVG1 LHI95B	RGVG1 AID95C	RGVG1 AID95B	YVG1 PE00A	WVG1 EB00A	dbrG1 EBG1	ybrG1 EBG2	yg1 EBG2	yrG1 EBG2	wG1 EBG3	bWG1 EBG4	H8 LEI95	H8 AID95	H8 PE00
SiO_2	47.58	46.86	47.49	67.05	67.05	53.65	62.57	66.01	68.04	74.68	83.75	28.54	19.92	39.35
TiO_2	0.89	0.87	2.74	0.8	0.69	0.24	0.73	0.16	0.59	0.68	0.71	0.46	0.31	0.2
Al_2O_3	12.42	14.31	15.28	11.48	9.66	15.35	12.41	5.63	9.66	8.64	5.53	7.46	4.97	7.88
Fe_2O_3	6.76	7.12	12.41	4.7	2.92	10.13	2.86	2.22	7.38	3.79	1.9	3.87	2.61	6.71
MuO	0.12	0.1	0.16	0.09	0.07	<dl	<dl	<dl	<dl	<dl	<dl	0.04	0.03	0.28
MgO	4.9	5.43	6.51	1.91	1.8	1.01	2.46	9.13	0.47	1.09	2.17	2.9	2.21	1.18
CaO	17.73	18.81	10.39	4.44	10.97	17.63	11.5	11.3	3.36	2.35	1.67	26.57	35.22	22.29
Na_2O	1.66	1.16	3.1	4.82	1.08	1.08	2.83	1.66	2.4	1.68	1.74	0.12	<dl	1.36
K_2O	1.45	1.63	1.17	1.35	4.08	0.04	4.03	3.44	6.35	6.84	2.33	0.89	0.62	1.09
P_2O_5	0.52	0.41	0.41	0.97	0.66	0.79	0.38	0.38	1.72	0.11	0.1	0.09	0.08	0.46
LOI	5.14	3.13	0.26	2.25	0.89	<dl	<dl	<dl	<dl	<dl	<dl	29.03	33.8	19.08
Total	99.17	99.83	99.92	99.86	99.87	99.92	99.77	99.93	99.97	99.86	99.9	99.97	99.77	99.88

fragmentation of coarse ejecta masses could have triggered the hot air turbulence, thus producing at a small scale the expected effects of impact air blast (Kring 1997, Kring and Durda 2002).

3.3 Spatial Variability of the Primary 4 kyr BP Signal and Secondary Disturbances

The in situ 4 kyr BP burnt soil surface occurs as discrete patches over a few square meters, laterally merging to scattered burnt traces. The FIa/FIb micro-facies succession appears as relict micro-fabrics in the form of discrete carbonaceous micro-aggregates and heated biogenic aggregates (Fig. 7a,c). Compared to the complete two-fold burnt facies, the 4 kyr BP exotic particles are similar, but less abundant. The fresh aspect of the exotic debris suggests an intact 4 kyr BP signal, with minimal bias by post-depositional disturbances. Local variability of the dust assemblage reflects the erratic dispersion of the impact-debris, possibly linked to heterogeneities in the initial ejecta melt (Kring and Durda 2002).

The correlation between morphological changes of the 4 kyr BP exotic assemblage and the drastic alteration of the associated micro-fabrics are expressed by two types of distortion of the original signal by post-depositional processes. These refer to distinctive palaeo-environmental implications.

The first type displays a low amount of diagnostic particles, even when sorting a larger volume of soils, a lack of a continuum of particles from coarse to very fine sizes, and no stratigraphical discontinuity (e.g. Fig. 2, Tr. A3: II; Fig. 3, Tr. C: IIa). The rather fresh aspect of the grains, as well as the recognition in thin sections of the rare heated micro-aggregates, and the evidence of the intense bioturbation suggest the occurrence of an in situ reworking mostly by biological activity along the soil development over a short period of time.

The second type shows markers of the 4 kyr BP event in later deposits, in particular coarse vesicular glassy grains (Fig. 4, Tr. B1: IIa and Fig. 12g). Reworking is indicated by the absence of the diagnostic fine particles in the surrounding matrix and the lack of the distinctive micro-fabrics formed by the interaction of the intruding flow glass within the host materials. Considering their fragility, the source deposits with a nearly intact signal would occur in the local surroundings.

4 Recognition of Soil-Landscape Disturbances Following the 4 kyr BP Event

Although its characteristics vary according to the local climate and geomorphic factors, the impact-induced soil destabilisation appears as a unique combination of processes in contrast to the range of soil changes encountered during the previous six thousand years (Courty 1994, Courty et al. 1995).

4.1 The Post-Event Soil Record in the Eastern Khabur Basin (North-East Syria)

In the loamy lowlands of the N-E Syrian flood plain, the post-event pedo-sedimentary unit displays a spongy porosity typical of structural collapse, weak bioturbation, slight depletion of the calcitic fine mass, and lack of secondary carbonates. Coarse slaking crusts and dusty clay coatings rich in fine silt-sized clusters of carbonaceous spheroids, mineral spherules, and glass shards are common (Fig. 10a,c). Abraded heterogeneous particles typical of the 4 kyr BP assemblage are abundant. This high-energy erosion of the burnt soil surface indicates exceptional heavy rains immediately following the impact (Fig. 2, Tr. A1: IIa).

Water-transport and accumulation in micro-depressions of the exotic micro-debris suggests the rainfall increase to not have lasted more than a few months. The anomalous carbonate depletion not expected to occur under an arid Mediterranean climate expresses significant acidification, possibly triggered by the impact-induced changes in atmospheric chemistry (Toon et al. 1997). Acid rain would possibly result in from the production of nitrous oxides and other chemicals due to the injection in the upper atmosphere of carbon-rich aerosols (Toon et al. 1997). The incorporation of black carbon into the clay coatings (Fig. 10b) also suggests acidification by the carbonaceous aerosols remaining in suspension at great altitude before being washed by rainwater.

In the overlying unit, the upper part of the pedo-sedimentary horizon displays a gradual change to a fine textured deposit (Fig. 2, Tr. A1: IIb) with a vughy to fissural porosity, along with coarse-textured intercalations integrated to the dense fine mass (Fig. 10d). This trend attests for an overall decrease of heavy rainstorms, the persistence of fine rains favouring silting and water saturation, and of low evaporation as indicated by the lack of secondary carbonates.

Just above the marked decrease of coarse textural features, the excremental fabric with channel-like porosity and the fine-textured homogeneous micro-fabric indicate recurrent windstorms and persistent low evaporation with the re-establishment of a seasonal contrast (Fig. 10e).

In the overlying pedo-sedimentary unit, the decrease of silting, the development of a total excremental micro-fabric and re-activation of the carbonate redistribution (Fig. 2, Tr. A1: IV) indicate progressive stabilisation of soil-landscapes under a marked seasonal contrast, broadly similar to the pre-event conditions (Fig. 10h). Based on the chronological data, the impact-linked climate anomaly did not last more than one hundred years.

In shallow basins delineated by calcrete, the pedo-sedimentary layers directly linked to the 4 kyr BP event are sealed by lenses of weakly sorted, loose carbonate nodules (Fig. 10). The sheet-flow erosion of the calcrete that had remained stable since the Late Pleistocene attests for the exceptional violence of the runoff generated by the heavy rain showers immediately following the 4 kyr BP event (Fig. 2, Tr. B1: IIIa).

In the more arid areas, the 4 kyr BP event marks the contact between six thousand years of regular flooding and the establishment of endoreism. The few metres

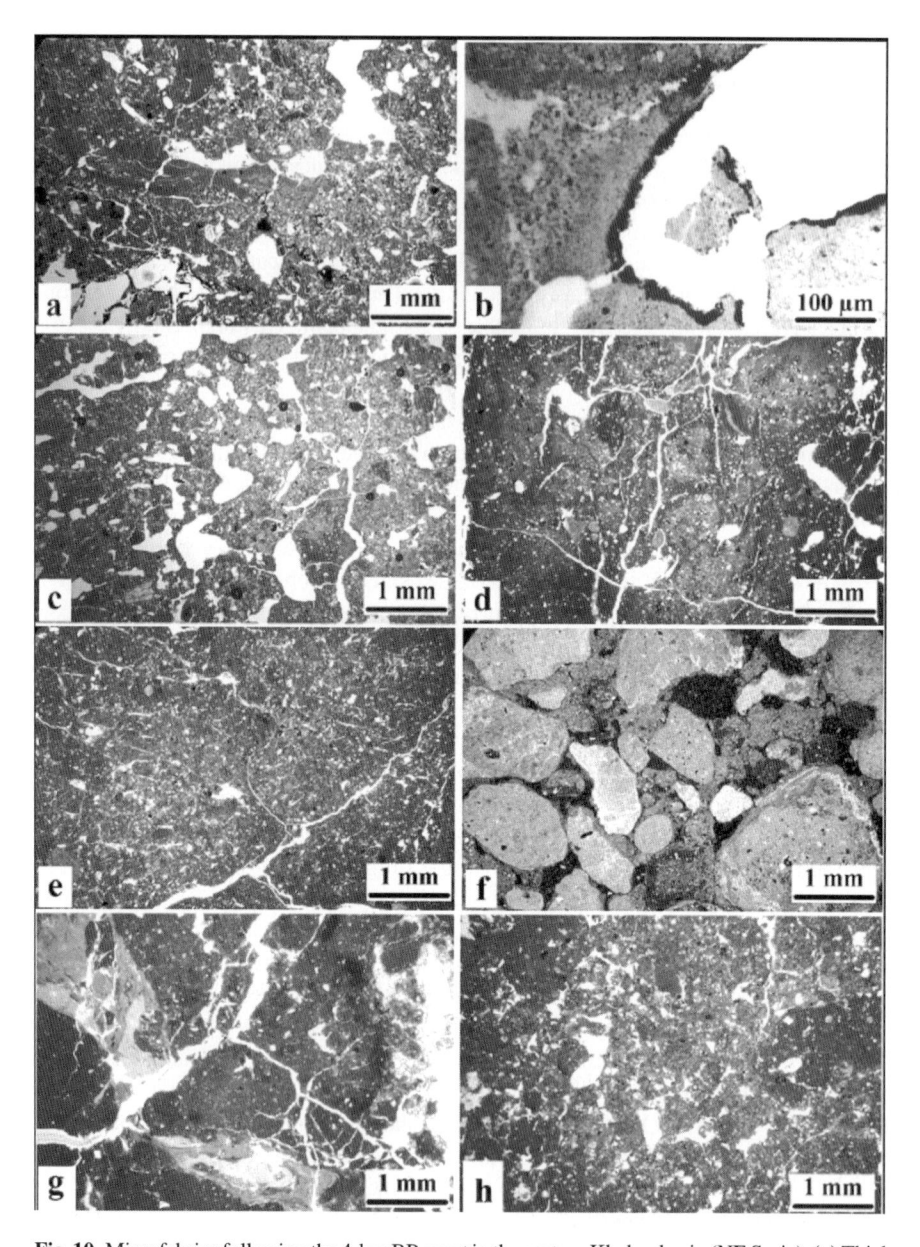

Fig. 10 Microfabrics following the 4-kyr BP event in the eastern Khabur basin (NE Syria). (**a**) Thick coarse textured slaking crusts. Tr. A1, IIb bottom, MPol, PPL. (**b**) Coarse clay coatings with rain-washed black carbon. Tr.A1, IIb bottom, MPol, PPL. (**c**) Spongy porosity, coarse clay coatings and intercalations. Tr.A1, IIb bottom, MPol, PPL. (**d**) Fissured to vughy microstructure with abundant intercalation and slaking crusts. Tr.A1, IIb bottom, MPol, PPL. (**e**) Fine textured micro-fabric. Tr.A1, IIb bottom, MPol, PPL. (**f**) Water-transported calcrete (see Fig. 2, tr. B1, IIIa). (**g**) Unusual thick clay coatings (see Fig. 2, tr. C1, IIb). (**h**) Well-developed excremental microfabric and diffuse calcitic impregnation. Tr.A1, III, MPol, PPL

thick succession of weakly bioturbated fine textured Entisols developed on the flood deposits with diffuse calcitic coatings merge to a fine-loamy Natraqualf with abundant textural features (Fig. 2, Tr. C1: IIa to IVc; Fig. 10g). Sudden salt accumulation well-expressed by the marked pH increase from 8.4 in the lower Entisols to 9.7 in the upper Natraqualf reflects the sudden disorganisation of drainage, most probably due to the massive re-mobilisation of dust with the maintenance of exceptional air turbulence.

4.2 The Post-Event Soil Record in the Vera Basin (Spain)

In the Vera basin, the filling of shallow dry valleys in the lower uplands (Fig. 3, Tr. B) above the 4 kyr BP burnt soil surface displays a heterogeneous micro-fabric. It comprises loosely-packed, heated reddish brown aggregates similar to the underlying unit, with disorganized earthworm pellets, abundant limestone fragments derived from the close outcrops, fragmented coarse-textured slaking crusts, and reworked pedogenic calcitic nodules (Fig. 11b).

The characteristics of the reworked pedogenic carbonates indicate an origin from the previously-developed Calciorthids and Camborthids formed on the uplands during the Holocene climatic optimum (Fig. 11a) at a period when the entire soil-landscape was broadly stable (Courty et al. 1995). The post-event pedo-sedimentary unit (Fig. 3, Tr. B: IIb) results from the severe erosion by high-energy runoff of the soil surface that recorded the 4 kyr BP event. Above, the sharp contact with a massive unit formed of fine-textured schistaceous sediments, with a flow structure and a vughy porosity, attests for liquefaction of the Precambrian schists by an exceptional mudflow immediately following the 4 kyr BP event (Fig. 11b,c). This severe mass movement of previously stable outcrops (Fig. 11a,b), attests for the considerable soil loss in the uplands due to the 4 kyr BP event (Fig. 3, Tr. B: IIb). The severe soil destabilisation is traced along the slopes of the midlands, down to the upper flood plain of the Aguas, although undisputable traces of the 4 kyr BP burnt surface are rare (Fig. 3, Tr. C: IIa).

In the overlying pedo-sedimentary units, the coarse texture, the micro-fabric heterogeneity, the poorly-developed structure together with the occurrence of slaking crusts attest for soil landscape instability up to the present time (Fig. 3, Tr. B: III to IVc; Fig. 11d). In contrast to the overall stability during the few thousand years, the severe soil loss linked to the 4 kyr BP event seems to have induced irreversible geomorphic changes in this highly dissected landscape.

The lack of interaction between the burnt soil layer and the overlying mudflow unit suggests a significant reduction of moisture retention of the heated soil surface, possibly due to the formation of hydrophobic compounds under fire (Almendros et al. 1990, Giovannini et al. 1983, Hubbert et al. 2006). The temporary decrease of soil wettability might have increased the risk of soil erosion as observed under wildfire conditions (Scott and Van Wyk 1990).

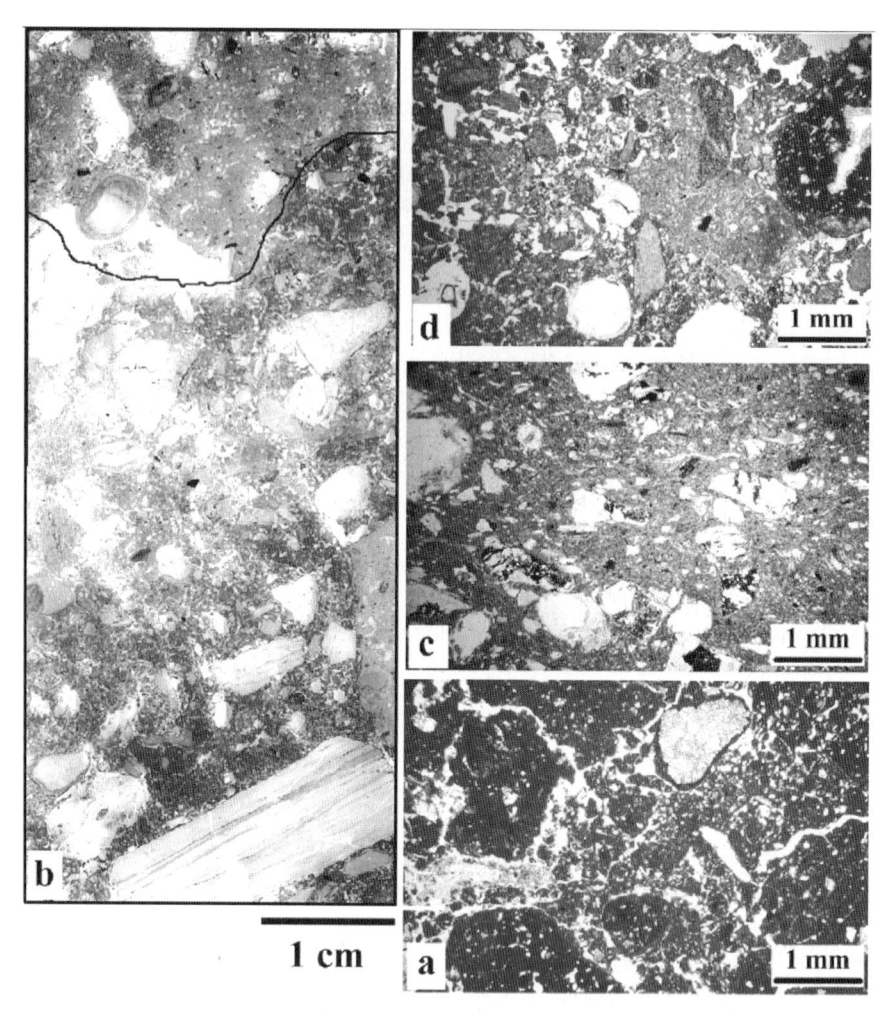

Fig. 11 Rambla Añofli, lower Aguas valley (Vera basin, Spain). Microfabrics related to the 4 kyr BP event (See Fig. 3, tr. B.). (**a**) well structured pre-event stable soil. Tr.B, I, MPol, PPL. (**b**) Sharp contact between the 4 kyr BP eroded surface soils and the subsequent schistaceous mudflow. Tr.B, IIb/III, MPol, PPL. (**c**) The fine textured schistaceous mudflow. (**d**) Limestone clasts, coarse textured slaking crusts and weakly developed excremental fabrics indicating maintenance of soil instability following the 4-kyr BP event. Tr.B, IVb, MPol, PPL

4.3 The Post-Event Soil Record in the Lower Moche Valley (Perou)

In the Moche valley, the burnt soil surface with the 4 kyr BP signal occurs within a thin blanket of structureless fine sand derived from aeolian reworking of the surrounding sand dunes (Fig. 4, Tr. A: II; Fig. 12b). The episode of sand remobilisation

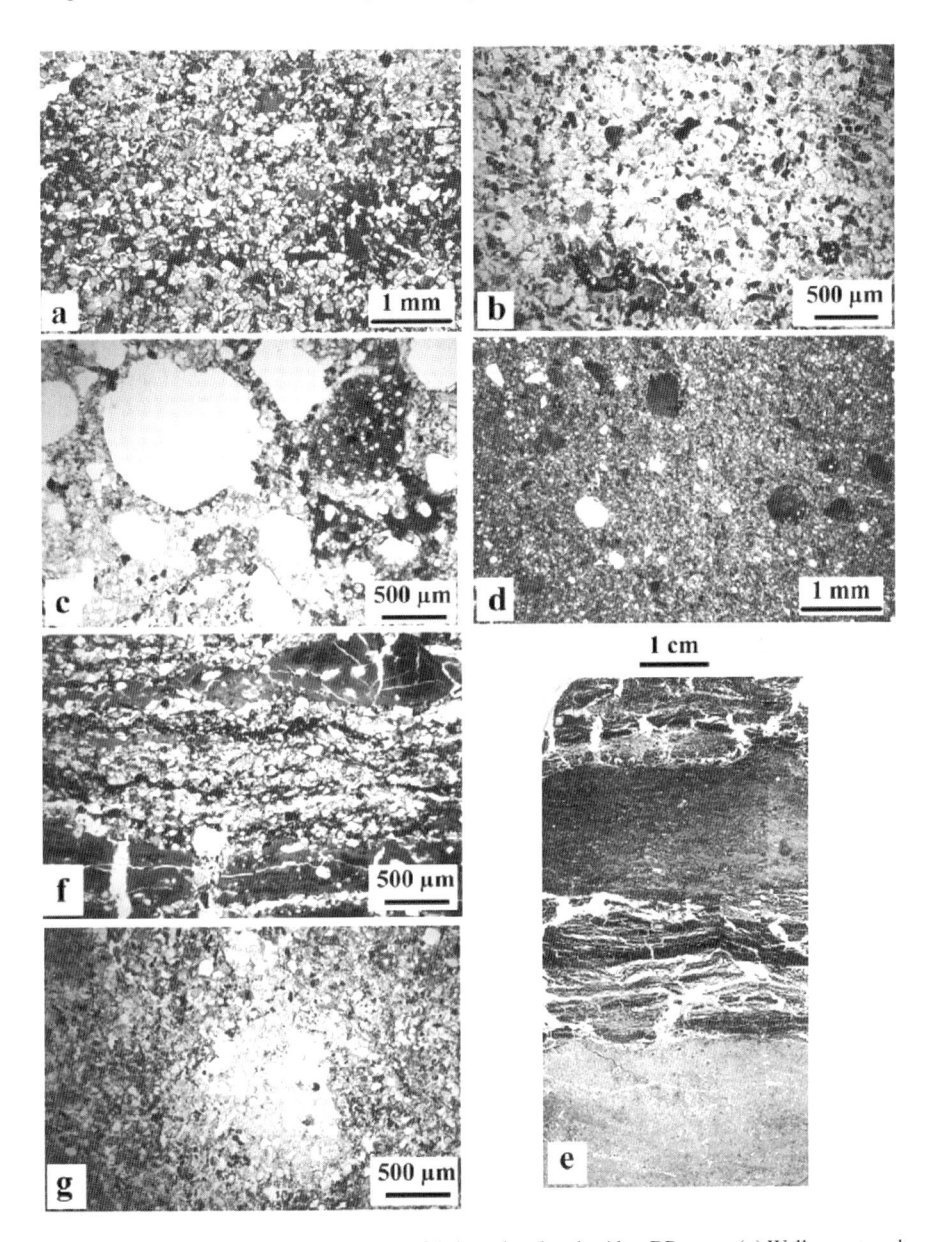

Fig. 12 Lower Moche valley (Peru). Microfabrics related to the 4 kyr BP event. (**a**) Well-structured pre-event soil with diffuse calcitic impregnation, sesquioxide hypocoatings. Tr. A, I, MPol, PPL. (**b**) Relict 4 kyr BP burnt soil surface within aeolian sands. Tr. A, II, MPol, PPL. (**c**) Impact glass beads within the heated host soil surface. Tr. A, II, MPol, PPL. (**d**) Massive fine textured flooded unit. Tr. A, III, MPol, PPL. (**e**) Finely laminated burnt organo-mineral deposits. Tr. B1, I, MPol, PPL. (**f**) Detailed view of (**e**) the clay and charred layers. (**g**) Impact glass beads in secondary position. Tr. B1, IIa bottom, MPol, PPL

is, however, not distinctive as compared to the recurrent instability of sand dunes recorded throughout the Holocene, linked to the inherent fragility of the coastal ecosystem under a hyper-arid climate.

The first and major environmental disturbance (Fig. 4, Tr. A: III; Fig. 12d) is recorded by the sharp depositional contact over the structureless fine sand of a massive fine-textured unit resulting from extensive flooding. This unit contrasts with the previous torrential discharge in braided channels of coarse sands, abundant gravels, and pebbles of local origin (Fig. 4, Tr. A: II; Fig. 12a). In the overlying massive clay-rich unit the abundance of illitic clay with a silt fraction derived from schistic materials indicates a substantial contribution from far-traveled components.

The resemblance of the clay-rich unit to the upstream valley deposits in the Andes at an altitude of about 4000 m incites to invoke a significant rainfall increase in the upper catchment valleys, immediately following the 4 kyr BP event. The related downslope mass movement of the soil caused the development of exceptional floods in the lower valleys, along with the erosion of the landscapes that had remained stable for more than five thousand years.

The following 1500 years (approx. 3700–2200 yr uncal. BP) were dominated by cyclical episodes of high energy alluvial reworking of the local sands and low energy regular flooding with deposition of micro-stratified fine textured alluvium (Fig. 4, Tr. B: I).

The micro-stratified sequence shows the alternation of the finely laminated sandy silt, deposited by gentle flooding, and finely laminated organo-mineral deposits rich in charred remains and burnt aggregates (Fig. 12e,f). This evolution suggests the maintenance throughout the year of the marshy conditions in the lower valley allowed by regular flooding from the upstream basin, episodically interrupted by wildfires during recurrent drought. After this episode of geomorphic stability, a new phase of channel incision eroded most of the previously formed flood plain when the torrential alluvial regime with recurrent exceptional El Niño episodes was re-established (Fig. 4).

Our interpretation of the existence of more humid conditions in the lower Moche valley at around 4 kyr BP seems to be in agreement with the increased precipitation known to have occurred from about 4000 yr BP throughout South America as indicated by the higher lake levels (Marchant and Hooghiemstra 2004).

5 Discussion

The synchronous occurrence in distant regions of the unique 4 kyr BP exotic assemblage with similar anomalous soil micro-fabrics suggests the possibility of a common trigger with widespread effects. The glass components have no equivalent in volcanic or anthropogenic by-products but have characteristics resembling impact glass.

In the four sites studied, mixing of the exotic micro-debris with remobilised dust derived from the local soils attests for violent air turbulence directly linked to the fallout of the impact-ejecta. The heterogeneity of the impact-products dispersion explains the spatial variability at local to regional scales of the dust-accumulation and the related micro-fabrics. The timing of the dust event related to the impact-pulverisation process at the ground surface appears abrupt and rather short from the succession of the pedo-sedimentary micro-fabrics, as predicted by models (Kring and Durda 2002). The maximum phase of dust remobilisation occurred at the paroxysm of the 4 kyr BP event and persisted for some years due to the fragility of the soil-landscapes before the regeneration of the soil cover was completed. The occurrence of glass shards typical of the 4 kyr BP exotic assemblage (Cullen et al. 2000) confirms that the spike from core M5–422 in the Arabian Gulf can be assigned to the distinctive 4 kyr BP dust. The absence of a similar dust spike in core 905 offshore near Somalia, considered to provide a high resolution record of dust flux during the Holocene (Jung et al. 2004), would support a spatially variable dust re-mobilisation.

The regional manifestation of the dust event would possibly depend upon the availability of the dust sources over the destabilised lands, the directions of the windstorm tracks from land to ocean, as well as spatial variability of the impact-ejecta dispersion. The misinterpretation of the overlapping radiometric dates for core M5–422 would explain how a real short-term dust event was transformed by Cullen et al. (2000) into an erroneous 300 years long mega-drought. In addition, the lack of detailed characterization of the dust assemblage does not exclude reworking by bottom currents of the original 4 kyr BP micro-debris. Vertical dispersion of the micro-debris would be easily confused with a dust event of long duration, unless the integrity of the dust record is securely established.

The sequence of pedo-sedimentary events following the 4 kyr BP abrupt event expresses: (i) direct effects of the impact-ejecta delivery processes, in particular the burning, the air blast and the atmospheric dust loading and (ii) indirect consequences of the dust-fallout-related disruption of soil-landscape stability.

The important variations in intensity and duration of the environmental disturbances between the three regions studied reflect their intrinsic resilience. Therefore, the sequence of pedo-sedimentary events identified in the three regions studied does not show at around 4 kyr BP any other environmental change than the ones initially triggered by the 4 kyr BP impact-ejecta event.

In contrast to the widely accepted occurrence of a rapid climate change during the 4200–3800 yr BP time interval (Mayewski et al. 2004), we propose that the widespread pronounced environmental changes recorded at about 4 kyr BP correspond to the direct and indirect consequences of the 4 kyr BP impact-ejecta fallback. This distinctive event might have exerted an influence on the global climate, in particular through injection in the upper atmosphere of volatile components and carbon-rich aerosols, known to induce cloud condensation nuclei effects and albedo changes (Toon et al. 1997).

6 Conclusions

This comparative study of the 4 kyr BP event in different regions demonstrates that soil micro-fabrics are reliable archives to reconstruct the timing of environmental perturbations induced over wide areas. Similar to the leading role played by high resolution records from deep sea, lake, and ice cores, soil sequences where rapid burial has allowed the exceptional preservation of a high quality signal appear to be of major interest. These peculiar situations offer a unique opportunity to discriminate the instantaneous soil reactions directly triggered by unusual phenomena such as a cosmic collision or any other internal cause, from the more long-term indirect effects on soil landscapes.

As illustrated by data from NE Syria, the possibility to capture high quality signals, in a wide diversity of local conditions in cumulative soil landscapes, is essential to appreciate both the spatial diversity inherent in the involved phenomena, and the spatial variability of soil responses with respect to the local factors. These ideal situations provide interpretative keys to approach the more fragmentary records available from the great majority of regions. These relict signals would be easily confused with human-induced perturbations (i.e. the 4 kyr BP signal in West France), or with a gradual climate change (i.e. the 4 kyr BP signal in the Moche valley) or with a recurrent erosional crisis (i.e. the 4 kyr BP signal in the Vera basin).

Although of different quality, data retrieved from a large variety of regions should help to finally understand a global event throughout its spatial complexity.

Acknowledgements Our deep thanks are addressed to those involved in this project who contributed to solving the puzzle of the 4 kyr BP event: Hélène Cachier, Tyrone Daulton, François Guichard, Eric Leroy, Jean-Louis Pastol, Bernhard Peucker-Ehrenbrink, Greg Ravizza, Franck Poitrasson, Gérard Sagon, Urs Schärer, Alex Shukolyukov, and Michael Walls. The detailed analytical characterization of the impact signal will be presented elsewhere.

We are greatly indebted to all those who have facilitated our access to the materials used in this Chapter: Harvey Weiss and Bertille Lyonnet for NE Syria; Vincente Lull and his team for the Vera basin; Claude Chauchat and Santiago Uceda for the Moche valley Peru; and Claude Burnez and Catherine Louboutin at Ebeon (France).

Nicolas Fedoroff and Stéphane Gaffié are gratefully acknowledged for their support in the field and in the laboratory.

References

Alley RB, Meese DA, Shuman CA, Gow AJ, Taylor KC, Grootes PM, White JWC, Ram M, Waddington ED, Mayevski PA, Zielenski GA (1993). Abrupt increase in snow accumulation at the end of the Younger Dryas event. Nature 362: 527–529

Almendros G, Gozales-Vila FJ, Martin F (1990) Fire-induced transformation of soil organic matter from an oak forest: An experimental approach to the effects of fire on humic substances. Soil Sci 49 (3): 158–168

Bond GC, Showers W, Elliot M, Evans M, Lotti R, Hajdas I, Bonani G, Johnson S (1999) The North Atlantic's 1–2kyr climate rhythm: Relation to Heinrich events, Dansgaard/Oeschger cycles and the little ice age. In: Clark PU, Webb RS, Keigwin LD (eds.) Mechanisms of Global Climate Change at Millennial Time Scales. Am Geophys Union Geophys Monogr, Washington DC 112: 35–58

Bullock P, Fedoroff N, Jongerius A, Stoops G, Tursina T, Babel U (1985) Handbook for Soil Thin Section Description, Waine Research Publications, Wolverhampton, UK

Burnez C, Fououeré P (1999). Les enceintes néolithiques de Diconche (Saintes, Charente-Maritime). Poitiers, 1999 (Mémoire XV)

Courty MA (1994) Le cadre paléogéographique des occupations humaines dans le bassin du Haut Khabour (Syrie du nord-est). Premiers résultats. Paléorient 20(1): 21–59

Courty MA (1998) The soil record of an exceptional event at 4000 B.P. in the Middle East. In: Peiser BJ, Palmer T, Mailey ME (eds.) Natural Catastrophes During Bronze Age Civilisations. Archaeological, ecological, astronomical and cultural perspectives. BAR Inter Series 728: 93–108

Courty MA (2001) Evidence at Tell Brak for the Late ED III/Early Akkadian Air Blast Event (4 kyr BP). In: Oates D, Oates J, McDonald H (eds.) Excavations at Tell Brak, Vol. 2: Nagar in the Third Millennium B.C. McDonald Institute Monographs, Cambridge, UK

Courty MA (2003) Tracing the 4 kyr BP dramatic event in soils and archaeological sediments across the Near East. In de Merocheji P, Thalman JP, Margueron JC (eds.) Actes du 3ème Congrès International d'Archéologie du Proche Orient, Paris

Courty MA, Goldberg P, Macphail RI (1989) Soil, Micromorphology and Archaeology. Cambridge Manuals in Archaeology, CUP, Cambridge

Courty MA, Fedoroff N, Jones MK, Castro P, McGlade J (1995) Environmental dynamics. In: Van der Leeuv S (ed.) Understanding the Natural and Anthropogenic Causes of Soil Degradation and Desertification in the Mediterranean Basin. The Archaeomedes project. Final Report of the contract EV5V-0021 for the Directorate General XII of the Commission of the European Union

Courty MA, Fedoroff M, Greenwood P, Grice K, Guichard F, Mermoux M, Schärer U, Shukolyukov A, Smith DC, Thiemens MH (2005) Formative processes throughout lands and seas of the 4-kyr BP impact signal. SEPM Research Conference. In: The Sedimentary Record of Meteorite Impacts. Springfield, Missouri, USA

Courty MA, Deniaux B, Crisci A, Fedoroff M, Greenwood P, Grice K., Mermoux M, Schärer U, Shukolyukov A, Smith DC, Thiemens MH (2006) High resolution soil records of impact-ignited fires at 4 kyr BP in various regions of the Near East, West Europe and Peru. Geophys Res Abs 8:33–78

Cullen HM, de Menocal PB, Hemming S, Hemming G, Brown FH., Guilderson T, Sirocko, F (2000) Climate change and the collapse of the Akkadian Empire. Evidence from the deep sea. Geology 28: 379–382

de Menocal P (2001) Cultural responses to climate change during the late Holocene. Science 292: 667–673

Gaffie S, Bruand A, Courty MA (2001) Analyse de la microstructure d'un horizon de surface enterré sous des matériaux archéologiques du Bronze Ancien en Syrie (2200 BC). CR Acad Sci Série II 332: 153–160

Giovannini G, Lucchesi S, Cervelli S (1983) Water-repellent substances and aggregate stability in hydrophobic soil. Soil Sci 135(2): 110–113

Graup G (1999) Carbonate-silicate immiscibility upon impact melting: Ries crater, Germany. Meteorit Planet Sci 34: 425–438

Heimann R (1982) Firing technologies and their possible assessment by modern analytical methods. In: Olin JS, Franklin AD (eds.), Archaeological Ceramics, Smithsonian Institution Press, Washington, DC

Hubbert KR, Preisler HK, Wohlgemuth PM, Graham RC, Narog MG (2006) Prescribed burning effects on soil physical properties and soil water repellency in a steep chaparral watershed, southern California, USA. Geoderma 130: 284–298

Humbert L (1972) Atlas de pétrographie des systèmes carbonatés. Technip, Paris

IUSS Working Group, WRB (2006) World reference base for soil resources. 2nd edition. World Soil Resources Reports 103. FAO, Rome

Jung SJA, Davies GR, Ganssen GM, Kron D (2004) Stepwise Holocene aridification in NE Africa deduced from dust-borne radiogenic isotope records. Earth Planet Sc Lett 221: 27–27

Kieffer SW (1977) Impact conditions required for formation of melt by jetting in silicates. In: Roddy, DJ, Pepin RO, Merrill RB (eds.) Impact and Explosion Cratering. Pergamon, New York

Kring DA (1997) Air blast produced by the Meteor Crater impact event and a reconstruction of the affected environment. Meteorit Planet Sci 32: 517–530

Kring DA, Durda DD (2002) Trajectories and distribution of material ejected from the Chicxulub impact crater: Implications for post-impact wildfires. J Geophys Res 107(E6) 10 1029/2001JE0011532

Marchant R, Hooghiemstra H (2004) Rapid environmental changes in African and South American tropics around 4000 years before present: A review. Earth-Sci Rev 66: 217–260

Mayewski PA, Rohling EE, Stager JC, Karlen W, Maasch KA, Mekk LD, Meyerson EA, Gasse F, van Kreveld S, Holmegren K, Lee-Thorp J, Rosquist G, Rack F, Staubwasser M, Shneider RR, Steig EJ (2004) Holocene climate variability. Quaternary Res 62:242–255

Ott d'Estevou Ph, Montennat C, Alvadao C (1990) Le bassin de Vera – Garrucha. In: Montenat C (ed) Documents et travaux de l'IGAL, Paris

Pavlik RS, Holland HJ, Payzant EA (2001) Thermal Decomposition of Zircon Refractories. J Am Ceram Soc 84 (12): 2930–2936

Scott DF, Van Wyk DB (1990) The effect of wildfire on soil wettability and hydrological behaviour of an afforested catchment. J Hydrol 121: 239–256

See TH, Wagstaff J, Yang V, Horz F, MacKay GA (1998) Compositional variation and mixing of impact melts on microscopic scales. Meteorit Planet Sci 33(4): 937–948

Soil Survey Staff (1993) Keys to Soil Taxonomy. US Department of Agriculture Natural Conservation Service. 9th edition, Washington DC.

Thompson L, Mosley-Thompson N, Davis ME, Henderson KA, Brecher HH, Zagorodnov, VS, Mashiotta T, Lin PN, Mikhalenko V, Hardy DR, Beer J (2002) Kilimanjaro ice core records: Evidence of Holocene climate change in Tropical Africa. Science 298: 589–593

Toon OB, Zahnle K, Morrison RP, Turco R, Covey C (1997) Environmental perturbations caused by asteroid impacts. Rev Geophys 35: 41–78

Wang S, Zhou T, Cai J, Zhu J, Xie Z, Gong D (2004) Abrupt climate change around 4 ka BP: Role of the thermohaline circulation as indicated by GCM experiment. Adv Atmos Sci 21(2): 261–295

Weiss H (2001) Beyond the Younger Dryas. Collapse as adaptation to abrupt climate change in ancient West Asia and the Eastern Mediterranean. In: Bawden G, Reycraft RM (eds.) Environmental disaster and the Archaeology of Human Response. Maxwell Museum of Anthropology, Albuquerque, NM. Anthropological papers 7: 75–98

Weiss H, Courty MA, Wetterstrom W, Meadow R, Guichard F, Senior L, Curnow A (1993) The origin and collapse of Third Millennium North Mesopotamian Civilisation. Science 261: 995–1004

Weiss H, Bradley RS (2001) What drives societal collapse? Science 291:609–610

Zazo C, Goy JL, Dabrio CJ, Bardaji T, Somoza L, Silva PG (1993) The last Interglacial in the Mediterranean as a model for the present Interglacial. Global Planet Changes 7: 109–117

Clay Illuviation in a Holocene Palaeosol Sequence in the Chinese Loess Plateau

He Xiubin, Bao Yuhai, Hua Lizhong and Tang Keli

Abstract The loess-palaeosol sequence on the Loess Plateau of China has been subjected to sophistic genetic processes such as dust accumulation, soil erosion, deposition and soil formation. These processes should be carefully distinguished when we clarify the soil genesis and classification, and interpret physical and chemical indicators to evaluate the bio-climate changes. Heavy minerals, pollen, and soil properties were studied in a typical Holocene loess profile in the north of the Loess Plateau, consisting of a palaeosol underlain by the Malan loess (late Pleistocene) and overlain by modern loess. The palaeosols, which were developed during ca. 8800–4400 ^{14}C years BP., consists of an upper humus-rich AB horizon over a clay-rich Bt horizon. The highest content of clay coatings was about 7.9% (by area percentage) appears at a depth of 140 cm, the lower part of the clay-rich horizon. The humus-rich horizon at a depth of 80 cm was intensely weathered in terms of heavy mineral analysis, and the higher pollen content and diversity also indicate a stronger weathering bioclimatic environment at that time. The Calcitic layer overlying the clay coating suggested that the carbonate material was derived from the overlying

He Xiubin
Institute of Mountain Hazards and Environment, CAS, Chengdu, Sichuan 610041, China; Institute of Soil and Water Conservation, CAS, Yangling, Shaanxi 712100, China, e-mail: xiubinh@imde.ac.cn

Bao Yuhai
Institute of Mountain Hazards and Environment, CAS, Chengdu, Sichuan 610041, China; Graduate University of Chinese Academy of Science, Beijing 100039, China, e-mail: byhcw@126.com

Hua Lizhong
Institute of Mountain Hazards and Environment, CAS, Chengdu, Sichuan 610041, China; Graduate University of Chinese Academy of Science, Beijing 100039, China, e-mail: hualizhong2008@yahoo.com.cn

Tang Keli
Institute of Soil and Water Conservation, CAS, Yangling, Shaanxi 712100, China, e-mail: kltang@ms.iswc.ac.cn

S. Kapur et al. (eds.), *New Trends in Soil Micromorphology*,
© Springer-Verlag Berlin Heidelberg 2008

modern loess. This phenomenon may imply that the clay translocation could occur within 60 cm, indicating a strong eluviation-illuviation process that took place during the development of the palaeosol. The results give valuable information on the genesis of the profile and classification of the palaeosol, and for the interpretation of the proxy indicators of bio-climate changes.

Keywords Holocene · loess-palaeosol · clay translocation · the loess plateau · China

1 Introduction

The over-100-meter thick loess-palaeosol sequence on the Loess Plateau of China has been a focus of geo-scientists' attention in the past decades, because the loess sequence holds implications for bio-climatic changes in the Quaternary and anthropological history. The top surface of the loess layer has been recognized as Holocene loess by geologists (Liu 1985, An et al. 1990, Porter and Zhisheng 1995), which is of special importance in loess-based paleo-climatic studies as it usually serves as the basis for interpreting the eco-environmental conditions under which the older sequences were developed (Liu et al. 1996). The top 40–60 cm soils are considered as human-made material such as application of plaggen manure, irrigation with sediment-rich water and tillage activities as indicated by the compacted layers and some cultural relics in the south profiles (Zhu 1962, STCRG 1991, Guo and Fedoroff 1990). New evidence shows human-made material only account 10–25% of the topsoil in the southern region, but little in the northern area (Tang 1981, He et al. 2002). For this reason, Chinese soil scientists recognized the Holocene profile as agricultural loess soil, namely black loam soil or dark loessial soils, classified as the Caliche Soils in Chinese system (STCRG 1991), Mollisols in Soil Taxonomy (Soil Survey Staff 1996) and Chernozems in WRB (FAO/ISRIC/ISSS 1998).

There have been a number of studies on the loess sequence in northwestern China, with alternations of loess and palaeosols, to document Quaternary glacial/interglacial episodes or fluctuations of winter/summer monsoons (e.g. An et al. 1990, Porter and Zhisheng 1995). The climatic record in the loess-palaeosol sequence has been found generally comparable to those of the oxygen isotope record of deep-sea sediments (e.g. Shackleton and Opdyke 1977 Kutzbach and Street-Perrott 1985). The records based on loess events seem also to agree with the variations of solar radiation resulting from perturbations of the Earth's orbit, the so-called Milankovitch cycles (Ding et al. 1994, Bronger and Catt 1998, Liu et al. 1999). However, the details and extent of such climatic fluctuations remain unclear and the model does not fully explain the geological, biological and historical evidence.

Recently, with advances in dating methods and the use of a variety of palaeo-environmental techniques such as soil micromorphology, mineralogy, and pollen analysis, a number of studies have systemically addressed the genetic complexity of the loess-palaeosol sequences (Courty and Fedoroff 1985, Liu 1985, Bronger and Heinkele 1989, Fedoroff et al. 1990, Bronger 1991, Bronger et al. 1998, Kemp

1985a,b and 1998a,b). The Holocene loess-palaeosol profile is of special importance in loess-based palaeo-climatic studies, as it usually serves the basis for interpreting the eco-environmental conditions under which the older loess-palaeosol sequences were developed (Liu et al. 1996, Bronger and Heinkele 1989, Sun et al. 1991, Moore et al. 1991, Yang and Yuan 1991, Guo and Fedoroff 1990, Liu et al. 1996, Xiangjun et al. 1997, Moore 1998, Kemp 1998a,b).

In this study, a dated Holocene loess-palaeosol sequence from the north of the Loess Plateau, the Ansai profile, is analyzed with a variety of techniques with the aim of further elucidating the sedimentary history and pedogenetic processes with particular attention to clay translocation processes.

2 Setting

The Loess Plateau is located in the Upper and Middle Reaches of the Yellow River at an altitude between 1000 and 2000 m and is affected by the prevailing northern monsoon in the winter and southern monsoon in the summer time. The annual precipitation ranges between 200 and 700 mm, decreasing from southeast to northwest, and the vegetation consists of temperate forests of temperate forest-steppe, temperate steppe, temperate desert-steppe, and temperate steppe-desert from southeast to northwest (Fig. 1).

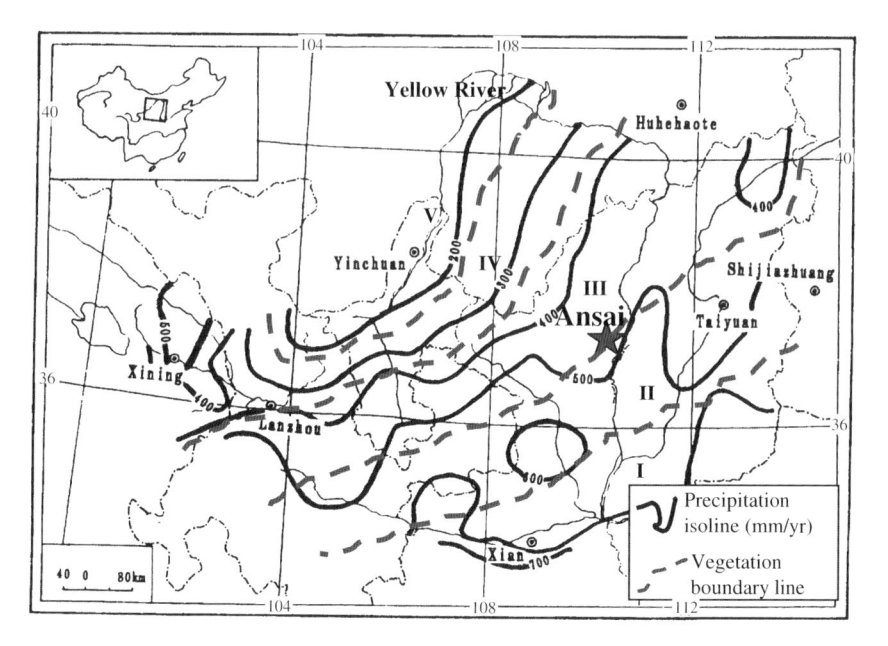

Fig. 1 Location of the Loess Plateau and the Ansai section (the red star): I. Broadleaved deciduous forest; II. Forest steppe; III. Steppe; IV. Desert-steppe; V. Steppe-desert (after Xiangjun Sun et al., 1997)

The Ansai section discussed in this paper is located at the transition from forest-steppe to steppe. Soils have been cultivated for agricultural production. Field edges and gully slopes are covered by herbs, such as Bothriochloa ischaemum, Stipa bungeana, Lespideza dahurica and Cleistogenes chinesis.

3 Methods

A typical Holocene loess-palaeosol profile of about 280 cm thick, which includes a 60-cm-thick surface layer of fresh loess (Lx), followed by a 180-cm thick palaeosol (So) and a 40-cm-thick parent material. Samples for thin sections, heavy mineral analysis, pollen analysis, and soil physico-chemical analyses were taken at 10 or 20 cm intervals according to the pedogenetic horizons recognized in the field (Table 1 and Fig. 2). Two samples for ^{14}C dating were taken from depths of 60–80 cm and 200–220 cm.

Soil texture (carbonate free samples), total organic carbon and calcium carbonate content of bulk samples were accomplished by routine methods (Head 1980). Heavy minerals in the 0.25–0.01 mm fraction were identified under a binocular microscope; some minerals were identified with energy dispersive X-ray analysis (EDAX). About 400 grains were distinguished in each sample.

The following ratios were employed to reflect the weathering intensity:

$$K1 = rm \text{ unstable group/relatively stable group}$$

$$K2 = rm \text{ unstable group/opaque mineral group}$$

$$F = K1 + K2$$

where the unstable group includes hypersthene, augite, hornblende, anthophyllite, and biotite.

The relatively stable group includes epidote, chlorite, apatite, garnet, and muscovite. The opaque mineral group includes stable magnetite, limonite, and hematite. When the value of F is lower, the content of unstable minerals is also lower, and the stronger is the degree of weathering and pedogenesis, assuming that the loess materials did not change during the Holocene. Thus the F value is interpreted as an index of the degree of weathering (Liu 1985, He et al. 1997).

Radiocarbon dating was carried out by the National Key Laboratory of Quaternary Geology and Loess in Xi'an. The material used for dating was soil organic matter (insoluble fraction in NaOH).

Undisturbed samples for micromorphological analysis were hardened with methylacrylate (He 1998) and a LEICA LABORLUX 12 microscope was used for petrographic analyses. The area percentages of clay coatings in thin sections were measured by the Quantiment 500, coupled with a microscope-based computer image processing system developed by the Leica Company (He 1997).

Table 1 Lithostratigraphy and soil properties of the loess-palaeosol sequence

Sampling depth (cm)	Munsell code	Genetic horizons	14C dating (yr BP)	Total organic C (%)	CaCO$_3$ (%)	Clay coating[a] (%)	Clay content[b] (%)	Sand content[b] (%)
0–20	7.5YR7/3	Lx		0.5	8.2	nd	17.4	1.2
20–40	7.5YR6/3			0.5	12.3	nd	16.1	2.0
40–50	7.5YR6/3			0.4	10.6	nd	22.3	1.4
50–60	7.5YR5/3			0.5	11.5	nd	21.7	1.1
60–80	2.5YR4/8	AB	4370 ± 80	0.8	11.2	3.0	14.1	9.7
80–100	2.5YR4/7	Bt1		0.8	3.3	3.5	16.2	8.0
100–120	2.5YR4/7			0.7	2.2	5.9	17.5	8.0
120–140	5YR4/6	Bt2		0.5	0.1	7.9	17.6	6.5
140–160	5YR4/6			0.5	0.1	3.8	14.9	7.0
160–180	5YR5/6			0.4	0.2	3.2	11.2	7.2
180–200	7.5YR5/4	BC		0.3	0.8	1.3	10.5	9.1
200–220	7.5YR5/4		8790 ± 120	0.3	0.7	nd	10.8	9.2
220–240	7.5YR6/4			0.3	0.8	nd	9.9	9.6
240–260	10YR8/3	C		0.1	1.5	nd	6.3	10.3
260–280	10YR8/3			0.1	1.7	nd	8.0	10.6
280–300	7.5YR7/4	Lo	10,000 ± 200[c]	0.2	12.4	nd	31.8	nd
300–320	7.5YR7/4			0.2	11.3	nd	27.0	nd

[a] Statistical area percentage in thin sections;

[b] Particle-size analysis: Clay: <0.005 mm; Sand: 1.00–0.25 mm;

[c] From Liu (1985);

nd — too little to be determined.

Lx — the fresh loess cover layer

So — the palaeosol layer of the Holocene profile

Lo — the Malan Loess profile

Schematic profile	Depth (cm)	Genetic horizons		Description
	0	Lx		Fresh loess
	50			
			AB	Red-brown, organic horizon
	100			
			Bt	Argillic horizon with strong prismatic structure
	150	So		
	200			
			BC	Greyish-black horizon with calc areous concretions and nodules
	250			
			C	Pale-yellow sandy loess
	300	Lo		Slity loess with vertical joints

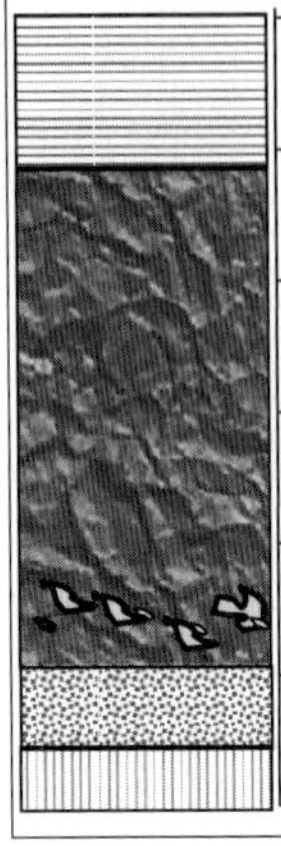

Fig. 2 The Ansai palaeosol and the stratigraphy of the loess-palaeosol sequence where Lx stands for the fresh loess cover layer; So sands for the palaeosol layer of the Holocene profile; and Lo stands for the Malan Loess profile (Liu, 1985; Tang et al., 1991)

Samples of 200 g of loess were used for pollen analysis after pretreatment with concentrated hydrochloric acid (36% HCl) for two hours, along with subsequent addition of K_2CO_3 (7%) and heating to the boiling point for 3–5 min. The pollen grains were extracted by using a heavy liquid (potassium iodide) (Ke 1997). The percentage of each taxon was calculated using the total of the land based plants as the sum of the pollen. Although pollen is reasonably abundant (152–483 grains per sample) in most of the samples, the C horizon of the palaeosol provided only 60–110 grains per sample.

4 Results

4.1 Chronostratigraphy

Because of the complicated processes of soil erosion and loess deposition, it is difficult to distinguish the genetic horizons according to the horizon designation criteria. There are many controversial descriptions on the classification and horizon designation (Zhu 1962 and 1994, Kemp 1985a,b, Bronger and Heinkele 1989, Tang et al. 1991, He et al. 2002).

The soils used in this study at Ansai, Northern Shaanxi of China, includes three parts, the top covered fresh loess (Lx), the middle palaeosol (So), and the later Pleistocene Malan loess (Lo). The middle palaeosol (So) can be pedogenetically sub-divided into the AB, Bt, BC and C horizons, which formed above the Malan Loess (Lo). Soil organic matter from the AB horizon (60–80 cm) gave the ^{14}C date of about 4, 370 ± 80 yr BP (Lab. No. XLQ-23), and that at a depth of 200–220 cm of 8, 790 ± 120 yr BP (Lab. No. XLQ-24). The boundary between the Holocene loess and Malan loess was recognized by texture; the coarse, sandy loess of the Holocene overlies the silty loess of the Malan (Liu 1985, He et al. 1997) (Table 1). The age of the top soil of the latter is about 10, 000 ± 200 yr BP by radiocarbon dating in the other sections (Liu 1985).

4.2 Lithostratgraphy and Micromorphology

The surface loess horizon (Lx: 0–60 cm) is characterized by a gray-yellow color, weakly-developed structure and fine sand texture, with a sharp boundary to the palaeosol below (Table 1). Under the microscope, a great deal of secondary carbonate was found either in dispersed form or along the walls of voids, roots or pores. A Dark brown color, relatively coarse texture (Table 1), and strongly to moderately and weakly-developed structure from top to bottom were characteristic of the palaeosol profile (60–280 cm; Table 1, and Fig. 2). The amount of organic carbon was highest in the AB horizon (60–80 cm) and gradually decreased with depth (Table 1 and Fig. 6).

The content of $CaCO_3$ was relatively high (11.24%) at the top and sharply decreased with depth. Investigation in thin section showed that the carbonates were all in secondary form (Fig. 3a,b) and sometimes appeared as carbonate coatings overlying the clay coatings (Fig. 3c). A few primary carbonate grains appear only at the bottom of the C horizon. The red-brown spots of Fe-Mn features and the area percentage of clay coatings on thin section are highest in the middle of the B horizon (120–140 cm), coinciding with the maximum clay content (<0.005 mm).

The micromorphological characteristics of the B horizon of this buried palaeosol are: The soil-matrix of the argillic horizon consists of a great deal of reddish-brown oriented clay, and the mineral grains connected with the clay forming a pluctoamictic fabric (Kubiëna 1938, FitzPatrick 1984) (Fig. 3d,e). The clay matrix has mainly striated orientation (Tang 1981), which can be found either in the dispersed form or around skeleton grains, or connected with voids and canals, presenting a reddish/yellow-brown color. The cutans surrounding the grains or voids are 10–100 µm thick.

4.3 Pollen Analysis

Pollen diagrams from previous work on Holocene palaeosol horizons (Sun et al. 1991, Xiangjun et al. 1997) are dominated by herbs (up to 90%) such as Artemisia, Chenopodiaceae and Compositae, representing steppe or possibly forest-steppe with a semi-arid climate. The possibility of warm-temperate broad–leaved deciduous forest on the Loess Plateau deduced from the pollen data is still controversial (Liu 1985, Tang et al. 1991, He et al. 1999). The pollen diagram from Ansai is also dominated

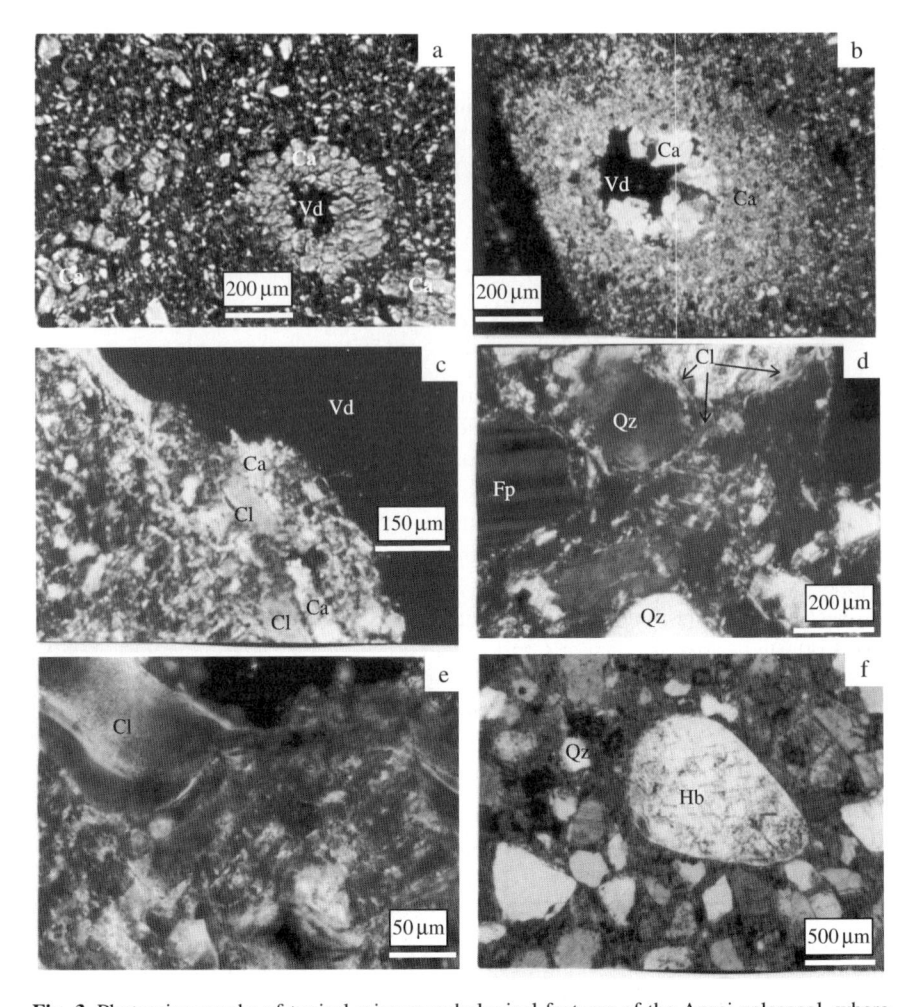

Fig. 3 Photomicrographs of typical micromorphological features of the Ansai palaeosol, where Ca-calcite; Cl-clay coating; Fp-feldspar; Hb-hornblende; Qz-quartz; and Vd-voids or cracks: **(a)** secondary crystalline calcitic coating around voids, crossed polarize light (XPL), at a depth of 30 cm; **(b)** calcitic coating around voids and crystalline grains of calcite in the voids (XPL), at a depth of 50 cm; **(c)** calcitic coatings overlying the micro-laminated clay coatings on the wall of a crack(XPL), at a depth of 90 cm; **(d)** clay cutans surrounding the mineral grains (XPL), at a depth of 120 cm; **(e)** enlarged portion of (d) (XPL); and **(f)** well preserved mineral grains (XPL), at a depth of 170 cm

by herbaceous pollen (including small shrubs), which accounts for 80–97% with Artemisia, which is the most abundant taxon through the profile (60–75%). The pollen diagram can be divided into five zones (Fig. 4).

Zone 1:   Top horizon of the Malan loess (Lo) from 280–320 cm. Herb pollen denom-
          inates, accounting for more than 90%. Artemisia is most abundant (>80%),
          followed by Chenopodiaceae and Compositae. Tree pollen percentages are

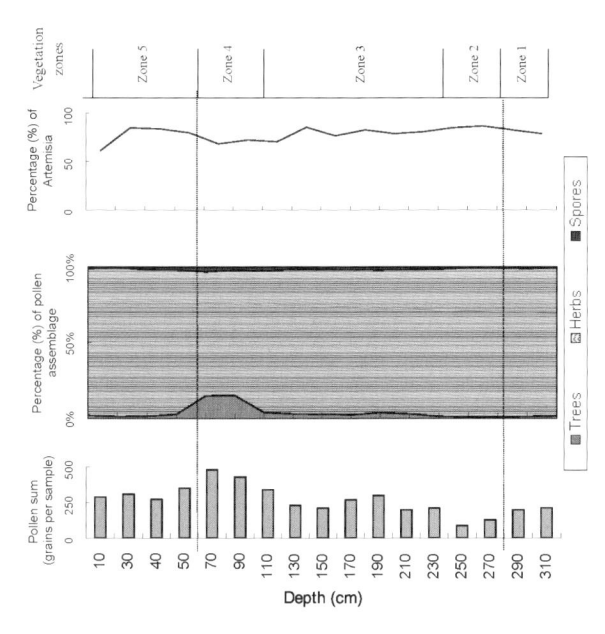

Fig. 4 Simplified pollen diagram from the Ansai loess-palaeosol sequence

low (1–2%). Pinus is the main type, while a few grains of Quercus and Betula appear.

Zone 2: Parent material (C) of the palaeosol from 240–280 cm. This zone is similar to Zone 1 in terms of the pollen assemblage. However, the pollen content is at a minimum (pollen sum c. 100), Artemisia attains maximum values and the tree pollen percentages are slightly lower than before.

Zone 3: The B and BC horizons of the palaeosol from 120–240 cm. This zone is characterized by an increase in both pollen content and taxon diversity, and a slight increase in tree pollen percentages (5%). Besides Pinus, Betula and Quercus, other trees such as Corylus, Ulmus and Cupressae are also found. Artemisia remains as the main contributor to the herbaceous pollen proportion although Cyperaceae, Ranuculaceae and Liliaceae comprise a considerable part (c. 10%). Ferns, such as Polypodiaceae and S. Sinensis, appear in small amounts (1–2%).

Zone 4: The AB and Bt horizons of the palaeosol from 60–120 cm. This zone is characterized by a sharp increase in tree pollen percentages, the diversity of taxa and pollen content values. Quercus is the dominant tree other than Pinus. Tsuga and Acer are present in appreciable amounts. Furthermore, the appearance of Podocarp, Castanea, Celtis and Rhus reflect the subtropical and warm-temperate environment of south China (south beyond the Qinling Mountain) and are generally considered as indicator species for climate (Wu 1980). Although mesic plants, such as Cyperaceae, Ranuculaceae, Polygonum and Cruciferae, contribute substantial proportions, Artemisia is still the dominant herb component, along with algae such as Concentricysitis.

Zone 5: Surface loess (Lx)-fresh- from 10–60 cm. This zone exhibits an irregular decrease in tree pollen to less than 2% at the surface and is dominated by Pinus and Quercus. Mesic herbs and fern spores also decrease significantly and many disappear.

4.4 Heavy Mineral Analysis

In almost all samples, heavy minerals such as hypersthene, augite, hornblende, anthophyllite, biotite, epidote, chlorite, apatite, garnet, muscovite, and opaque minerals can be found. These minerals account for 95% of the bulk amount of heavy minerals in loess. The dominant heavy minerals are opaque minerals, epidote and hornblende, which account for >70% of the total heavy minerals.

The degree of weathering, as indicated by the F index, is sensitive (Fig. 5). Maximum values of the F index, and hence minimum weathering, is characteristic of the C horizon of the paleaosol. Weathering increases up to the palaeosol with minimum F values in the AB horizon, coinciding with the peaks in total organic C and tree pollen percentages. The largely unweathered nature of the modern loess is indicated by consistently high values of the F index.

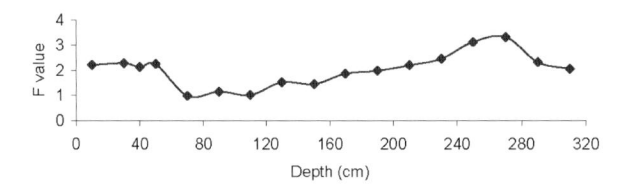

Fig. 5 F values of heavy minerals from the Ansai loess-palaeosol sequence

5 Discussion

5.1 Paleopedosedimentary Processes

During the Holocene, the Loess Plateau was subjected to a complete sequence of climatic changes from dry-cool to humid-warm and back to dry-cool (Liu 1985). At the time of the Opium period of the middle Holocene, There was a luxury forest on the Loess Plateau (Tang et al. 1991). This is confirmed by the present pollen analysis which was characterized by an increase in both pollen content and taxon diversity, and a sharp increase in tree pollen percentage in this period.

The palaeosols of the Holocene loess profile in northern China are generally recognized as the result of a humid warm climate in the middle Holocene (Liu 1985, An et al. 1991), while soil scientists described them as agricultural loess soils and termed them as Lutu soils (Ustalfs) in the south, heavy black loam soils

(Argiustolls) in the middle and light black loam soils (Haplustepts) in the north (Zhu 1962, He et al. 2002). A number of studies indicate that these profiles may be of post-depositional origin of earlier Holocene/Late Pleistocene loess and thus may include partially weathered material, which is difficult to recognize (Liu 1985, He 1997, Kemp 1998a,b and 2001).

The loess deposition and soil development were simultaneous with alternations of intensity of each process in response to bio-climatic changes. The soil formation processes may proceed at the overlying surface or disturb the stratigraphic information of the pedosedimentary features, and subsequent erosion may remove all the marks of these processes (Tang et al. 1991, He and Tang 1999). Such phenomena were also reported in other paleosols as well (e.g. Buol et al. 1997, Catt 1986, Birkeland 1999, Hall 1999, Mason and Kuzila 2000).

Results of this study suggest that the accumulation of sandy loess in the early Holocene occurred with no or very weak pedogenesis. This dry-cool stage is represented by the C horizon, which is characterized by a coarse texture, extremely low contents of $CaCO_3$, humus and pollen, and weakly weathered mineral grains. As the content of the medium to coarse sand fraction (1.00–0.05 mm) is almost the same throughout the palaeosol profile, and quite different from the layers above and below (Table 1), the palaeosol could have been developed either during or after the accumulation of this relatively coarse sandy loess.

Although the coarse sandy loess began to accumulate before $8,790 \pm 120$ yr BP, and possibly shortly after 10,000 yr. BP, the preservation of pollen stratification in the palaeosol profile supports the conclusion that loess accumulation continued during soil development to at least $4,370 \pm 80$ yr BP, which can be regarded as the maximum age for the cessation of soil development and for palaeosol formation (Matthews 1985).

5.2 Mechanical Eluviation-Illuviation of Clay Particles in the Palaeosol Profile

Many papers have discussed the clay-rich or argillic horizon in loess profiles, but the genesis of such argillic horizons is still disputed (Brewer 1968, Mermut and Pape 1971 and 1973, Zhu 1994, Bronger et al. 1998). Considering the pH (7.8) (Zhu 1962, Liu 1985), it has been suggested that the clay-rich horizon was not formed by mechanical eluviation but formed in situ (Mermut and Pape 1971 and 1973, Zhu 1994). In this study, however, the most strongly weathering horizon according to the lowest F values combined with highest contents of organic matter, is 60 cm above the clay-rich horizon (Fig. 6), which contains numerous oriented clay coatings indicative of illuviation with a reddish brown or yellow-brown color, and relatively high F values.

The fresh surface of quartz and feldspar grains and weakly weathered unstable heavy minerals are well preserved in the clay-rich zone (Fig. 3d,f). These phenomena imply that the clay probably accumulated through mechanical eluviation-illuviation

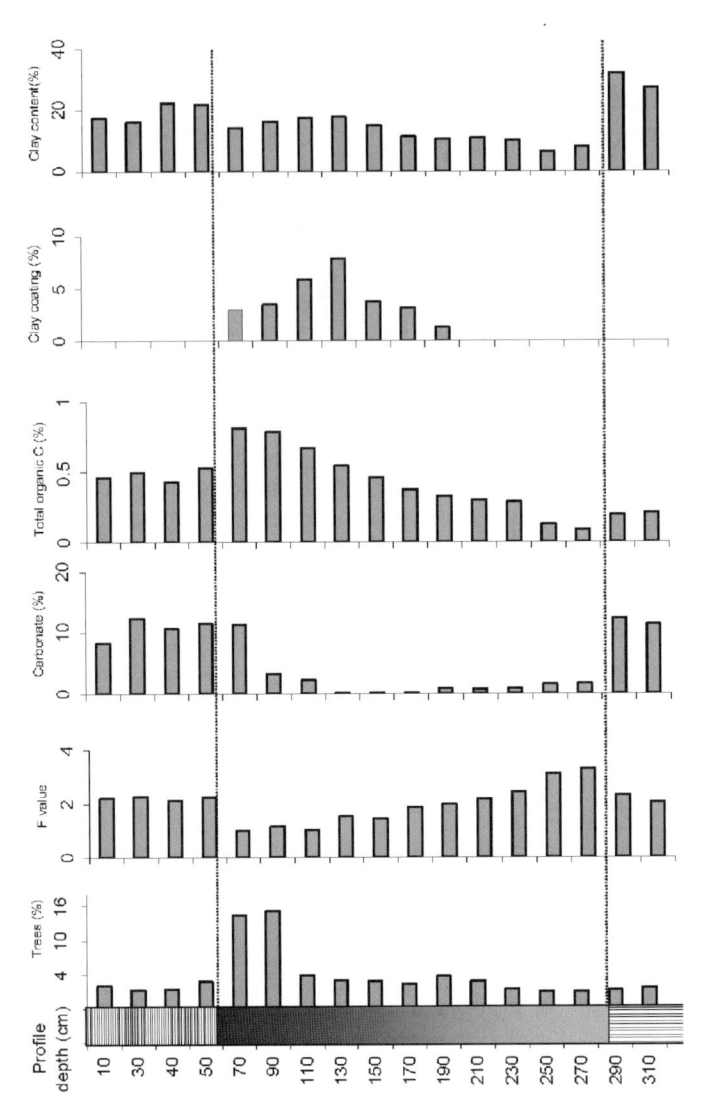

Fig. 6 Comparison of different bio-climatic indicators from the Ansai loess-palaeosol sequence

processes when the pH was lower. Thus, the palaeosol may be a type of Brown soil (FitzPatrick 1984, Courty and Fedoroff 1985, Catt 1986, Stoops and Eswaran 1986).

5.3 Bio-Climatic Stages of Profile Development

After a long period (possibly 100,000 yrs) of cool-dry climate when the Malan loess was deposited (Liu 1985), the climate entered an early-Holocene cool-dry phase, which initiated about 220 cm of sandy loess deposition (depth of 60–280 cm). Then the climate became warm and humid, as a culmination in the development of a warm-temperate forest during the Holocene Optimum. Pedogenesis, characterized by enhanced physico-chemical weathering and biological activity, took place alongside a reduced rate of loess accumulation.

Continued loess deposition is inferred to have been the primary cause of the pollen stratification, and a declining rate of loess deposition may have contributed to the corresponding increase in weathering. Finally, in the late-Holocene, there was still a certain amount of precipitation (about 400 mm) similar to the present time (Tang et al. 1991). Thus, biological activity indicated by the root pores and earthworm channels, and weathering indicated by the accumulation of secondary carbonate, were still prevailing. This led to the redistribution of the more soluble salts. The top palaeosol, the AB, and even the B horizon, therefore, have a certain amount of carbonate, derived from the overlying loess, leading to the weak alkaline pH of <7. This model of palaeosol formation is summarized in Fig. 7.

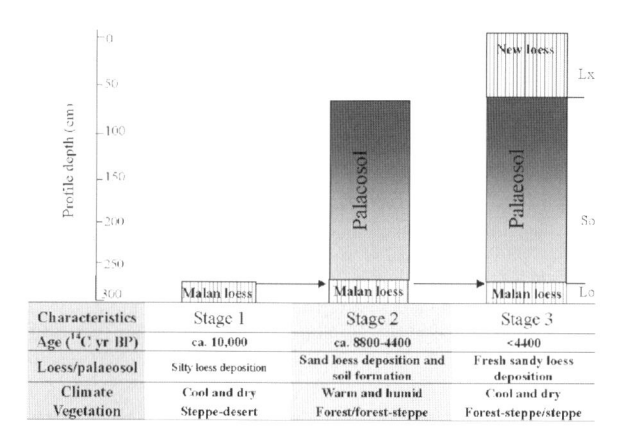

Fig. 7 The sequence of formation stages of the Holocene loess-palaeosol sequence

6 Conclusion

The palaeosol of the Holocene loess profile was developed between at least 8790 yr BP and 4370 yr BP. Yellowish/reddish-brown clay coatings were identified in the B horizon. Comparison of different bio-climate indicators, based on mineralogy, micromorphology, and palynology suggest that the oriented clay cutans were developed by the mechanical translocation of clay, and that the argillic horizon developed by strong weathering during the Holocene Optimum, and the subsequent eluviation-illuviation of clay minerals. The distance of vertical translocation of clay is up to 60 cm. The proxy indicators of bio-climate changes, such as grain size distribution, magnetic susceptibility and some chemical indicators, may have been greatly influenced by such soil formation processes. Thus the bio-climatic situation may be misinterpreted, especially when high time resolution is considered.

The development process of the Holocene loess profile can be divided into the following stages: (a) coarse, sandy loess deposition with a dry and cool climate; (b) strong soil formation in a humid and warm climate with reduced loess deposition; and (c) renewed loess deposition during the return to a dry and cool climate, overprinting new features on the buried soil profile by vertical transportation of carbonate and bio-turbation. Thus, the single palaeosol represents not only the period of humid and warm climate, as was reported by previous researchers, but also bear the imprint of several different bio–climatic environments. Further investigation should be undertaken when attempting to classify the palaeosol and clarify the pedogenesis.

Acknowledgements This work was supported by the Knowledge Innovation Programs from Chinese Academy of Sciences (KZCX3-SW-146) and the National Natural Science Foundation of China (No. 90502007). The authors express sincere thanks to Prof. Zhou Mingfu for 14C Dating, Prof. Zhao Jingpo for assistance in thin section preparation, Ms. Ke Manhong for assistance in pollen analysis, and to Prof. Mermut and one anonymous referee for critical comments and corrections on an early draft of the manuscript.

References

An Z, Kukla G, Porter, SC (1991) Late Quaternary dust flow on the Chinese loess plateau. Catena 18: 125–132

An Z, Liu T, Lu Y (1990) The long-term palaeomonsoon variation recorded by the loess-palaeosol sequence in central China. Quaternary International 7/8: 91–95

Birkeland PW (1999) Soils and Geomorphology, 3rd ed. Oxford University Press, New York

Brewer R (1968) Clay illuviation as a factor in particle-size differentiation in soil profiles. Trans. 9th International Congress of Soil Science, Adelaide, IV, pp. 489–499

Bronger A (1991) Argillic horizons in modern loess soils in an ustic soil moisture regime: comparative studies in forest-steppe and steppe areas from Eastern Europe and the Unite States. In: Stewart, B.A. (Ed.), Advance in Soil Science 15, Springer-Verlag, New York, pp. 41–90

Bronger A, Catt JA (1998) Summary outline and recommendations on palaeopedological issues. Quaternary International 51/52: 5–16

Bronger A, Heinkele T (1989) Micromorphology and genesis of palaeosols in the Luochuan loess section, China: pedostratigraphical and environmental implications. Geoderma 45: 123–143

Bronger A, Winter R, Sedov S (1998) Weathering and clay mineral formation in two Holocene soils and in buried palaeosols in Tadjikistan: towards a Quaternary palaeoclimatic record in central Asia. Catena 34: 19–34

Buol SW, Hole R, McCracken J, Southard RJ (1997) Soil Genesis and Classification. 4th ed. Iowa State University Press, Ames, IA

Catt JA (1986) Soils and Quaternary Geology. Oxford Science Publication, Oxford

Courty MA, Fedoroff N (1985) Micromorphology of the recent and buried soils in a semi-arid region of Northwest India. Geoderma 35: 287–332

Ding Z, Yu Z, Rutter NW (1994) Towards an orbital time scale for Chinese loess deposits. Quaternary Sciences Reviews 13: 3970

FAO/ISRIC/ISSS (1998) World Reference Base for Soil Resources

Fedoroff N, Courty MA, Thompson ML (1990) Micromorphological evidence of palaeoenvironmental change in Pleistocene and Holocene palaeosols. In: Douglas, L.A. (Ed.), Soil Micromorpholoy: A Basic and Applied Science. Elsevier, Amsterdam, pp. 653–665

FitzPatrick E (1984) Micromorphology of soils. Chapman and Hall, London

Guo ZT, Fedoroff N (1990) Genesis of calcium carbonate in loess and in palaeosols in central China. Developments in Soil Science 19: 355–359

Hall RD (1999) A comparison of surface soils and buried soils: Factors of soil development. Soil Science 164: 264–287

He X (1997) Application of image processing technique in quantitative study of soil micromorphology. Chinese Journal of Soil Science 28: 110–112 (in Chinese)

He X (1998) Consolidation of thin section with methylacrylate. Chinese Journal of Soil Science 29: 95–96 (in Chinese)

He X, Tang K (1999) Potentiality of vegetation construction on the Loess Plateau. Soil and Water Conservation in China 204: 31–35(in Chinese)

He X, Tang K, Lei X (1997) Heavy mineral record of the Holocene environment on the Loess Plateau in China and its pedogenetic significance. Catena 29: 323–332

He Xiubin (1999) Pedogenesis of loess profile and evolution of erosional environment on the Loess Plateau since 0.2 Ma. BP. Soil Erosion and Water Conservation 5: 92–95

He X, Tang K, Tian J, Matthews JA (2002) Paleopedological investigation in three agricultural loess soils on the Loess Plateau of China, Soil Science 167: 478–491

Head KH (1980) Manual of soil laboratory testing. Pentech Press, London

Ke M-H (1997) A method of pollen-spore analysis in loess. Journal of Botany 39: 672–676 (in Chinese)

Kemp RA (1985a) Distribution and genesis of calcitic pedofeatures within a rapidly aggrading loess-palaeosol sequence in China. Geoderma 65: 303–316

Kemp RA (1985b) Soil Micromorphology and the Quaternary. Quaternary Research Association Technical Guide 2. QRA, Cambridge

Kemp RA (1998a) Micromorphology of loess-palaeosol sequences: a record of palaeoenvironmental change. Catena 35: 179–196

Kemp RA (1998b) Role of micromorphology in palaeopedological research. Quaternary International 51/52: 133–141

Kemp RA (2001) Pedogenic modification of loess: signicance for palaeoclimatic reconstructions. Earth Science Reviews 54: 145–156

Kubiëna W (1938) Micropedology. Collegiate Press, Ames, IA

Kutzbach JE, Street-Perrott FA (1985) Milankovitch forcing of fluctuations in the level of tropical lakes from 18 to 0 kyr B.P. Nature 317: 130–134

Liu T (1985) Loess and the Environment. China Ocean Press, Beijing

Liu T, Ding Z, Rutter N (1999) Comparison of Milankovitch periods between continental loess and deep sea records over the last 2.5 Ma. Quaternary Science Reviews 18: 1205–1212

Liu T, Guo Z, Wu N, Lu H (1996) Prehistoric vegetation on the Loess Plateau: steppe or forest? Journal of Southeast Asian Earth Sciences 13: 341–346

Mason JA, Kuzila MS (2000) Episodic Holocene loess deposition in central Nebraska. Quaternary International 67:119–131

Matthews JA (1985) Radiocarbon dating of surface and buried soils: principles, patterns and prospects. In: K.S. Richards, S. Ellis and R.R. Arnett (Eds.), Geomorphology and Soils, Allen and Unwin, London, pp. 269–288

Mermut AR, Pape T (1973) Micromorphology of "In Situ" Formed Clay Cutans in Soils. Leitz Scientific and Technical Information, West Germany 2, 147–150

Mermut AR, Pape T (1971) Micromorphology of Two Soils from Turkey with Special Reference to "In Situ" Formation of Clay Cutans. Geoderma 5, 271–281

Moore PD (1998) Did forest survive the cold in a hotspot? Nature 391, 124–125

Moore PD, Webb JA, Collinson ME (1991) Pollen Analysis. Blackwell Science, London

Porter SC, Zhisheng A (1995) Correlation between climate events in the North Atlantic and China during the last glaciation. Nature 375: 305–308

Shackleton NJ, Opdyke ND (1977) Oxygen isotope and palaeomagnetic evidence of early northern hemisphere glaciation. Nature 270, 216–219

STCRG (Soil Taxonomic Classification Research Group, Institute of Soil Science, Chinese Academy of Sciences) (1991) Chinese Soil Taxonomic Classification. Science Press, Beijing (in Chinese)

Stoops G, Eswaran H (1986) Soil micromorphology. Van Nostrand Reihold Company, New york

Sun J, Ke M, Zhao J, Wei M, Sun X, Li X, Liu S, Chang P (1991) Palaeo environment of the last glacial stage in Loess Plateau. In: Sun J. and Zhao J. (eds.), Quaternary of the Loess Plateau, pp. 154–184. Science Press, Beijing (in Chinese)

Tang K (1981) Micromorphology and genesis of palaeosol in Wugong section. Kexue Tongbao 26(3): 177–179

Tang K, Zhang P, Wang B (1991) Soil erosion and eco-environment changes in Quaternary. Quaternary Research 4: 49–56 (in Chinese)

Soil Survey Staff (1996) Keys to Soil Taxonomy. 6th ed. US Department of Agriculture, Washington DC

Wu Z (1980) Vegetation of China. Science Press, Beijing

Xiangjun S, Changqing S, Fengyu W, Mengrong S (1997) Vegetation history of the loess plateau of China during the last 100,000 years based on pollen data. Quaternary International 37: 25–36

Yang Q, Yuan B (1991) Environment on the Loess Plateau Regions and Its Evolution. Science Press, Beijing (in Chinese)

Zhu X (1962) Lutu Soil. Agriculture Press, Beijing (in Chinese)

Zhu X (1994) Discussion of clay coating of palaeosols on the Loess Plateau. Journal of Soil Science 31(4): 6–9 (in Chinese)

Soil Microstructure and Solution Chemistry of a Compacted Forest Soil in a Sub-Boreal Spruce Zone in Canada

J. M. Arocena, Z. Chen and P. Sanborn

Abstract This study was conducted to identify and quantify changes to soil micro-structure and solution chemistry in order to understand the long-term effects of machinery-induced compaction on the soil. The study area near Log Lake, northeast of Prince George, BC, is one of three installations in the Sub-boreal Spruce Zone of the Long-Term Productivity Study of BC Ministry of Forests and Range. Selected plots representing major treatment combinations from the foregoing experiments were used for the current study. Soil thin sections from undisturbed (control) and compacted soils were prepared to study soil microstructure. Soil solution was extracted from field-moist samples using the immiscible displacement-centrifugation technique.

Results showed that compaction reorganized soil substances and changed soil pore space characteristics. Porous types of soil microfabrics (granic, granoidic, plectic) that are common in control (uncompacted) soils, are replaced by dense types of soil microfabrics (such as banded or porphyric fabrics) in compacted soils. The microscopic porosity in compacted soils appeared lower than in control samples. Compaction decreased soil macroporosity and reduced the connectivity of soil pores as well. Void types that have less connectivity, such as vughs and planar

J. M. Arocena
College of Science and Management; Canada Research Chair – Soil and Environmental Sciences University of Northern British Columbia 3333 University Way, Prince George, BC Canada V2N 4Z9, e-mail: arocenaj@unbc.ca

Z. Chen
Environmental Monitoring and Evaluation Branch Alberta Environment, 9820–106 Street, Edmonton, AB Canada T5K 2J6

P. Sanborn
College of Science and Management, University of Northern British Columbia 3333 University Way, Prince George, BC Canada V2N 4Z9

S. Kapur et al. (eds.), *New Trends in Soil Micromorphology*,
© Springer-Verlag Berlin Heidelberg 2008

voids, are increased in compacted soils, likely at the expense of voids that have better connectivity (compound packing voids). An increase in proportion of isolated vughs relative to overall soil microscopic pore space was also observed. This is probably related to formation of "relict" macropores, resulting from soil compaction. The decrease in soil microscopic pore space and pore connectivity might have restricted nutrient movement or enhanced precipitation of organic salts. Soil solution chemistry investigation revealed a decrease in Ca and K in soil solution to such a degree that might be sufficient to curtail vigorous plant growth.

Keywords Compaction · porosity · packing void · planar voids · forest soils · soil microstructure

1 Introduction

Forest practices such as site preparation and harvesting can compact soil and remove organic matter (OM) from the surface soil (Mo et al. 1995, Busse et al. 1996, Huang et al. 1996, Rab 2004). Soil compaction from mechanized agriculture or forestry operations results mainly from wheel traffic (Kooistra and Tovey 1994, Bruand et al. 1997). Soil compaction and loss of OM alter soil microstructure and consequently the rates of nutrient and water supply to plant roots. The degree of compaction and loss of organic matter from the forest floor have direct influence on the weathering rates of minerals, nutrient mineralization rates and consequently plant growth (Zabowski et al. 1996, Worrell and Hampson 1997). However, Kranabetter et al. (2006) reported that after 12 years, in situ rates of N mineralization were uniform irrespective of compaction and OM removal treatments.

Poor seedling germination was reported for compacted soil (Johnson et al. 1991), and reductions in calcium (Ca^{2+}) and magnesium (Mg^{2+}) concentrations of 60 and 55% respectively, in the 0–10 cm depth interval of the mineral soil, following the removal of forest floor (Brais et al. 1995). The long-term reduction in the concentration of K^+ in soil solution, due to forest floor removal and compaction treatments, could be as high as 88% (Arocena 2000). Although the mechanism may be complex, numerous studies (Greacan and Sands 1980, Tuttle et al. 1985, Corns 1988) concluded that soil compaction results in decrease in forest productivity.

Another study, although not of forest soils, reported that compaction increased the points of contacts between particles and induced stronger absorption of water at high matric potential (Assouline et al. 1997). However, these studies are retrospective in nature and do not provide the forward-looking perspective which is becoming increasingly necessary (Powers et al. 1990). Recently, Rab (2004) reported that effects of compaction remain in soils even after ten years. Collection of additional data on the relationship between compaction and nutrient dynamics is pertinent to the understanding of long-term effects of compaction on soil nutrient characteristics.

Soil compaction induces structural modifications that can influence plant growth. Compaction adversely affects soil microstructures, reduces rates of water infiltration and increases runoff. It restricts root penetration, root growth and thus plant productivity in compacted soils (Yao and Wilding 1994). Bouwman and Arts (2000) reported that at a compaction load of 8.5t per m^2, roots did not penetrate deeper than 20cm, while Li et al. (2004) found a decrease in microbial N content upon OM removal and decrease in microbial C:N ratio with compaction. Soil compaction reorients soil particles, changes pore size distribution, increases bulk density, and causes pores to collapse (Bresson and Zambaux 1990). Crawford et al. (1995) indicated that precise description of soil microstructure is necessary to relate the soil-moisture release curve to the real soil structure.

The objectives of this study were to: (1) describe and compare soil microstructure, (2) estimate the alteration of microscopic pore space and types of microscopic pores, and (3) determine the changes in soil solution chemistry, in order to assess the impacts of soil compaction and soil organic matter removal on soil productivity. The results might be useful to the understanding of the long-term effects of soil compaction and organic matter removal caused by machinery on forest productivity.

2 Materials and Methods

2.1 Site Description and Sample Collection

The experimental site was part of the British Columbia Ministry of Forests and Range Long-term Site Productivity study in the Sub-Boreal Spruce zone, and is located at Log Lake, 60km northeast of Prince George, BC. The description of the site and selected properties of the soils are given in Table 1 and in Arocena and Sanborn (1999).

Table 1 Selected site characteristics and the soil properties (after Holcomb 1996, Trowbridge et al. 1996)

Characteristics	Description
Site Characteristics	
Location	54° 21'N 122° 37'W
Elevation (masl)	780
Slope and aspect	0–3% south
Vegetation	Dominant – Douglas fir and white spruce
	Minor – Lodgepole pine and sub-alpine fir
Parent material	Morainal Blanket
Soil Properties	
Soil classification	Orthic Humo-Ferric Podzol[1]
	Cryorthod[2]
Humus form	Hemimor
pH (CaCl$_2$) of Ae horizon	4.2
CEC of Ae horizon (cmol$_c$ kg^{-1})	4.4
Clay content (g kg^{-1})	77
Coarse Fragments (g kg^{-1})	390

Table 1 (Continued)

Characteristics	Description
1-Year Post-Treatment BD^3 (Mg m^{-3})	
C_0OM_0	1.52 (1.53)^4
C_1OM_1	1.87 (1.45)^4
C_2OM_2	1.71 (1.50)^4
1-Year Post-Treatment Forest Floor	Mass (Initial Mass) (kg m^{-2})
C_0OM_0	9.3 (8.3)^5
C_1OM_1	9.6 (8.4)^5
C_2OM_2	0.0 (7.2)^5

[1]Soil Classification Working Group (1998).
[2]Soil Survey Staff (1999).
[3]Total Bulk Density=Total Weight/Total Volume (for 0–20 cm soil depth).
[4]values in bracket represent initial BD.
[5]values in bracket represent initial forest floor biomass.

The operational application of the compaction and organic matter removal treatments were described previously in Holcomb (1996), Trowbridge et al. (1996) and Powers (2006). In brief, organic matter was removed when the sites were dry and, when necessary, manual labor was used to avoid compaction. The original experimental design had three levels of soil compaction (C) and organic matter (OM) removal. The three levels of C-treatments are C_0 – no compaction, C_1 – light compaction, and C_2 – heavy compaction. Light and heavy compaction treatments are defined as 40 and 80%, respectively, of the difference between the hypothetical growth-limiting maximum and pre-harvest conditions. The three levels of OM-treatments are OM_0 – boles (tree trunk) only removed, OM_1 – boles, crowns removed, and OM_2 – boles, crowns, plus forest floor removed (Holcomb 1996, Powers 2006).

For current soil microstructure investigation, we selected one representative plot for each of the following treatment combinations: C_0OM_0, C_1OM_1, and C_2OM_2. In addition, soils from an adjacent un-harvested stand (control site) were also collected. The micromorphological study was intended to further investigate the changes caused by soil compaction and organic matter removal. It was time-consuming and therefore needed to focus on selected treatment combinations. Intact samples were collected in situ using Kubiena boxes to protect soil microstructure. A total of six samples were collected from each of above treatments and the undisturbed site for the preparation of soil thin sections from the Ae horizon.

All nine compaction-OM removal treatments were used for the soil solution chemistry investigations.

2.2 Soil Thin Section Preparation and Description

Soil thin sections were prepared following the methods of Brewer (1976) and Brewer and Pawluk (1975). The thin sections were 5×8 cm in dimension and about 30 μm in thickness. Soil thin section description was carried out following the concepts and terminology in Brewer and Pawluk (1975) for microfabrics, Brewer (1976) for plasma fabrics, type of voids and pedofeatures and Bullock et al. (1985) for microstructure. Recognition of the different types of voids were conducted following the guidelines given in Bullock et al. (1985) using a Zeiss petrographic microscope. Abundance of solid substances or abundance of microscopic pore types was observed following the micromorphological illustrations by Bullock et al. (1985). Microscopic light manipulation (plain light, partially polarized light, and cross-polarized light) was performed to further identify above substances and voids in the thin sections. The abundance of overall miscroscopic pore space of thin sections was estimated as 100% minus solid space.

Briefly, the different types of voids are defined as (1) compound packing – voids resulting from packing of compound individuals, such as peds, which do not accommodate each other; (2) planar voids- voids that are planar in appearance according to the ratio of principal axes, the edges of corresponding planes are generally accommodating; (3) vughs – largely isolated voids, spherical to elongate, or irregular, and not normally connected to voids of comparable size, and (4) chamber – voids that are near spherical, connected by channels, with smoothed walls, and no cylindrical shape (Brewer 1976, Bullock et al. 1985). The distribution of the different types of voids is estimated as percentage of that pore space on the thin section.

2.3 Soil Porosity

Soil porosity was calculated from the bulk density data from (Holcomb 1996, Trowbridge et al. 1996) as quoted in Table 1. The following calculation formula from Danielson and Sutherland (1986) was used:

$$\text{Soil porosity} = (1 - \text{bulk density/solid density}) \times 100\% \tag{1}$$

2.4 Sample Collection for Soil Solution Extraction and Chemical Analyses

Field moist samples were collected from the Ae horizon (0–12 cm) at nine monthly intervals between June 1997 and September 1998, except August 1998 and when the soil was frozen (November 1997-March 1998). Samples were collected in 50-mL centrifuge tubes containing 20 mL of tetrachloroethylene (TCE). The TCE was used to extract the soil solution with the immiscible displacement-centrifugation technique (Wolt 1994). Centrifugation was carried out at 14,000 rpm at 4 °C for 15 min. The soil solution was filtered using a 0.45-μm filter and stored in polyethylene bottles at 4 °C prior to analysis. Soil solutions were analyzed for

Ca, Mg, K, Si, and Al using inductively coupled plasma-atomic emission spectrometry at the Central Equipment Laboratory, University of Northern British Columbia.

The availability indices for the K^+ and Ca^{2+} (and Mg^{2+}) were calculated using the following molecular ratios:

$$K_{AR} = \frac{[K^+]}{\sqrt{[Ca^{2+}] + [Mg^{2+}]}} \text{ (Brady and Weil 1996)} \qquad (2)$$

$$Ca_{AR} = \frac{[Ca^{2+}]}{[Al^{3+}]} \text{ (Cronan and Grigal 1995)} \qquad (3)$$

where, [ ] denotes mmol metal L^{-1} soil solution.

2.5 Data Analyses

As the estimates of soil microscopic pore space are no more than semi-quantitative, only the major trends of the data will be presented.

For soil solution composition, correlation matrices were constructed among the concentrations and ratios of Ca, Mg, K, and Al, as well as the moisture content. ANOVA (multivariate-repeated measures design) was conducted to compare the concentrations of Ca, Mg, K, and Al in the soil solution from various treatments where the homogeneity of variance was tested with the Sen and Purin test. The Least Significant Difference and Tukey's HSD tests were used to determine the differences between means of significantly different treatments.

3 Results

3.1 Microstructure

The soil fabric of the surface layer of Control and C_0OM_0 samples is dominated by the presence of granic and granoidic fabric units (Fig. 1a) of about 200–3000 μm in size (Table 2). These units are arranged in such a way that the intergranular spaces between these units are mostly compound packing voids with an overall microscopic void space around 65% in these samples. These granic and granoidic units have silasepic plasmic fabric. There is also a minor component of matriplectic-matrigranic fabric (Fig. 1b) for C_0OM_0 treatment. The fabrics for surface soils of C_1OM_1 and C_2OM_2 soils are largely porphyric (Fig. 1c), formed with dense matrix materials (Table 2). The overall microscopic void space is about 35% and dominated by vugh type voids of about 180–1200 μm in size (Fig. 1d). Planar voids are also observed in some parts of soil thin sections. The plasmic fabrics for C_1OM_1 and C_2OM_2 soils are similar to Control and C_0OM_0 samples and are dominated by silasepic and skelsepic fabrics. It

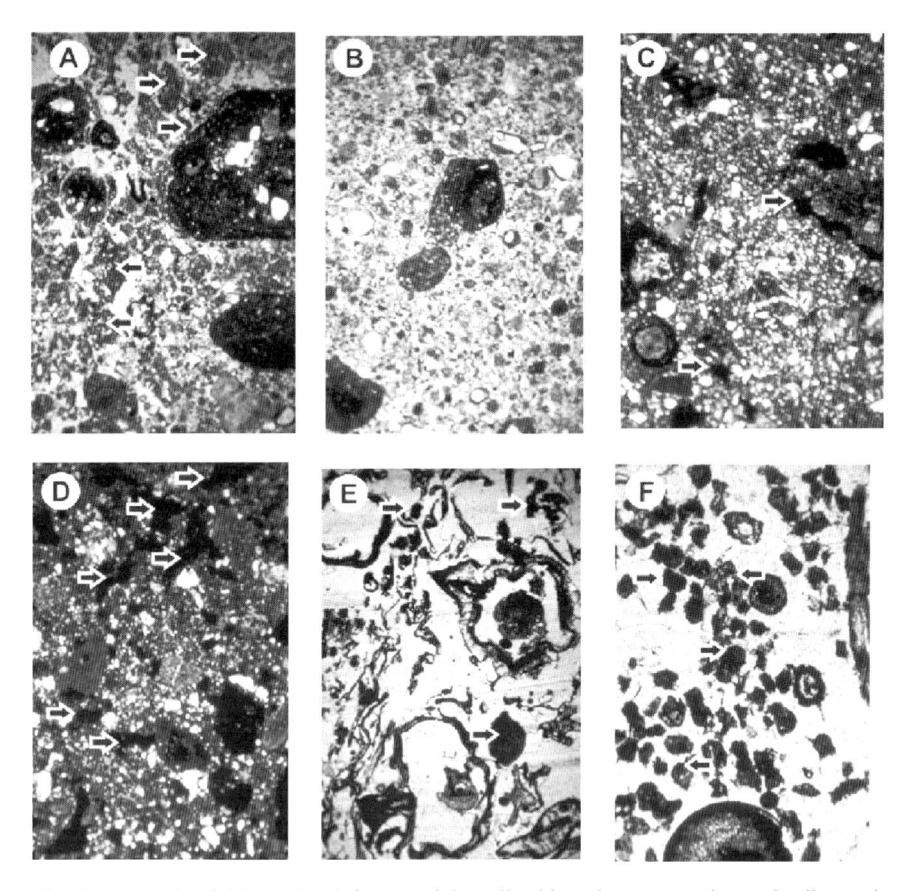

Fig. 1 Select microfabric and pedofeature of the soil subjected to compaction and soil organic matter removal treatment: (**a**) granic ($\rightarrow$) or granoidic ($\leftarrow$) fabric and compound packing voids, PPL. 7.2 mm, (**b**) matriplectic-matrigranic porphyric microfabric, PL, FL = 7.2 mm, (**c**) porphyric fabric with isolated vughs ($\rightarrow$) XPL, FL = 7.2 mm, (**d**) irregular vughs (185–2,370 μm) with total microscopic porosity ~30–40% XPL, FL = 7.2 mm, (**e**) humi-phytogranic microfabric, PL, FL = 7.2 mm, and (**F**) phyto-humic granic, granoidic fabric with fecal pellets of collembola ($\rightarrow$) and enchytraeid ($\leftarrow$), PL, FL = 3.3 mm. Legend – XPL-cross-polarized light, PL-polarized light, FL = frame length

should be noted that there are small regions in C_1OM_1 and C_2OM_2 soils that show soil fabrics other than porphyric, such as humi-phytogranic (Fig. 1e) and banded fabrics (Table 2).

The pedofeatures observed in all soil samples are dominated by diffuse to sharp Fe and Mn nodules with diameter ranging between 300 and 5000 μm. Generally, Fe and Mn nodules are the dark and rounded soil materials. There are also pedofeatures from biological activities such as humigranic fecal materials likely from enchytraeids and casts from mesofauna (Fig. 1f). Some of the biological casts are showing disintegration or fusion into other casts.

Table 2 Description of selected thin section of soils subjected to compaction and organic matter removal treatments from the Log lake site

Sample identification and thin section description

Control – Old Growth Site 2B

Zone 1

Fabric[1] – Few equant to oblate phytogenic units that range in size of $0.37 \times 10^3 - 1.1 \times 10^3$ µm are present. Frequent matrifragmoidic units are also present. They are 0.79×10^3 to 2.3×10^3 µm in size and have an oblique orientation.

Plasma fabrics[2] – silasepic.

Voids – total void space is about 70%. The compound packing voids are dominant. Minor amount (~5%) of short irregular planes and irregular orthovughs (5–10%) are present.

Pedofeature – Very few fecal pellets of humic or moder composition occurs close to the phytogenic units or in voids. These fecal pellets are 0.096×10^3 to 0.14×10^3 µm in size and are moderately infected by fungi. The shape of the fecal pellets varies from irregular to well defined and are likely a mixture of collembola and enchytraeid excrement.

Zone 2

Fabric[1] – Dominant (~50–70%) banded, dense, matrifragmoidic units, with horizontal to sub-horizontal orientation occurs in this zone. The upper boundary of these platy aggregates is darker and better defined than the lower boundary. The platy aggregates are 1.1×10^3 to 7.9×10^3 µm in length and 0.24×10^3 to 1.8×10^3 µm in thickness. The lower portion of this zone grades to banded matriporphyric fabric due to presence of a denser s-matrix.

Plasma fabrics[2] – Silasepic.

Voids – Total porosity is about 40–45%. The dominant voids in upper portion is horizontal to subhorizontal, discontinuous planes in length of 1.2×10^3 to 3.9×10^3 µm and account about 30% of the area. The intra-aggregate voids are irregular orthovughs and vesicles. They account for about 10–15% of the area. In the lower portion of this zone where the banded porphyric fabric dominates, vughs and vesicles become the major void forms and account about 25% of the area, while the irregular planar voids occur in about 15% of the area.

Pedofeature – Trace amount of disintegrated, humigranoidic casts in 0.18×10^3 to 0.55×10^3 µm occur randomly in very limited area.

Zone 3

Fabric[1] – Few matriplectic units of approximately 0.18×10^3 to 1.5×10^3 µm in size occur randomly throughout this zone. The matrigranoidic units are largely 0.18×10^3 to 1.8×10^3 µ in size, with the upper size of 3.2×10^3 µm. The finer aggregates (<0.18×10^3 µm) are distributed in the inter-aggregate space of the larger aggregates.

Plasma fabrics[2] – Silasepic.

Voids – The total void space is about 65–70% of this zone. The dominant packing voids are compound packing voids. Minor amount (~2%) of planes and vughs are also present.

Pedofeature – (a) Few (~5%) diffuse to sharp Fe/Mn nodules are present. They are 0.39×10^3 to 2.0×10^3 µm and occur randomly. (b) Trace amount (~<1%) of humigranitic fecal pellets occurs in very limited area. They are likely from enchytraeid worms.

Table 2 (Continued*)*

Sample identification and thin section description

C_0OM_0 – Plot 1 -Site 1C

Fabric[1] – Few (~10–15%) coated skeleton grains of 0.19×10^3 to 3.2×10^3 µm in size occur in inter-aggregate space of common granular aggregates of 0.79×10^3 to 3.9×10^3 µm in size. The large granular aggregates occur randomly. The finer granular aggregates ($< 0.79 \times 10^3$ µm) also fill the inter-aggregate space of the large aggregates. In two small areas (about 10% together), a matriplectic-matrigranic fabric component is present. In the middle bottom location, a matriplectic-matrigranoidic-porphyric mixed intergrade fabric component occurs at a small area of about 10% of the thin section.

Plasma fabrics[2] – silasepic throughout.

Voids – Total void space is about 60%. Compound packing voids are dominant and account for about 55% of the area of thin section. Minor amount of interconnected vughs occupies about 5% of the area.

Pedofeature – Very few (~5 %), equant, diffuse to sharp, Fe/Mn nodules occur as clusters. Their size ranges from 0.79×10^3 to 3.9×10^3 µm.

C_1OM_1- Plot 6 – Site1A

Zone 1

Fabric[1] – The upper 0–0.7 cm is covered with a thin layer of phyto-humigranic substances. The phyto-fragments are horizontally distributed and are 0.18×10^3 to 9.1×10^3 µm in size. The humic substances are strongly disintegrated. Below this thin organic layer, mineral matrix is densely distributed and forms a porphyric fabric. Very few charred organic substances are present with a horizontal orientation within the mineral matrix. Very few (~2%) phytofragments are present in mineral substances.

Plasma fabrics[2] – Dominantly silasepic, with a very minor, weak skelsepic component.

Voids – Total void space is about 30%. The dominant voids (~25% in area) are isolated irregular vughs that are 0.18×10^3 to 0.90×10^3 µm in size. Very few

(~5%) root channels and horizontal planes are also present.

Pedofeature – Very few (~2%) disintegrated mesofaunal casts in size of approximately 96 to 0.24×10^3 µm appear in random patches, in root channels or mixed with mineral substances in upper portion of this zone. Very few Fe nodules (~ <1%) in size of about 0.40×10^3 to 1.6×10^3 µm are present.

Zone 2

Fabric[1] – The major change in zone 2 as compared with zone 1, is a significant increase in abundance of the brownish black Fe/Mn nodules. In the lower portion of zone 2, there is a sharp decrease in abundance of these Fe/Mn.

Plasma fabrics[2] – the same as zone 1.

Voids – total porosity is about 30%. Dominant voids (about 30% in area) are irregular vughs in size of 0.28×10^3 to 2.0×10^3 µm. Very few (about 2% in area) vesicles are also present.

Pedofeature – Frequent (~15–30%) brownish black Mn/Fe nodules with diffuse to sharp boundary, and mainly in size of 0.18×10^3 to 2.4×10^3 µm occur throughout the upper to middle portions of zone 2. Some of the large nodules are up to about 4.3×10^3 µm in size. The frequency of these nodules decreases to about 5% in area in lower zone 2.

Table 2 (Continued)

Sample identification and thin section description

C$_2$OM$_2$ – Plot 9 – Site 3C

Fabric[1] – The skeleton grains, Fe/Mn nodules, very few phytofragments are located on the dense soil matrix. The soil matrix is dense porphyric microfabric. The size of soil skeleton grains ranges from 0.79×10^3 to 2.0×10^3 μm. The equant to acicular phytofragments are 0.37×10^3 to 5.5×10^3 μm in size. A minor matripletic-matrigranoidic fabric component is present at the right edge of the thin section (to a depth of 0–2.5 cm), and at the lower 5.2–6.7 cm regions. The minor fabric component accounts for about 10% in area of the thin section.

Plasma fabric[2] – Dominantly silasepic, with a very minor skelsepic component. The minor matriplectic-matrigranoidic fabric region has the same plasma fabric as the porphyric region.

Voids – The total void space is about 35%. In the dominant fabric region, the dominant voids are irregular orthovughs of 0.18×10^3 to 1.2×10^3 μm. Very few vesicles in size of about 0.25×10^3 μm are also present. Few joint-planes in length of about 12×10^3 μm and width of 0.39×10^3 to 0.80×10^3 μm are also present. Very few root channels, in length of 1.2×10^3 to 1.6×10^3 μm, are also present. The large planar voids account for about 20% of the voids.

Vughs, vesicles, and root channels occupy about 15% of the area. In the minor fabric region, the total porosity is about 50%, dominantly compound packing voids.

Pedofeature – (a) Few (~15%), equant to oblate Fe/Mn nodules generally with size of 0.39×10^3 to 2.8×10^3 μm, are present in clusters. The large Fe/Mn nodules are up to 5.1×10^3 μm in size. These nodules have sharp boundary. (b) Very few organic casts of mesofauna, 0.048×10^3 to 0.058×10^3 μm in size are present inside of some acicular phytofragments. The irregular boundary of these casts suggests that they are likely from enchytraeids. Very few (~ <2%) disintegrated castes are present in soil matrix. In the minor fabric region, some equant to irregular fecal pellets with mull composition, are likely from enchytraeids as well. Their size ranges from 0.048×10^3 to 0.096×10^3 μm.

[1] Granic – Unaccomodated, typically loosely packed, discrete units without coatings or bridges between units; Granoidic – Similar to granic but the units are not discrete, appearing to be fused at their contacts; Plectic – The matrix coats the framework members and broadens and extends to form bridges between them; Porphyric – The coarser particles occur in a groundmass of finer materials. (After Brewer 1976, Bullock et al. 1985).

[2] Silasepic – Soil material has wide range of particle size with relatively high proportions of silt size grains so that domains are difficult to recognize; Skelsepic – Plasma separations with striated orientation occur on the surface of sand grains. (After Brewer 1976).

3.2 Soil Microscopic Pore Space and Soil Porosity

The overall median microscopic void space from six thin section observations ranged from 40 to 60% (Table 3). There appear to be a decrease from about 60% in the Control plots to about 40% in C_2OM_2 plots, corresponding to a reduction of about 35% (Table 3). The soil porosity values calculated from the one-year post treatment bulk density data reported by (Holcomb 1996, Trowbridge et al. 1996) were 42.6% for C_0OM_0, 29.4% for C_1OM_1 and 35.5% for C_2OM_2. The trend is consistent with the microscopic pore space changes reported here. Compaction also altered the distribution of microscopic pore types. The amount of compound packing voids appears to be the dominant type of voids in Control and C_0OM_0 plots (Table 2). Table 3 indicated a reduction of compound packing voids in C_1OM_1 and C_2OM_2 plots, compared to the Control and C_0OM_0 plots. The median estimates of vughs are high in the C_1OM_1 treatment (85%) and C_2OM_1 treatment (70%), whereas the chamber type of voids are lowest among the different types of voids.

3.3 Soil Solution Chemistry

Concentrations of Ca, Mg, K and Al in the soil solutions and the effects of compaction and OM removal treatments on the composition of the soil solution are given in Fig. 2 and Table 4. The effect of soil compaction on the Ca content is apparent from the significant reduction of Ca concentration, from 11 mg L^{-1} in C_0 plots to 8.3 mg L^{-1} in the C_2 plots (Table 4). The magnesium content in the soil

Table 3 Range (and median) total microscopic pore space, distribution of different types[1] of pores and soil porosity in soils subjected to compaction and organic matter removal treatments, Log Lake site. ($n=6$)

Treatment[2]	Total microscopic porosity	Packing	Planar	Vugh	Chamber	Soil porosity[4]
	Range (Percentage)[3]					
Control	50–70	50–100	0–20	0–50	0–0	–
	(60)	(90)	(5)	(5)	(0)	
C_0OM_0	45–70	0–100	0–45	0–80	0–10	42.6
	(60)	(80)	(0)	(20)	(0)	
C_1OM_1	30–100	0–100	0–50	0–100	0–0	29.4
	(40)	(0)	(15)	(85)	(0)	
C_2OM_2	30–60	0–60	0–80	10–100	0–15	35.5
	(40)	(0)	(30)	(70)	(0)	

[1]% of total microscopic porosity.
[2]Organic Matter (OM) treatments: OM_0=boles (tree trunk) only removed; OM_1=boles+crown removed; OM_2=boles, crown+forest floor removed; Compaction (C) treatments: C_0=no compaction; C_1=intermediate compaction; C_2=heavy compaction.
[3]rounded to nearest 5 or 10 %; median values are in parentheses.
[4]Soil porosity is calculated from the reported post-treatment soil bulk density data in Table 1.

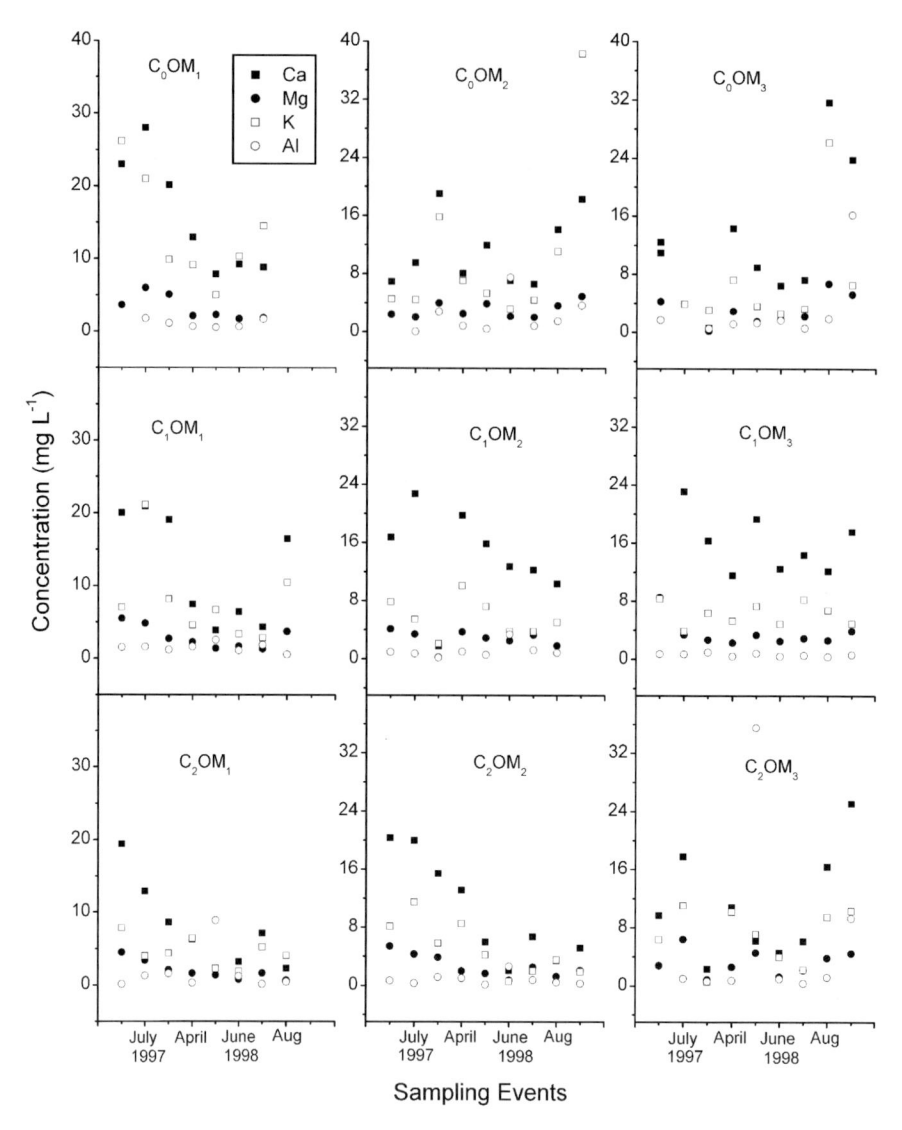

Fig. 2 Scatter plots of the concentrations of Ca, Mg, K and Al in soil solution extracted from the Ae horizon of soils subjected to forest floor removal and compaction treatments from the Log Lake site. (No detectable values for C_0OM_0, C_2OM_2 and C_2OM_2 on Sept 98)

solution ranged from 2.0 mg L^{-1} in the C_2OM_0 to 3.7 mg L^{-1} in the C_1OM_2 treatment and appears not to have been affected by the compaction and OM-removal treatments (Table 4). The mean K concentration ranged from 4.6 mg L^{-1} in the C_2OM_0 treatment to 13.7 mg L^{-1} in the C_0OM_0 treatment, whereas the compaction treatment significantly lowered the mean K content in the soil solution from

Table 4 Mean contents (and standard deviation) of Ca, Mg, K and Al (mg L^{-1}) and molecular ratios of $\dfrac{K}{\sqrt{(Ca+Mg)}}$ and Ca/Al in soil solution extracted from Ae horizon of soils subjected to compaction and organic matter removal, Log Lake site

	OM$_1$	OM$_2$	OM$_3$	Compaction mean
Calcium	$n=6$	$n=6$	$n=6$	$n=18$
C$_0$	14 (6.4)	9.9 (4.9)	8.3 (4.8)	11^b (5.4)
C$_1$	10 (7.4)	13 (6.2)	22 (17.0)	15^b (11.4)
C$_2$	7.8 (6.1)	11 (6.8)	6.6 (3.1)	8.3^a (5.4)
OM-Mean ($n=18$)	11 (6.5)	11 (5.7)	12 (11.6)	
Magnesium	$n=6$	$n=6$	$n=6$	$n=18$
C$_0$	2.8 (1.3)	2.8 (0.87)	2.1 (1.4)	2.6 (1.1)
C$_1$	2.5 (1.6)	2.8 (1.4)	3.7 (2.4)	3.0 (1.8)
C$_2$	2.0 (1.3)	2.7 (1.7)	2.4 (1.3)	2.4 (1.3)
OM-Mean ($n=18$)	2.4 (1.3)	2.8 (1.2)	2.7 (1.7)	
Potassium	($n=7$)	($n=7$)	($n=7$)	($n=21$)
C$_0$	13.7^a (7.4)	6.4^a (4.3)	4.9^b (3.0)	8.4^b (6.2)
C$_1$	7.7^a (6.3)	5.7^b (2.8)	6.3^a (1.7)	6.6^{ab} (3.9)
C$_2$	4.6^b (2.1)	5.8^b (3.8)	5.9^b (3.9)	5.4^a (3.2)
OM-Mean ($n=21$)	8.6 (6.5)	6.0 (3.4)	5.7 (2.9)	
Aluminum	$n=5$	$n=5$	$n=5$	$n=15$
C$_0$	0.95 (0.50)	2.5 (3.0)	1.0 (0.49)	1.5 (1.7)
C$_1$	1.6 (0.58)	1.3 (1.2)	0.63 (0.25)	1.2 (0.83)
C$_2$	2.4 (3.6)	1.2 (0.91)	7.6 (15.6)	3.8 (8.7)
OM-Mean ($n=15$)	1.7 (2.0)	1.6 (1.8)	3.1 (8.7)	
K/(Ca+Mg)	$n=6$	$n=6$	$n=6$	$n=18$
C$_0$	0.48 (0.22)	0.28 (0.13)	0.28 (0.16)	0.35 (0.19)
C$_1$	0.25 (0.10)	0.23 (0.07)	0.23 (0.05)	0.24 (0.07)
C$_2$	0.23 (0.07)	0.20 (0.10)	0.24 (0.14)	0.22 (0.10)
OM-Mean ($n=18$)	0.32 (0.17)	0.24 (0.10)	0.25 (0.12)	
Ca/Al	$n=5$	$n=5$	$n=5$	$n=15$
C$_0$	9.5^{ab} (3.7)	7.4^{ab} (7.2)	5.0^a (3.4)	7.3 (4.9)
C$_1$	4.1^a (4.0)	9.7^{ab} (6.5)	17^b (3.4)	10 (6.7)
C$_2$	9.3^{ab} (11.6)	9.8^{ab} (9.3)	5.4^{ab} (4.9)	8.2 (8.3)
OM-Mean ($n=15$)	7.7 (7.1)	9.0 (7.0)	9.1 (6.5)	

Organic Matter (OM) treatments: OM$_0$=boles (tree trunk) only removed; OM$_1$=boles+crown removed; OM$_2$=boles, crown+forest floor removed; Compaction (C) treatments: C$_0$=no compaction; C$_1$=intermediate compaction; C$_2$=heavy compaction.
Within each element or ratio, means superscripted with different letters or numbers are significantly different ($p<0.05$).

8.4 mg L^{-1} in the C$_0$ to 5.4 mg L^{-1} in the C$_2$ treatments. Generally the mean K content was lower in the C$_1$, C$_2$ and OM$_2$ treatments compared to the C$_0$ and OM$_1$ treatments. The mean Al concentrations were not affected by any treatments and the concentrations of the cations followed the order of Ca>K>Mg>Al.

The values for K$_{AR}$ ranged from 0.20 to 0.48 (Table 4). Soil compaction and OM removal had no significant effects on K$_{AR,-}$ while the Ca$_{AR}$ values ranged from 4.1 in the C$_1$OM$_1$ and 17 in the C$_1$OM$_2$ treatment (Table 4). The interaction between compaction and OM treatments significantly affected the C$_{AR}$ values.

4 Discussion

4.1 Soil Microstructure

The change from porous microfabrics (granic and granoidic) at Control and C_0OM_1 plots, to dense microfabrics (porphyric) found in C_1OM_1 and C_2OM_2 plots, provided micromorphological evidence of soil compaction. This is a result of granular-shaped aggregates in Control and C_0OM_0 plots being replaced by dense soil matrix in C_1OM_1 and C_2OM_2 plots. The trend in change of soil microscopic pore space is consistent with soil porosity data generated from soil bulk density data of related previous studies (Holcomb 1996, Trowbridge et al. 1996). Similar results were presented by Bresson and Zambaux (1990) for forest soils in Normandy, where soil compaction caused the disappearance of granular aggregates and promoted the formation of a dense matrix.

The shift in the soil fabric is probably due to the collapse of large pores under the compaction pressure, such as large packing voids and compound packing voids, during the compaction treatment. The collapse of the large pores changes the pore structural configuration of the soil following compaction. Among the several factors affecting soil compaction is the soil water content. In a laboratory simulation of soil compaction, Bresson and Zambaux (1990) concluded that the degree of soil compaction from a pressure of 250 kPa was significantly more severe for wet samples compared to dry soils. This is due to the reduction of the internal friction angle between microaggregates. In field conditions, the collapse of the soil pores can also result from cultivation.

Livingston et al. (1990) reported a reduction in the mean equivalent diameter of the pores in cultivated soils under hardwood forest, and Aguilar et al. (1990), reported the loss of intrapedal porosity upon plowing of soils planted to olives in Spain. The mechanical stress from compaction causes re-organization of the macropores into dominantly horizontal orientation (Yao and Wilding 1994), which favours the formation of planar voids in highly compacted soils.

The effect of compaction on soil fabric was not uniformly expressed in the soil matrix. The presence of several types of fabrics in some small portions in C_1OM_1 and C_2OM_2 soils suggests that there are some variations in the compactability of the soils. This is likely the result of the anisotropism (or non-uniformity) in the soil, which is highly pronounced in the area studied, due to the morainal nature of the parent material. Bresson and Zambaux (1990) attributed similar results to the presence of gravel in a dominantly silty Fragiudalf in Normandy, France. The presence of 35% coarse fragments in our area, might have contributed to the non-uniformity of compaction effects on the soil fabrics.

Variations in the compactability of the soils due to structural anisotropism, might be beneficial to survival of plants under stressed conditions. It is interesting that little difference in soil response is observed under the two rates of compaction forces in this micromorphological investigation and related bulk density data in Table 1 from others (Holcomb 1996; Trowbridge et al. 1996). Thus the lower compaction rate used to establish C_1OM_2 plots may be strong enough to cause significant damage to soils.

Soil porosity is one of the physical properties of soil that is significantly affected by compaction. The decrease in microscopic soil pore space and is expected from the increase in the bulk density (28%) of the C_1OM_1 treatment. This translates to a theoretical decrease from 50 to 36% in total porosity of the soils. Similar results were reported earlier both in agricultural and forest soils by Bruand et al. (1997) and Carr (1987a,b). Physical stress from compaction crushed a significant portion of the macropores (e.g., compound packing voids) and caused the reduction in total porosity (Bresson and Zambaux 1990, Bruand et al. 1997).

Crushing of compound packing voids by compaction could explain the significantly lower amount of packing voids in the C_2OM_2 compared to Control and C_0OM_1 plots. The significantly higher amount of planar voids in the C_2OM_2 plot compared to Control and C_0OM_0 plots is also the result of the physical crushing of macropores upon soil compaction. In addition to the general decrease in macropores, Bruand et al. (1997) suggested that compaction also results in the formation of "relict" pores that are isolated remnants of macropores. Bruand et al. (1997) utilized mercury porosimetry to determine the amount of "relict" macropores. The significantly high amounts of vugh type of voids observed in the C_1OM_1 plot might be due to "relict" macropores. These micromorphological changes in distribution of pore types revealed a reduction in physical connectivity of soil pore space. Thus these micromorphological observations contributed to our understanding to change in bulk physical measurements, such as bulk density and soil porosity.

4.2 Soil Solution Chemistry

The dominance of Ca in the soil solution could be due to the breakdown of anorthite ($CaAl_2Si_2O_8$), releasing Ca into the soil solution. Acidity in the Ae horizon enhances the breakdown of anorthite and other minerals. Organic acids normally come from the overlying forest floor. However, in the black spruce forests of Alaska, Van Cleve and Dryness (1983) reported elevated levels of Ca^{2+} and Mg^{2+} in the soil solution after the mechanical removal of the forest floor. This suggests that a statistical interaction between OM removal and soil compaction factors may occur. The complicated biochemical and physiochemical mechanisms deserve further research. The abundance of Ca ions is also related to the 75–85% Ca occupancy of the exchange complex, as is common in soils containing 2:1 expanding-type clay minerals (Van Cleve and Dryness 1983, Trowbridge et al. 1996). The decrease in the Ca content in heavily compacted soils is consistent with the findings of Lipiec and Stepniewski (1995).

In terms of Ca availability to plants, the C_{AR} values are above 1.8 and indicate an environment conducive to plant growth (Cronan and Grigal 1995). The apparent effects of compaction and OM treatments on C_{AR} values might be associated with the reduction in porosity and the interaction of the decrease in acidity due to the removal of OM and to physical processes such as wetting-drying, and freeze-thaw cycles. In 1998, the mean soil pH measured in OM_2 plots (pH=5.04) was significantly higher (p<0.05) than OM_1 plots at pH 4.86 and OM_0 plots at pH 4.87 (Kranabetter et al. 2006).

The decrease in the K^+ concentration from C_0 to C_2 treatments is consistent with Lipiec and Stepniewski (1995). Wolkowski (1991), as cited in Lipiec and Stepniewski (1995), reported that compaction reduced K^+ concentration by 4–11%, however, the reduction was not large enough to reduce the shoot growth of maize. Our data agree with Wolkowski (1991 cited in Lipiec and Stepniewski 1995) where compaction did not change the K_{AR} values.

There is inconsistency in the explanation of the likely cause of the decrease in the concentrations of Ca^{2+} and K^+ in the soil solution. According to Barber (1984), the plant uptake of Ca^{2+} is largely through a mass-flow mechanism in soils. The reduced connection of soil pores may restrict migration of solutes from micropores to medium-sized pores where soil solutions are likely more accessible to plant roots. If nutrient migration is retarded, the ionic exchange at the solid-aqueous interface could also be compromised in highly compacted soils. Some of the solutes can be blocked in isolated pore spaces, such as vughs, in compacted soils and subsequently become not accessible to plant roots and the immiscible displacement by TCE.

Ishaq et al. (2001) reported a 24% reduction in K uptake of wheat grown in compacted soil. Lipiec and Hatano (2003) argued that "greater root–soil contact" coupled with high unsaturated hydraulic conductivity increased water movement towards the root, hence increasing water (and nutrient) uptake by plants in compacted soil.

The poor growth of trees in the compacted soil contradicted Lipiec and Hatano (2003). In compacted soil, we think that the decreased Ca^+ and K^+ concentrations in the soil solution might be related to the removal of ions by processes such as precipitation of organic salts (e.g., Ca-oxalate). Perhaps, the low Ca^+ and K^+ in the solution of the compacted soil is the result of the low exchange capacity (~4 cmol$_c$ kg^{-1} soil, Arocena and Sanborn 1999). This could be due to the silicate clays, which will be oriented on the walls of the dominant micropores left after compaction and most likely not contributing to the exchange capacity of the compacted soil. Another possibility may be that formation of the porphyric fabric makes some exchange sites and some solutes isolated in closed micro-pore space, and makes them not accessible to the extraction process. In the uncompacted soil, the dissolution of the minerals provides Ca^+ and K^+ in to the solution in addition to the contribution from the exchange sites. The sand fraction containing Ca and K-minerals are the significant parts of the materials enclosing the macropores.

5 Conclusion

The application of physical stress of compaction on forest soils alters the soil microstructure. It can be concluded from the study that compaction of an Orthic Humo-Ferric Podzol (Cryorthod) near Prince George, BC, changes soil microstructure from a dominantly granic and granoidic to dense and phosphoric soil fabrics. However, the effect of compaction is not uniformly expressed in the soil, because of the inherent anisotropy (or non-uniformity) of the soil associated with the morainal type of parent material. The soil compaction treatment decreased total porosity from about 60% in Control (and C_0OM_0) plots to about 45% in compacted (C_1OM_1 and C_2OM_2) plots. The main effect of compaction is the destruction of the macroporosity of the soil and increase in the amount of planar voids at the expense of compound packing voids. The high amounts of vughs in the C_1OM_1 plot might be related to the "relict" macropores formed upon soil compaction.

In terms of the relevance of the results to plant growth, the decrease in total soil porosity might be sufficient to curtail vigorous plant growth. Specifically, compaction result in decreases in porosity, as well as in the Ca^+ and K^+ concentrations of the soil solutions. The latter may also be related to the reduced connectivity of soil pores that hinders migration of solute in compacted zones and/or inorganic processes such as precipitation of organic salts. Our results showed that soil compaction influences both physical and chemical properties of the soil.

Acknowledgements We would like to thank Forest Renewal BC for providing financial support to undertake this study. Support from Natural Sciences and Engineering Research Council and the Canada Research Chair program are also acknowledged.

References

Aguilar J, Fernandez J, Ortega E, de Haro R, Rodriguez T (1990) Micromorphological characteristics of soils producing olives under nonploughing compared with traditional tillage methods, In: Douglas LA (ed.) Soil Micromorphology: A Basic and Applied Science, Development in Soil Sci. 19. Elsevier, Amsterdam

Arocena JM (2000) Cations in solution from forest soils subjected to forest floor removal and compaction treatments. For Ecol Manage 133:71–80

Arocena JM, Sanborn P (1999) Mineralogy and genesis of selected soils and their implications for forest management in central and northeastern British Columbia. Can J Soil Sci 79:571–592

Assouline S, Tavares-Filho J, Tessier D (1997) Effect of compaction on soil physical and hydraulic properties: Experimental results and modeling. Soil Sci Soc Am J 61:390–398

Bouwman LA, Arts WBM (2000) Effects of soil compaction on the relationships between nematodes, grass production and soil physical properties. Appl Soil Ecol 14: 213–222

Barber SA (1984) Soil nutrient bio-availability: a mechanistic approach. John Wiley and Sons Inc, New York

Brady NC, Weil RR (1996) The nature and properties of soils. 11th ed. Prentice-Hall Inc., New York

Brais S, Camiré C, Paré D (1995) Impacts of whole-tree harvesting and winter windrowing on soil pH and base status of clayey sites of northwestern Quebec. Can J For Res 25:997–1007

Bresson LM, Zambaux C (1990) Micromorphological study of compaction induced by mechanical stress for a Dystrochreptic Fragiudalf. In: Douglas LA (ed.) Soil Micromorphology: A Basic and Applied Science, Development in Soil Sci. 19. Elsevier, Amsterdam

Brewer R (1976) Fabric and mineral analysis of soils. 2nd ed. R.E. Krieger Publishing Co., Huntington, New York

Brewer R, Pawluk S (1975) Investigation of some soils developed in hummocks of the Canadian sub-Arctic and southern Arctic regions. I. Morphology and micromorphology. Can J Soil Sci 55:301–319

Bruand A, Cousin I, LeLay C (1997) Formation of relict macropores in clay-loamy soil by wheel compaction. In: Shoba S, Gerasimova M, Miedema R (eds.), Soil Micromorphology: Studies on Soil Diversity, Diagnostics and Dynamics. Van Gils B.V., Wageningen

Bullock P, Fedoroff N, Jongerius N, Stoops G, Tursina T (1985) Handbook for soil thin section descriptions. Waine Research Publications, Albrighton, Wolverhampton, UK

Busse MD, Cochran PH, Barrett JW (1996) Changes in ponderosa pine site productivity following removal of understory vegetation. Soil Sci Soc Am J 60:1614–1621

Carr WW (1987a) The effect of landing construction on some forest soil properties. A case study. FRDA Report no. 003. Govt. of Canada and British Columbia Ministry of Forests and Lands

Carr WW (1987b) Restoring productivity on degraded forest soils: Two case studies. FRDA Report no. 002. Govt. of Canada and British Columbia Ministry of Forests and Lands

Corns IGW (1988) Compaction by forestry equipment and effects on coniferous seedling growth on four soils in the Alberta foothills. Can J For Res 18:75–84

Crawford JW, Matsui N, Young IM (1995) The relation between the moisture-release curve and the structure of soil. Eur J Soil Sci 46:369–375

Cronan CS, Grigal DF (1995) Use of calcium/aluminum ratios as indicators of stress in forest ecosystems. J Environ Qual 24:209–226

Danielson RE, Sutherland PL (1986) Porosity, In: Klute A (ed.) Methods of Soil Analysis, Part I, Physical and Mineral Methods. 2nd ed. ASA and SSSA, Madison, WI

Greacan EL, Sands R (1980) Compaction of forest soils. A review. Aust J Soil Res 18:163–189

Holcomb RW (1996) The long-term soil productivity study in British Columbia. FRDA report, ISSN 0835–0752; 256

Huang J, Lacey ST, Ryan PJ (1996) Impact of forest harvesting on the hydraulic properties of surface soil. Soil Sci 161:79–86

Ishaq M, Ibrahim A, Hassan M, Saeed M, Lal R (2001) Subsoil compaction effects on crops in Punjab, Pakistan: II. Root growth and nutrient uptake of wheat and sorghum. Soil Till Res 60:153–161

Johnson CE, Johnson AH, Huntington TG, Siccama TG (1991) Whole-tree clear-cutting effects on exchangeable cations and soil acidity. Soil Sci Soc Am J 55:502–508.

Kooistra MJ, Tovey NK (1994) Effects of compaction on soil microstructure. In: Soane BD, van Oowerberk C (eds.) Soil Compaction in Crop Production. Elsevier Sci., BV. Amsterdam

Kranabetter JM, Sanborn P, Chapman BK, Dube S (2006) The contrasting response to soil disturbance between lodgepole pine and hybrid white spruce in sub-boreal forests. Soil Sci Soc Amer J 70:1591–1599

Li Q, Allen HL, Wollum AG (2004) Microbial biomass and bacterial functional diversity in forest soils: effects of organic matter removal, compaction, and vegetation control. Soil Biol Biochem 36: 571–579

Lipiec J, Hatano R (2003) Quantification of compaction effects on soil physical properties and crop growth. Geoderma 116:107–136

Lipiec J, Stepniewski W (1995) Effects of soil compaction and tillage systems on uptake and losses of nutrients. Soil Till Res 35:37–52

Livingston SJ, Norton LD, West LT (1990) Effects of long-term cultivation on aggregate stability, organic carbon distribution, and porosity of two soil series. In: Douglas LA (ed.) Soil Micromorphology: A Basic and Applied Science, Development in Soil Sci. 19. Elsevier, Amsterdam

Mo J, Brown S, Lenart, M (1995) Nutrient dynamics of a human-impacted pine forest in a MAB reserve of subtropical China. Biotropica 27:290–304

Powers RF (2006) Long-term soil productivity: genesis of the concept and principles behind the program. Can J For Res 36:519–528

Powers RF, Alban DH, Miller RE, Tiarks AE, Wells CG, Avers PE, Cline RG, Loftus NS Jr, Fitzgerald RO. (1990). Sustaining productivity in North American forests: problems and prospects. In: Gessel SP, Lacate DS, Weetman GF, Powers RF (eds.). Proceedings of the seventh North American forest soils conference Vancouver, British Columbia

Rab MA (2004) Recovery of soil physical properties from compaction and soil profile disturbance caused by logging of native forest in Victorian Central Highlands, Australia. For Ecol Manage 191:329–340

Soil Classification Working Group (1998) The Canadian system of soil classification. Agric. And Agri-Food Can. Publ. 1646 (revised), National Research Council Canada, Ottawa

Soil Survey Staff (1999) Soil Taxonomy a basic system of soil classification for making and interpreting soil surveys. USDA Agric. Handbook 436, 2nd ed., US Govt. Printing Press, Washington DC, USA

Trowbridge R, Kranabetter M, Macadam A, Battigelli J, Berch S, Chapman W, Kabzems R, Osberg M, Sanborn P (1996) The effects of soil compaction and organic matter retention on long-term soil productivity in British Columbia. Establishment Report, Experimental Project No. 1148. BC Ministry of Forest, Victoria, March 1996

Tuttle CL, Golden MS, Meldahl RS (1985) Soil surface removal and herbicide treatment: effects on soil properties and loblolly pine early growth. Soil Sci Soc Am J 49:1558–1562

Van Cleve K, Dryness CT (1983) Effects of forest-floor disturbance on soil-solution nutrient composition in black spruce ecosystem. Can J For Res 13:894–902

Wolt J (1994) Soil solution chemistry: Applications to environmental science and agriculture. John Wiley and Sons, New York

Worrell R, Hampson A (1997) The influence of some forest operations on the sustainable management of forest soils-a review. Forestry 70:61–85

Yao L, Wilding LP (1994) Micromorphological study of compacted mine soil in east Texas. In: Ringrose-Voase AJ, Humphreys GS (eds.) Soil Micromorphology: Studies in Management and Genesis. Development In Soil Sci 22. Elsevier, Amsterdam

Zabowski D, Rygiewicz PT, Skinner MF (1996) Site disturbance on clay soil under a radiata pine. Plant and Soil 186:343–351

Index

Printing: Krips bv, Meppel, The Netherlands
Binding: Stürtz, Würzburg, Germany